ENCYCLOPAEDIA OF
HORTICULTURE

ENCYCLOPAEDIA OF HORTICULTURE

A. Edwards

ANMOL PUBLICATIONS PVT. LTD.
NEW DELHI - 110 002 (INDIA)

ANMOL PUBLICATIONS PVT. LTD.
H.O.: 4374/4B, Ansari Road, Daryaganj
New Delhi - 110 002
Ph.: 23278000, 23261597
B.O.: No. 1015, Ist Main Road, BSK III Stage
III Phase, III Block
Bangalore - 560 085 (India)
Visit us at: www.anmolpublications.com

Encyclopaedia of Horticulture

First Published, 2006

ISBN 81-261-2918-2

PRINTED IN INDIA

Printed at Arora Enterprises, Delhi.

Preface

Basically, horticulture is the art or science of cultivating gardens. In fact, man has had great relation with horticulture since ancient times. But in a systematic way it emerged on the scene very late. In terms of education and training it was considered as part and parcel of agriculture. But now horticulture is a fullfledged subject, being taught at various agricultural universities and colleges. Since horticulture has proved an important component of agrarian economy at State level in India, separate departments of horticulture are created to look after the horticultural developments at field level.

This encyclopaedia on horticulture encompasses all terms pertaining to horticulture and allied science. Terms are defined and explained authoritative in simple language. A multidisciplinary and inter-disciplinary approach has been followed to compile this encyclopaedia giving priority to the interests of its readers and professional users.

A veritable mine of information, this work will prove a dependable source book to students, teachers, agricultural scientists and others in the field.

The compiler/editor expresses his deep sense of gratitude to all scholars whose works are cited or and expressions are borrowed. He is equally grateful to his colleagues and friends for assistance while compiling the material. Finally, he is thankful to the publishers for bringing out this book in superb way.

—*Editor*

A

Abandoned wells. Abandoned *drainage wells* and abandoned water wells on vacant farmsteads are of particular concern for agriculture. Abandoned wells can present both safety risks and a direct conduit by which groundwater can be contaminated by surface runoff. A number of states have incentive and/or regulatory programs to cap or seal abandoned wells.

ABC soil. Soil having a distinctly developed profiles including A, B and C horizons.

Abelia. Genus of evergreen and deciduous shrubs, in the Caprifoliaceae (honeysuckle) family, some of which are hardy in southern England. The Chinese *Abelia chinensis* and *A. schumannii*, and the Mexican *A. floribunda* are cultivated for their summer and autumn flowers.

Aberdeen-angus. A breed of black, hornless beef cattle, developed using interbreeding from a mixture of breeds. They have a long, deep body.

Abiology. Means science of non-living things.

Abnormal seedings. The seedings lacking the capacity for continued development into normal healthy plants. Examples include seedlings with weak or malformed parts such as stubby roots, split hypocotyl, absence of primary leaves and/or terminal bud.

Abomasum. It is the fourth stomach of Ruminants which is lying close to the Omasum or third stomach, in which digestion gets completed by the supply of digestive juice. It is also known as Reed Stomach, Rennet Stomach, and sometimes called Maw.

Abortion antigen. An antigen which finds use in making the agglu-tination test.

Abortion. Means miscarriage, slippings, or premature delivery of partlally developed offspring. (Contagious Abortion)

Abortive. (1) Only partially developed, such as incomplete seed or frost nipped bud. (2) Undeveloped; rudimentary.

Abrasion. (1) A superficial injury caused to the skin or the mucous lining (2) The wearing away of part of the earth's surface by the action of water, wind or ice.

Abrasive. Substances which are able to facilitate the mechanical transmission of certain viruses. They tend to increase the num-ber of wounds and also the type of injury conducive to success-ful establishment of infection.

Abscisic acid (ABA, Abscisin, Dormin). (1) A plant hormone which has growth inhibiting action and promotes abscission. It is associated with the onset and maintenance of dormancy. (2) Naturally occurring plant growth inhibitor, promoting senescence, abscission and / or dormancy.

Abscission zone (Abscission layer). Zone where transverse division and separation of cells occur, leading to abscission of organ

Abscission. (1) Detachment of leaves, flowers, fruits or other plant parts. (2) The shedding of leaves, flowers, or fruits following the formation of the abscission zone. (3) The term used for the shedding of leaves, foliage branches, floral parts, fruits or bark by the layingdown of an abscission layer. (4) Also, refers to the liberation of a fungus spore by the disappearance of an adjoining layer or wall. (5) A natural dropping of leaves, flowers and other plant parts. (6) The falling of leaf, twig-tip, etc, from a clean-cut scar, by a self-healing wound.

Absolute drought. The term used for a period of atleast fifteen consecutive days, none of which has received as much as 0.25 milimeters of rain.

Absolute growth rate. Defined as the rate of increase in size of a growing plant (or a part of it) in a given time, under specified conditions.

Absolute humidity. Refers to the mass of water vapour present in a unit volume of air. It may be expressed as grains per cubic meter.

Absolute transpiration. The term used for the rate of water loss from a plant determined experimentally.

Absolute water requirement. Means the quantity of water in acre-inches per crop season which is absorbed by the crop together with the evaporation from the crop producing land. Also called 'consumptive use of water'.

Absolute weed. One of no use to farmers under any circumstances.

Absolute. Purified floral concrete. Impurities like plant waxes, albumin, colouring matter etc., are removed by separation techniques using absolute alcohol.

Absorbed water. Water which is held mechanically in a soil mass and possessing physical properties which is not substantially different from ordinary water at the same temperature and pressure.

Absorptiometer. An instrument used for measuring the absorption of nearly monochromatic radiation in the visible range by a gas or a liquid, and so is used to find out the concentration of the absorbing constituents in the gas or liquid.

Abstraction licence. A licence issued by a water authority for the abstraction of water from a water source for domestic or commercial use, after for irrigation purposes.

Abutilon. One of a group of 90 related species of tropical or semitropical ornamental plants. The Indian mallow or velvet leaf (*Abutilon theophrastus*) is one of the more common; it has bell-shaped yellow flowers and is the source of a jutelike fibre. Many of the species are pollinated by hummingbirds. (Genus *Abutilon*, family Malvaceae.)

AC soil. Soil which has a profile containing only A and C horizons but no clearly developed B horizon.

Acanthaceae. Family of dicotyledonous plants, mostly tropical herbs and shrubs with tubular, bilaterally symmetrical flowers which are often in racemes or clustered in leaf axils, and are showy and sometimes enclosed in colourful modified leaves (bracteoles). The fruits are capsules, often discharging the seeds explosively. The best-known genus is *Acanthus*. Other genera include *Aphelandra, Beloperone* (shrimp plant), *Dianthera, Fittonia, Hygrophila, Jacobinia, Odontonema, Ruellia, Schaueria, Thunbergia,* and *Whitfieldia*.

Acanthus. Herbaceous plant with handsome lobed leaves. Twenty species are found in the Mediterrranean region and Old World tropics, including bear's-breech (*Acanthus mollis*) whose leaves were used as a motif in classical architecture, especially on Corinthian columns. (Genus *Acanthus*, family Acanthaceae.)

Acaricides. Substances that are able to destroy acarids (ticks and mites). They find use in the treatment of acariasis.

Acarology. Deals with the study of acarina which is a large order of small to microscopic arachnids including ticks and mites.

Accelerated erosion. Refers to the erosion of soil material which takes place at a rate greater than that of normal erosion. It is caused due to the destruction of vegetative cover or due to some activity of man.

Accent plant. Attention-getting plant due to its color or form.

Accession. (1) Refers to a cultivar which is registered at a genetic resources centre (2) Any entry which has been newly received in a crop improvement programme.

Accessory bud. A bud located above or beside the main axillary bud.

Accessory fruit. A fruit developing from accessory floral parts like floral tube, receptacle etc., in addition to the ovary.

Accessory. In Floriculture, an item used in support of landscape or floral design such as figurine, bird bath, pool or accent rocks.

Acclimatization. (1) The process of adapting to a new environment. (2) The physiological adaptation of an animal or plant to changes in climate or environment, such as light, temperature, or altitude. (3) The term used for the process by which human beings, animals and plants get adapted to an unfamiliar set of climatic conditions.

Accredited herds scheme. Brucellosis (Accredited Herds) Scheme.

Accredited milk. Milk sold by a producer-retailer from a herd ac-credited as free from Brucellosis.

Accrescent. Enlarged and persistent. Especially of a calyx which becomes enlarged as the fruit ripens.

Accretion. (1) Refers to the process of growth in which material has been added to the outside in non-living matter (2) Also, refers to the gradual addition of new land to old by the deposition of sediment carried by the water of a system.

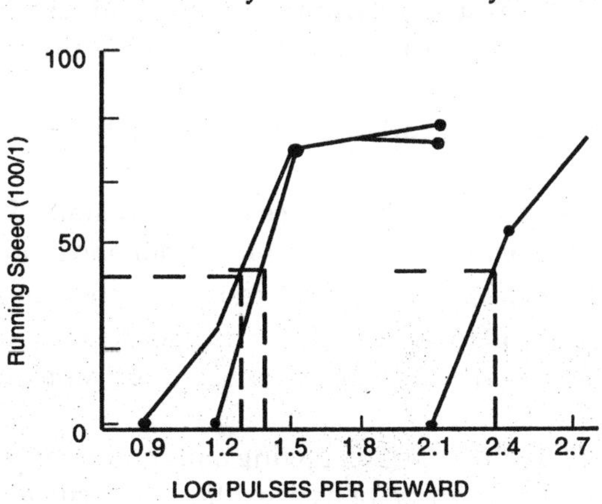

Fig.: Accumulation or Summation Curve

Accumulation or summation curve. Curve which is obtained by plotting settling time on the abscissa and the weight of the deposited particles on the ordinate. It is representing the total amount of accumulated material at various time intervals.

ACE. Agriculture in Concert with the Environment.

Acentric. Whole or part of a chromatid or chromosome which lacks centromere.

Acephalous. Lacking a head. Said of a style lacking a well developed stigma.

Acequia. An irrigation ditch or canal.

Acetabularia. It is the genus of tropical or subtropical marine algae of the division Chlorophyta.

Acetic acid bacteria. Bacteria which are able to produces acetic acid from alcohol.

Acetobacter. It is the genus of gram-negative, chemoorganotrophic, aerobic, ovoid or rod shaped non-spring bacteria. They are able to oxidize ethanol to acetic acid and acetic acid to carbon dioxide.

Acetomonas. It is the genus of gram-negative, aerobic, chemoor-ganotrophic, ovoid or rod shaped polarly flagellate (or non-motile) non-spring bacteria belonging to the family Pseudo-monadaceae.

Acetonemia. The term used for the metabolic disorder of cattle which causes an abnormal quantity of ketone bodies in the blood and urine. It is responsible for a distinctive flavour to milk. It is associated with poor quality of ration.

Achene. (1) A dry indehiscent, one seeded fruit in which the ovary wall remains free from the seed coat. (2) A small, dry, one-seeded Fruit formed from a single Carpel, which does not split open on ripening but rots gradually with seed germination. The pips on strawberries are achenes.

Achrodextrins. Dextrins which are produced when the hydrolysis of starch has carried out to the stage that it imparts no colour with iodine.

Achromatin. Non-staining basic substance of the nucleus excluding the chromatin.

Acid deposition/Acid rain. Abnormally acidic (low pH) precipitation (or dry deposition) resulting from emissions of sulfur and nitrogen compounds that transform during chemical processes in the atmosphere. Acid deposition can affect the chemistry of soils and acidify lakes, adversely affecting forests and fish. It does not adversely affect cropland. The Clean Air Act includes a program focused on controlling precursor emissions of acid deposition, primarily sulfur oxides from coal-fired electric utilities.

Acid fast bacteria. Bacteria keeping carbol fuchsin stain after addition of 25% sulphuric acid.

Acid fast organisms. Micro-organisms which, once effectively stained, cannot get decolorized when treated with mineral acids or mixtures of ethanol and acid.

Acid forming fertilizer. A fertilizer leaving acid residue in the soil when applied.

Acid scarification. Its purpose is to modify hard orimpermeable seed coverings. Soaking seeds in concentrated sulphuric acid is considered to be an effective method of doing this.

Acid soil. Soil with a pH of less than 7.0. Soil having a preponderance of hydrogen ions, and proba-bly of aluminium in proportion of hydroxylions. Specifically, it is a soil with a pH value less than 7.0.

Acinetobacter. It is the genus of gram-negative, aerobic, chemo-organotrophic, asporogenous, aflagellate bacteria of the family Neisseriaceae.

Aconite or monkshood or wolfsbane. Herbaceous plant belonging to the buttercup family, with hooded blue–mauve flowers, native to Europe and Asia. It produces aconitine, a poison with pain-killing and sleep-inducing properties. (*Aconitum napellus*, family Ranunculaceae.) There are about 100 species throughout the northern temperate regions, all hardy *herbaceous plants* containing poison. Summer aconite (*Aconitum uncinatum*) is a common North American flower. Winter aconite (*Eranthis hyemalis*) belongs to another genus of the buttercup family; it has yellow buttercuplike flowers with six petals which appear in February and March; the leaves follow later. *A. ferox*, the bikh or bish of Nepal, contains the fatal bikh poison in its root.

Acorn. The fruit of oaks, a thick walled nut with a woody cup-like base.

Acquired immunity. Means specific immunity which is acquired through exposure to the antigen. Acre. A measure of land. 4840.sq. yds. Equal to 4 Roods, or 10 square Chains. (Scottish acre, 6150.4 sq. yds; Irish acre, 7840 sq. yds-both obsolete.) Also equivalent to 0.405 Hectares. Originally, the amount of land that could be ploughed in a day by a team of oxen.

Acquired lands. Lands in federal ownership that were obtained by the federal government through purchase, condemnation, gift, or exchange. One category of public lands.

ACR. Acreage conservation reserve.

Acre foot of water. Defined as the amount of water that would cover an acre of land to a depth of one foot, provided that there occurs no seepage, evaporation and runoff losses.

Acre. (1) 1 acre = 43,560 sq. ft = 208.7 ft.^{2} = 0.405 hectares; or 640 acres = 1 sq. mile (called a section). (2) A measure of land totaling 43,560 square feet. A square acre is 208.75 feet on each side.

Acreage allotment. Under provisions of permanent commodity price support law, a farm's acreage allotment is its share, based on its previous production, of the national acreage needed to produce sufficient supplies of a particular crop. Under the FAIR Act of 1996, acreage allotments are not applicable to the contract commodities, peanuts or sugar. However, acreage allotments still apply to tobacco.

Acreage base (or base acres). A farm's average planted acreage for a specific crop over the previous five years (for wheat or feed grains) or three years (for cotton or rice), plus land not planted because of certain acreage reduction or diversion programs. Commodity acreage bases were eliminated by the FAIR Act of 1996.

Acreage conservation reserve. The cropland acreage diverted from production under the acreage reduction program.

Acreage diversion program. Historically, commodity programs included provisions to reduce commodity supplies by diverting acreage to non-crop uses. Examples include paid diversion, unpaid diversion, set-aside, and acreage reduction programs. The FAIR Act of 1996 eliminated authority for the USDA to implement annual acreage reduction programs. The Conservation Reserve Program pays farmers for the long-term conversion of fragile crop land to conserving uses and is not considered to be an acreage diversion program.

Acreage limitation. With respect to commodity policy, acreage limitation might refer to planting constraints under an acreage reduction program, set-aside, or paid land diversion. In relation to water policy, it is the maximum number of acres that may be irrigated with less than full-cost water from Bureau of Reclamation projects. Generally, the acreage limitation for individuals or legal entities representing 25 people or fewer is 960 acres; however, amounts vary depending on a landowner's legal status. Also referred to as ownership limitation, ownership entitlement, or non-full-cost entitlement.

Acreage Reduction Program (ARP). A no longer authorized annual cropland retirement program for wheat, feed grains, cotton, or rice in which farmers participating in the commodity programs (in order to be eligible for nonrecourse loans and deficiency payments) were mandated to idle a crop-specific, nationally-set portion of their base acreage during years of surplus. The idled acreage (called the acreage conservation reserve) was devoted to a conserving use. The goal was to reduce supplies, thereby raising market prices. Additionally, idled acres did not earn deficiency payments, thus reducing commodity program costs. ARP was criticized for diminishing the U.S. competitive position in export markets. The FAIR Act of 1996 did not reauthorize authority for ARPs. ARP differed from a set-aside program in that under a set-aside program reductions were based upon current year plantings, and did not require farmers to reduce their plantings of a specific crop.

Acre-foot. The volume of water that would cover one acre of land (43,560 square feet) to a depth of one foot, equivalent to 325,851 gallons of water. An acre-foot is the basic measure of agricultural water use. On average, irrigators apply almost 2 feet of water on each acre through the crop growing season; the amount ranges from 4 feet in the Southwest to a half foot in some eastern states. Water withdrawn for *irrigation* from ground and surface sources totals about 150 maf (million acre-feet) of water annually.

Acre-inch. A unit of measurement used in Irrigation works. The amount of water required to cover an acre to a depth of 1 in (25.4 mm per hectare) equivalent to 101 tons or 22460 gallons.

Acropetal. From a physiological base to a physiological apex

Actinomycin D. An antibiotic that is able to inhibit the elongation of RNA chains.

Action levels. As opposed to tolerances (which are established for pesticide residues occurring as a direct result of proper usage), action levels are set for inadvertent residues resulting from previous legal use or accidental contamination. At the action level set by the Environmental Protection Agency, Food and Drug Administration and USDA are required to take enforcement action against the contaminated food or agricultural commodity. The term is also used in regulatory programs.

Activated sludge. Refers to the thick or thin watery suspension of sewage and associated protozoans, micro-metazoans and bacte-ria in certain sewage treatment processes. When the suspension is constantly stirred and air is blown through it, the organisms convert much of the organic fraction into gases, salts, and inert organic compounds.

Active immunity. Refers to the specific immunity which is afforded by the body's own immunological defence mechanism follow-ing exposure to antigen.

Active ingredient. In any pesticide product, the component that kills, or otherwise controls, target pests. Pesticides are regulated by the Environmental Protection Agency primarily on the basis of active ingredients.

Actual Production History (APH). A measure of an individual farmer's annual production of a commodity over a multi-year period. The APH serves as the basis for the farmer's "normal" crop yield in the crop insurance program. When the actual crop yield deviates by more than a certain percentage from the APH, an insured producer is eligible for an indemnity (loss) payment.

Actuarially sound. The financial goal of any insurance program (including the federal crop insurance program) is to operate on an actuarially sound basis; that is, total premiums collected should more than offset total indemnities paid out.

Acuminate. (1) Coming to a point with concave, curved sides. (2) Having a sharp, gradually tapered point. (3) Tapering gradually to a sharp point, as the tips of certain leaves.

Acute. (1) The tip or base of a leaf blade whose margins come to a broad point. (2) With sides forming an angle of less than 90°; shape pointed.

Acute exposure. Term used for radiation exposure of short duration.

Acute toxicity. The ability of a substance to cause harmful effects soon after a single exposure or dose. Also, any sever poisonous effect resulting from a singe short-term exposure to a toxic substance.

Adaptability. The capacity of an organism or species to become suited to its environment

Adaptation. The fitness of a population in a given environment

Adaptive enzyme. An enzyme which is able to synthesize only in the presence of an inducing agent, such as its substrate.

Adaptive trial. A trial conducted to assess the adaptability of a technology or a variety to a particular region

Additive (Fertilizer). A substance which is added to fertilizer to improve its chemical or physical condition.

Additive (Silage). Any material which is added to forage at the time of ensling.

Additives. (1) Substances added to Feedingstuffs or Concentrates during manufacture, which are not principal nutrient sources. Examples include Vitamins, minerals, Trace Elements, certain Antibiotics, etc. (2) Substances added to pesticides to improve their efficiency and efficacy. They include Surfactants (wetters). Adjuvant Oils and drift retardants.

Adiantum. Genus of *ferns* of the family Polypodiaceae. There are about 200 species, usually found in damp tropical woods in the Americas. Some species are used medicinally; others are cultivated as conservatory plants. They include the *maidenhair* fern *A. capillus-veneris.*

Adjuvant oils. Vegetable-based and mineral oils which are used to improve the effectiveness and reliability of pesticide sprays. The oils reduce evaporative loss from droplets in flight and help maintain the chemical in liquid phase for extended periods on leaves, aiding adsorption.

Adjuvant. A material which is added to an insecticide to aid its action. Adjuvant may act in various ways like emulsifying agents, wetting agents, spreaders and stickers.

Adpressed. Lying close to another organ, but not fused to it.

Adsorption complex. The group of substances in the soil which are capable of absorbing other materials (such as applied fertiliz-ers). Organic and inorganic colloidal substances form the greater part of the adsorption complex. Non-colloidal materials, such as sand and silt, exhibit adsorption to a much lesser extent than the colloidal materials.

Adult phase. the stage of plants capable of producing fruit and flowers.

Adulterated food. Generally, impure, unsafe, or unwholesome; however, the Federal Food, Drug, and Cosmetic Act, the Federal Meat Inspection Act, and the Poultry Products Inspection Act contain separate language defining in very specific (and lengthy) terms how the term "adulterated" will be applied to the foods each of these laws regulates. Products that are adulterated under these laws' definitions cannot enter into commerce for human food use.

Adulteration. The adding of water to milk.

Advance deficiency payments. Initial payments (ranging from 30 to 50% of the total payment) made to crop producers when they signed up for federal commodity programs. If the total deficiency payment was eventually calculated to be less than the advance deficiency payment, the producer was required to refund the difference. The FAIR Act of 1996 replaces the target price/deficiency payment subsidy mechanism with production flexibility contract payments.

Advance recourse loans. Price-support loans made early in a crop year to farmers to enable them to hold their crops for later sale, usually within the marketing year. Farmers must repay the recourse loan with interest and reclaim their collateral.

Adventious bud. An extra or special bud not found in the axil of a leaf.

Adventitious embryony (Nucellar embryony). Type of apomixis where embryos arise from a cell or group of cells either in the nucellus (usually) or in the integuments in addition to the regular embryo

Adventitious. (1) Arising in an abnormal position. Adventitious roots develop from parts of plants other than roots, *e.g.*, from stem or leaf cuttings. Adventitious buds develop from parts of plants other than in the axils of leaves, *e.g.*, from roots. (2) Not in the usual place (buds that have remained undeveloped so that they are no longer evidently axillary, or that really originate elsewhere, as on a root). (3) Produced in an unusual or abnormal position, or at an unusual time of development or away from the natural habitat

Adze. A tool consisting of a thin arched blade with its edge at right-angles to the handle like a hoe, used for slicing off the surface of a piece of wood, *e.g.*, in farming work.

Aeciospore. A yellow, single-celled, binucleate spore of the rust fungus which is formed in a special cluster or cup-like structure called the accium.

Aecium. A cup like structure which is produced on certain plants by rust fungi, in which binucleate single-celled spores are found.

Aegillaceous. Clayey. Argillaceous soils have a high proportion of clay particles. Mudstones and marls are examples of argil-laceous rocks.

Aeolian soil. Wind deposited soil which is easy to keep in good tilth, as coastal sand dune, of not much agricultural value. It occues in parts of Rajasthan, South-east Punjab and Kutch.

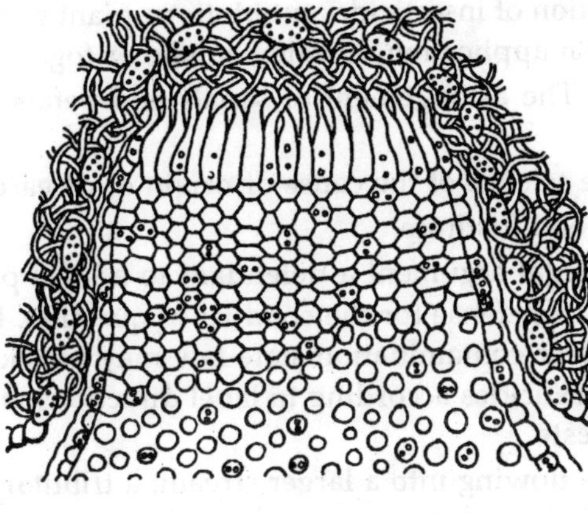

Fig.: Aecium

Aerate. Loosening or puncturing the soil to increase water penetration.

Aeration. (1) To increase the amount of air space in the soil by tilling or otherwise loosening the soil. Aeration is the process of increasing the soil's air content. An ideal soil will contain 50% air spaces, of which 25% will be air. A core aerifier will pull 1-2 inch plugs of soil from the ground. Plugs are deposited on the soil's surface where they will break down. Soil surrounding the plugged holes and the soil deposited on top will collapse and fill in. Turf soil should be aerated at least once a year on heavy compacted or clay soils. At a minimum recommended rate, turf should be aerated at least once every three to five years. DO NOT USE ANY AERIFYING EQUIPMENT THAT SIMPLY PUNCHES A HOLE IN THE SOIL. While a hole may be created, the "punched" soil merely compacts in the surrounding areas. Golf or football cleats, or shoes/sandals with nails pounded through the soles may create holes but only compact the surrounding areas. (2) Air is needed for composting bacteria to thoroughly break down plant material. Aeration can be accomplished by regular turnings of the compost pile, preferably weekly. A pitch fork or spading fork can be used to turn a pile. Plastic perforated pipes can be laid throughout the pile to increase air movement. If the temperature of a pile drops, the compost needs either moisture or aeration. (3) Supply of oxygen to the roots of plants and micro-organisms in the soil

Aerenchyma. Tissue of thin walled cells with large, air filled intercellular spaces, found in roots and stems of some aquatic and marsh plants

Aerial photograph. Refers to the photograph of the earth's surface which is taken from airborne equipment. It is sometimes called aerial photo or air photograph.

Aerial roots. Roots produced above ground, often used for climbing.

Aerobe. Any organism glowing in the presence of oxygen or air.

Aerobic respiration. Refers to enzymatic destruction of a substrate to release energy, by using elemental oxygen, evolving carbon dioxide and water.

Aeroceric acid. C26H44O3. Refers to a constituent of humus.

Aeroponics. A system of growing plants where the roots are suspended and bathed in a nutrient mist rather than a nutrient solution as in hydroponics

Aerosol. Type of formulation of insecticide, antibiotic or plant growth substances dispersed as very fine droplets on application, forming smoke or fog.

Aestivation (Estivation). The arrangement of sepals and petals in the flower bud before expansion.

Aetiology (Etiology). The science of the cause or origin of diseases.

AFFI. American Frozen Food Institute.

Affirmative defense. A legal argument a defendant to a legal preceding can raise which provides a protection or defense for the conduct in question. In actions for the costs of cleaning up groundwater contamination state law may provide an affirmative defense for farmers who use pesticides according to label directions and apply fertilizers based on the results of soil tests.

Affluent. Stream which is flowing into a larger stream; a tributary.

Afforestation. (1) Artificial establishment of a forest where tree crops are not previously grown or on deforested land. (2) The process of transforming an area into forest, usually when trees have not previously been grown there. (Deforestation, Reforestation)

After crop. A second crop from the same land in the same year.

After ripening. The term used for a period through which some seeds have to pass after ripening, before they will germinate.

After-cultivation. Harrowing, rolling and other cultivations carried out in a field after the crop has emerged.

Aftermath. Grass springs up again after the cutting of a crop of Hay, and can subsequently be taken as a second cut. Also known as eadish, eddish, rowen and lattermath.

After-ripening. The physiological changes that occur in a dormant seed to enable germination.

Agamospermy. Refers to the reproduction in which meiosis and fertilization get circumvented, so that the embryo (s) which develops in the seed has been usually indentical with its mater-nal parent in both chromosome number and genie content. It includes all types of apomictic reproduction in which embryos and seeds have been formed by asexual means. Agar. A complex sulphated galactan which is used as a base for several solid and semi-solid media used in microbiological stud-ies. It is formed by certain species of red algae.

Agar. A gelatinous substance obtained from certain species of red algae, widely used as a solidifying agent in aseptic culture.

Agaric. Any of a group of fungi of typical mushroom shape. Agarics include the field mushroom *Agaricus campestris* and the cultivated edible mushroom *A. brunnesiens*. Closely related is the often poisonous *Amanita*, which includes the fly agaric *A. muscaria*. (Genus *Agaricus*, family Agaricaceae.)

Agglomerated feeds. (1) Refer to the compacted or extruded form of individual ingredients. (2) Also, refers to a mixture of non-processed and/orindividually processed ingredients or a combi-nation of both.

Agglutination. The term used for a process in which cells or bacteria undergo cross linking with or gets attached to each other when the antigens on the surface of their cells undergo intersection with antibodies, which form bridges linking the antigen determi-nant sites of the different cells. Such reactions are used for identifying blood groups, bacteria etc.

Aggregate flower. A single flower heaped or crowded into a dense flower cluster.

Aggregate fruit. A fruit developing from a flower having several pistils that ripen together in a single mass

Aggregate Measure of Support (AMS). An indicator of the amount of domestic support for agriculture. As used in the Uruguay Round Agreement on Agriculture, the AMS refers to a measure of the gap between domestic and world prices multiplied by the quantity supported, plus any other commodity-specific transfers. Internal or domestic support reduction commitments in the Uruguay Round Agreement on Agriculture are expressed in terms of reductions in a total AMS covering all trade-distorting internal support measures for agriculture. An index that measures the monetary value of the extent of government support to a sector. The AMS, as defined in the Agreement on Agriculture, includes both budgetary outlays as well as revenue transfers from consumers to producers as a result of policies that distort market prices.

Aggregated. Joined together, confluent, as applied to bundle traces.

Agistment. A contract arising when one person, the agister, takes livestock belonging to another person to graze on his land for reward.

Agribusiness. Agriculturally related businesses that supply farm inputs (such as fertilizer or equipment) or are involved in the marketing of farm products (such as warehouses, processors, wholesalers, transporters, and retailers). Farms are not usually included when the term agribusiness is used. Producers and manufacturers of agricultural goods and services, such as fertilizer and farm equipment makers, food and fiber processors, wholesalers, transporters, and retail food and fiber outlets.

Agricultural chemicals. Used generally to refer to both pesticides and fertilizers and in some situations may include animal drugs.

Agricultural chemistry. Refers to that branch of the science of chemistry which deals with the composition and transformation of the plants and animals on which the economy of the farmer depends.

Agricultural Conservation Program (ACP). Administered by the Farm Service Agency, this largest and oldest conservation cost-sharing program paid farmers up to $3,500 per year as an incentive to install approved conservation practices. It was terminated in the FAIR Act of 1996 and replaced by a new Environmental Quality Incentives Program (EQIP).

Agricultural district. A planning term which defines an area in which farming is the preferred economic activity. Districts may be voluntarily created by landowners who receive benefits, usually in return for not developing the land for a certain number of years, or they may be designated in a local land use plan.

Agricultural diversification. A system of farming that encourages production of a variety of plants and animals and their products as opposed to monoculture or large-scale specialization. Advocates of diversification argue that it provides greater income stability. Specialized farms benefit from economies of size.

Agricultural economics. Branch of economics which deals with farm management and production.

Fig.: Agricultural Engineering

Agricultural engineering. (1) Deals with the application of knowl-edge, techniques and disciplines of various fields of engineering to the solution of problems which arise in the field of agriculture and rural living with the aim of reducing labour, improving agricultural productivity per worker, raising the standard of living of farmers and increasing the overall earnings per worker (2) Deals with the study in the design, construction and use of agricultural implements and buildings, soil and water manage-ment, use of electricity and processing of agricultural products.

Agricultural geology. A branch of applied geology which is con-cerned with the nature and properties of rocks and minerals of importance of soil formation.

Agricultural Marketing Service (AMS). A USDA agency that establishes standards for grades of cotton, tobacco, meat, dairy products, eggs, fruits, and vegetables. It also operates inspection and grading services and market news services, and provides supervisory administration for federal marketing orders.

Agricultural marketing. Includes all the services or functions which are involved in bringing produce from a fanner to the final consumer.

Agricultural pollution. Wastes, emissions, and discharges arising from farming activities. Causes include runoff and leaching of pesticides and fertilizers; pesticide drift and volatilization; erosion and dust from cultivation; and improper disposal of animal manure and carcasses. Some agricultural pollution is point source, e.g., large feedlots, which require permits under the Clean Water Act, but much is nonpoint source, meaning that it derives from dispersed origins, e.g., blowing dust or nutrients leaching from fields. As most pollution control programs have focused on particular categories of point sources, nonpoint and unregulated point sources account for an increasingly large proportion of remaining pollution. Based on state surveys, the Environmental Protection Agency concludes that agricultural sources account for over one-half the pollution impairing surface water quality in the U.S. The Clean Water Act mandates that states develop and implement management programs to control nonpoint sources of water pollution.

Agricultural science service. A division of the agricultural development and advisory service providing specialist advice to farmers on a range of scientific disciplines both directly and via field officers.

Agricultural Stabilization and Conservation Service (ASCS). A former USDA agency that was responsible for administering farm price and income support programs as well as some conservation and forestry cost-sharing program. Local offices are maintained in nearly all farming counties. Its functions were folded into the Farm Service Agency (FSA) as a consequence of USDA reorganization in 1994.

Agricultural zoning. A designation intended to protect farmland and farming activities from incompatible nonfarm uses. Agricultural zoning can specify many factors, such as the uses allowed, minimum farm size, the number of nonfarm dwellings allowed, or the size of a buffer separating farm and nonfarm properties.

Agriculture in Concert with the Environment (ACE). An Environmental Protection Agency program, administered cooperatively with USDA's Sustainable Agriculture Research and Education (SARE) program, to fund research projects that reduce the risk of pollution from pesticides and soluble fertilizers.

Agriculture. (1) Science of cultivating soil so as to produce economic crops and demanding great knowledge and skill in its scientific, commercial and artistic aspects. Very broadly the term is appli-cable to include both pastorcil and arable farming, even though the trend is some what towards the latter. (2) An activity of man aimed at the production of food, fibre, fuel etc., by the use of terrestrial resources. (3) The art, science and industry of managing the growth of plants and animals for human use.

Agrimony. Herbaceous plant belonging to the rose family, with small yellow flowers on a slender spike. It grows along hedges and in fields. (*Agrimonia eupatoria*, family Rosaceae.)

Agri-silviculture. Use of land for the concurrent production of agricultural and forest crops.

Agrobiology. Phase of the study of agronomy which deals with the relation of yield to the quantity of an added or available fertilizer element.

Agro-climatic regions. Broadly homogenous zones with respect to climate and edaphic factors.

Agroecology. Refers to the study of the relation of agricultural crops and environment.

Agro-ecology. The study of the relation of agricultural crops and environment

Agro-forestry. Self-sustaining land management system combining production of agricultural crops with that of tree crops with or without livestock.

Agrology. Refers to the study of applied phase of soil science and soil management.

Agronomy. The science of crop production and soil management. The branch of agriculture concerned with the theory and practice of field-crop production and the scientific management of soil.

Agrostology. (1) Deals with the study of grasses, their classification, management and utilization. (2) The science which deals with the study of grasses, their classification, management and utilization.

Ailanthus. Any of several trees or shrubs with compound leaves made up of pointed leaflets and clusters of small greenish flowers with an unpleasant smell. The tree of heaven (*Ailanthus altissima*), native to East Asia, is grown worldwide as an ornamental; it can grow to 30 m/100 ft in height and the trunk can reach 1 m/3 ft in diameter. (Genus *Ailanthus*, family Simaroubaceae.)

Ailantois. A membranous sac-like appendage for effecting oxygenation in the embryos of mammals, birds and reptiles.

Air blast sprayer. Sprayer having a large fan to produce a high speed, high volume air flow to break spray particles into small droplets and to carry the spray to the plant.

Air drainage. (1) Refers to the flow of cold air downhill. Freeze-sensitive crops are grown on hillsides so that on calm spring nights the cold air will drain down and away from the crop. (2) Air outlets and convection currents which prevent dead air and frost pockets.

Air dry soil. Soil losing its free moisture by evaporation or transpi-ration, and therefore having only hygroscopic moisture and chemically combined water.

Air layering. (1) Air layering is a process of propagating various houseplants involving the creation of roots on the plant's stem. Air layering is also used to reduce the height of some plants. Plants commonly air layered include rubber plants, dracaenas, scheffleras and figs. An upward cut or slit is made in the plant's trunk or stem, approximately 1/3 through. A toothpick is inserted to keep the cut open. Care must be made to prevent cutting the plant all the way through. The cut area is wrapped with moist peat moss or sphagnum moss, and then covered with a plastic wrap. Make sure the wrap covers all the moss; moss allowed to remain uncovered will dry and act like a wick, drying out the remaining moss. Secure the plastic wrap with tape or twist ties. Roots should form in a matter of weeks or months, depending on the type of plant. Maintain proper plant growing conditions. Once roots form, cut the plant below the root formation area and repot. (2) A specialized method of plant propagation accomplished by cutting into the bark of the plant to induce new roots to form; a process of producing a new plant by forming soil around a stem. (3) It is a method (asexual) of plant propagation by layering wherein roots are formed on the aerial part of a plant after the stem gets injured or girdled or slit at an angle. This portion is then kept in a moist rooting medium. It is commonly used for propagating a number of tropical and subtropical trees and shrubs. (4) Layering in above ground branches by removing a ring of bark and placing the rooting medium around.

Air plants. Plants not attached to the ground.

Air pollution. Contamination of the atmosphere by substances that, directly or indirectly, adversely affect human health or welfare. Air pollution results from human activities, both deliberate releases (as from smokestacks) and fugitive emissions (as dust blown from streets or fields), and from natural sources, including sea spray, volcanic emissions, pollen, etc. The Clean Air Act authorizes the Environmental Protection Agency to regulate air pollution (see National Ambient Air Quality Standards).

Air root-grafting. The grafting of roots on to a branch while it is still attached to the tree.

Air screen cleaner. Basic seed processing machine in which basic cleaning of all seeds is done by air screen cleaner.

Air space ratio. Refers to the ratio of volume of water which can be drained from a saturated toil under the action of force of gravity to total volume of voids.

Air void ratio. Refers to the ratio of the volume of air space to the total volume of voids in a soil mass.

Air-inarching. Technique of grafting a branch to the trunk of a tree.

Aitionomic. (1) Able to develop parthenocarpic fruits only in response to some stimulus external to the ovary. (2) The term used for the ability to develop parthenocarpic fruits only in response to some stimulus external to the ovary.

Akene. Dry, one seeded, small, indehiscent fruit of one carpel, seed like in appearance.

Alar. Trade name for daminozide, a plant regulator and therefore classed as a pesticide, that makes apples redder, firmer, and less likely to drop off trees before harvest. It was also used to a lesser extent on peanuts, tart cherries, concord grapes, and other fruits. Alar was suspended by the Environmental Protection Agency in 1989 following a controversy over allegations of cancer risk to children from residues of Alar and its breakdown product UDMH on apples and in apple products.

Alba/albus/album. White or whitish, usually applied to flowers

Albedo. Refers to the ratio of the amount of radiation reflected to the amount coming on a surface. The earth's mean albedo has been rearly 35 per cent.

Albido. The white tissue beneath the peel of citrus fruit.

Albinism. Means the condition in which the normal pigment is not present, such as in the skin, hair and eyes. In plants also lack of normal chlorophyll causes albinism.

Albino. An organism lacking normal pigmentation due to genetic factors. The condition is albinism

Albo-maculata. Spotted in white.

Albo-marginata. White edged or margined.

Albo-striata. striped or striated in white. Usually a variegated plant.

Albumens. A group of water-soluble Proteins which occur in many of the tissues and body fluids of animals; *e.g.*, egg albumen (egg white); serum albumen (in blood), lactalbumen (in milk), etc. Also called albumins.

Albuminoids. The simple proteins which are insoluble in all neutral solvents and in dilute acids and alkalies but are coagulated by heat. For example keratin, collagen etc.

Albuminous seed. Seed having endosperm at the time of germina-tion, which nourishes the growing seedling.

Albumins. A group of water-soluble Proteins which occur in many of the tissues and body fluids of animals; *e.g.*, egg albumen (egg white); serum albumen (in blood), lactalbumen (in milk), etc. Also called albumins.

Alchemilla. Genus in the family Rosaceae of about 200 perennial plants of northern temperate regions and the mountains of tropical Africa. *A. vulgaris,* with broadly lobed leaves, and yellowish-green flowers in loose clusters, is lady's mantle; *A. alpina,* Alpine lady's mantle; these and other species are found in Britain.

Alcohol. The family name of a group of organic chemical compounds that includes methanol, ethanol, isopropyl alcohol, and others. Ethanol is produced from crops or residues with a high carbohydrate content. Alcoholic beverages contain ethanol, and ethanol is blended with gasoline to produce gasohol. Most industrial ethanol produced in the U.S. is from corn wet-milling.

Aldrin. A persistent organochlorine insecticide, a derivative of chlo-rinated nupthalene, harmful to fish and under restricted use. Applied as a dust, spray or seed dressing.

Aleurone grains. Protein bodies which are present in certain oilseeds. They consist of ground substance having one or more of the crystalloids or protein crystals, calcium oxalate crystals, globoids believed to be phosphates with organic salts.

Aleurone layer. Layer of aleurone grains which is presents in the periphery of most seeds.

Aleuroplast. A plastid storing protein, often in the form of grains.

Alfalfa gate. In irrigation, it refers to a light sheet metal slide gate in a section of pipe for the control of water from sublaterals into fields and ditches.

Alfalfa meal. A feedingstuff comprising artificially dried and ground Lucerne, known in America as alfalfa.

Alfalfa weevil. A small, dark brown weevil that is problem in alfalfa.

Alfalfa. A valuable leguminous crop for forage or hay used in livestock feeding.

Algae singular alga. Highly varied group of plants, ranging from single-celled forms to large and complex seaweeds. They live in both fresh and salt water, and in damp soil. Algae do not have true roots, stems, or leaves. Marine algae help combat global warming by removing carbon dioxide from the atmosphere during photosynthesis. Algae were formerly included within the division Thallophyta, together with fungi and bacteria. Their classification changed with increased awareness of the important differences existing between the algae and Thallophyta, and also between the groups of algae themselves; many botanists now place each algal group in a separate class of its own. They can be classified into 12 classes, largely to be distinguished by their colour, including the green algae Chlorophyta, freshwater or terrestrial; stoneworts Charophyta; golden-brown algae Chrysophyta; brown algae Phaeophyta, mainly marine and including the kelps Laminariales and allies, the largest of all algae; red algae Rhodophyta, mainly marine and often living parasitically or as epiphytes on other algae; diatoms Bacillariophyta; yellow-green algae Xanthophyta, mostly freshwater and terrestrial; and blue-green algae Cyanophyta, of simple cell structure and without sexual reproduction, mostly freshwater or terrestrial. The life cycles of algae vary widely; some reproduce solely by vegetative means, others have a prolonged haploid phase, and a very restricted diploid phase. However, some advanced algae have their haploid generation reduced almost as much as it is in seed plants.

Algae. (1) Algae are microscopic plants of various colors — blue, green yellow, blue green, black, brown, red and orange. Most grow in the upper three inches of the soil and add a significant amount of organic matter to the soil. Algae are NOT the same as moss. As a group, algae seldom cause any problems with soil growing conditions or plants. (2) Simple, flowerless, unicellular or multicellular photosynthetic plants with unicellular organs of reproduction. They are aquatic plants, freshwater or marine, (*e.g.* seaweeds), or plants of damp situations, (*e.g.* damp walls, tree trunks, in soil, etc.)

Algicides. Chemical agents which are used for killing the algae (selectively).

Alismataceae. Family of about 13 genera and 80 species of plants, usually marsh plants or floating-leaved aquatics. They are monocotyledons with distinct sepals and petals, both three in number. The carpels are separate, and often numerous. The main British genera are *Alisma* (water plantain) and *Sagittaria* (arrowhead). The family is like Ranunculaceae in some ways, possibly indicating a common evolutionary origin of monocotyledons and dicotyledons.

Alkali flat. The term used for on alkaline, marshy area in an arid region in which one or more direct streams lead. It becomes a barren area of hard mud which gets covered with alkali when all the water is evaporated in the dry season, after heavy rainfall, it becomes a shallow muddy lake.

Alkali. A soil contains sufficient exchangeable sodium to interfere with the growth of most crops plants, either with or without appreciable quantities of soluble salts.

Alkaline soil. (1) A soil with a pH higher than 7.0 is an alkaline soil. pH is a measure of the amount of lime (calcium) contained in your soil. (2) Sweet Soil. Soil having a pH above 7.0 due to an excess of hydroxyl (OH-) ions, or having a high exchangeable sodium content, or both. Typical alkaline soils are those of the chalk and limestone hills, e.g., South Downs, Cotswolds and Craven Pennines. (Acid Soil, Neutral Soil).

Alkaline. A soil for which the pH reading of the saturated soil paste is higher than 7.

Alkaloids. (1) Group of nitrogen containing basic organic compounds which possess poisonous and medicinal properties e.g. atropine, morphine quinine and strychnine (2) Heterogenous group of basic nitrogen having substances that are produced by plants and possess potent pharmacological activities. They have been classified according to the type of heterocyclic group present.

Allelopathic. Plant that produces chemicals affecting other nearby plants' growth. Usually used to indicate a negative effect, inhibiting the growth of nearby plants. Black Walnut (*Juglans nigra*) trees inhibit the growth of many other plants.

Allelopathy. Any direct/indirect harmful effect that one plant has on another or mutually on each other through the production of chemical compounds that escape into the environment.

Alley. A narrow passage left across rows to facilitate harvesting and cultural operations.

Alleyways. Passage-ways between rows of hop plants in hop gardens.

Allochronic speculation. Refers to the production of new species during the passage to time, tending to form a gradation from one species to the next.

Allogamy. (1) Mode of reproduction in which a flower gets pollinated by pollen from another flower. (2) Cross fertilization in plants.

Allograft (Homograft). A graft between like species.

Allopatric Hybridisation. Refers to the hybridisation between in-completely differentiated species in a border zone, because of a premature break-down of a geographical barrier.

Allopatric. Refers to two or more related populations which are occupying mutually exclusive (but usually adjacent) geographi-cal areas and do not interbreed; this situation is often termed as contiguous allopatry; where two related populations get sepa-rated by a wide gap in which neither occurs, the situation is known as disjunct.

Allopolyploid. A polyploid organism to which two different species have each contributed one or more sets of chromosomes (e.g., probably cultivated wheats).

Allopolyploidy. Polyploidy involving the combination of chromosomes from two or more species.

Allotetraploid. An allopolyploid arising when an ordinary Hybrid between two species which contains a chromosome set from each parent, doubles its chromosome number.

Ordinary hybrids are usually sterile since their chromosomes cannot pair during meiosis. In an allotetraploid each chromosome can pair with its homologue, thus overcoming sterility, and immediately creat-ing a new species. Tobacco probably originated in this way. Allotetraploidy is only known in plants at present. Artificial production of allotetraploids is important in creating new agri-culture and horticultural varieties. Also called amphidiploid.

Allotment. In conjunction with commodity support programs, acreage allotments and marketing quotas serve to limit a farm's output or volume marketed. For federal lands grazing, an allotment is an area designated and managed for grazing of livestock. The Bureau of Land Management and the Forest Service stipulate the number of livestock and time period (season) of use for each allotment under their respective jurisdictions.

Allowable Sale Quantity (ASQ). A forestry term defined in law as the maximum amount of timber that can be sold every year, forever, from a national forest; in forest planning, the annual timber sale target for a national forest.

Allspice. Spice prepared from the dried berries of the evergreen pimento tree, also known as the West Indian pepper tree, (*Pimenta dioica*) of the myrtle family, cultivated chiefly in Jamaica. It has an aroma similar to that of a mixture of cinnamon, cloves, and nutmeg.

Alluvial deposit. Soil, sill, sand, gravel, or other material which has been deposited in place by the action of running or receding water.

Alluvial fan. The outspread sloping deposit of detril material which has been brought down by the action of water from neigh-bouring elevations upon a plain or open valley bottom and dropped where the stream gradient decreases abruptly, forming roughly a segment of a low cone with the apex at the point of debouchure.

Alluvial land. Areas of unconsolidated alluvium which are generally stratified and varying widely in texture, recently deposited by streams, and subject to frequent flooding. A miscellaneous land type.

Alluvial soil. Soil which is formed by the transportation and deposi-tion, by streams, of material which is carried to a considerable distance before getting deposited.

Alluvial. The term used for the material that gets transported and deposited by the running water.

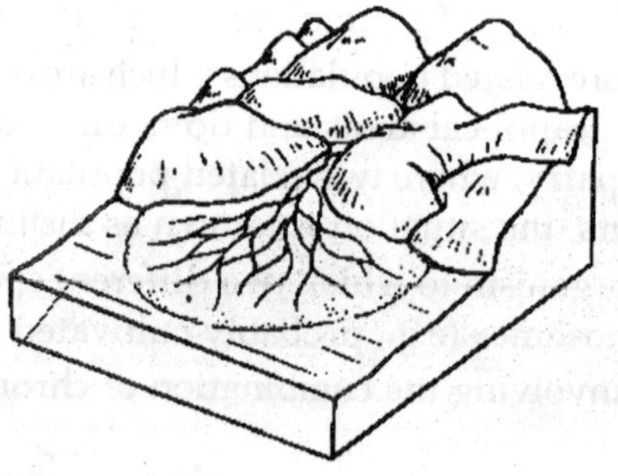

Fig.: Alluvial land

Alluvium. Material such as clay, silt and sand deposited by rivers and streams. (Colluvium).

Aloes wood or eaglewood. The heartwood of *Aquilaria ovata* and *Aquilaria agallocha*, found in tropical Asia. The innermost wood of these trees yields a fragrant resinous substance which is used in the East as a perfume and as a medicine.

Alpha amylase. An enzyme naturally occurring in wheat seed at germination, causing starch breakdown to sugar. If unduly active in milling wheat it reduces the quality of bread and cakes. Tested for by either the Hagberg Test or the Farrand Test.

Alphid. A plant house, also called greenfly. Aphids are small, soft-skinned, often green plant-bugs. They suck sap mainly from young leaves and shoots, causing reduced growth and vigour, and leave a sweet, sticky excretion (honeydew) which is eagerly eaten by ants and which occludes the leaf and shoot pores. Often found in large numbers on wild and cultivated plants. Many are pests, *e.g.*, Bean Aphis.

Alpina/alpinus/alpinum. Epithet meaning from an alpine area or just a very dwarf plant. It is also used to denote smaller species in a large genus.

Alpine garden. A garden comprised of plant species normally found growing at high elevations.

Alpine plant. (1) Tolerate the short growing season, cold, and wind of higher mountain elevations. They are typically low growing, small-leaf perennials. Snow cover depth often dictates the plant's growing height. (2) A plant suitable for rock gardens or one that normally grows in exposed mountain regions

Alpine. (1) A breed of goat native to the Swiss Alps, from which the British Alpine breed developed. (2) Any plant that is normally found in wild on mountains or in mountainous areas.

Alsike Clover. A species of clover, *Trifolium hybridum*, named after Alsike near Uppsala in Sweden, characterised by pale pink or white flowers, a smooth fairly upright stem, toothed leaflets with long pointed stipules, and a pod containing 1-3 small seeds. It tolerates adverse soil conditions better than Red Clover or White Clover, and is less susceptible to stem rot.

Alternate. The arrangement of leaves or buds one at a time along a stem at the nodes.

Alternate bearing (Biennial bearing). (1) The term used for the habit of bearing heavy and sparse crops in alternate years, inherent in mango and some other species of fruit trees, where terminals bearing fruit in an year put forth only leafy shoots next year. (2) The phenomenon of producing heavy bearing during 'on year' which is alternated by shy bearing during 'off year.':

Altitude (Elevation). Vertical distance of a level, a point or an object from mean sea level (MSL). *Ambient* (1) The environment of a particular time. (2) Set of climatic conditions existing during an experiment.

Alternate. (1) An arrangement of leaves, buds, and other organs borne singly at different heights on either side of the stem or axis. (2) Said of leaf scars or leaves occurring one at each node.

Alternative agriculture. A systematic approach to farming intended to reduce agricultural pollution, enhance sustainability, and improve efficiency and profitability. Overall, alternative agriculture emphasizes management practices that take advantage of natural processes (such as nutrient cycles, nitrogen fixation, and pest-predator relationships), improve the match between cropping patterns and agronomic practices on the one hand and the productive potential and physical characteristics of the land on the other, and make selective use of commercial fertilizer and pesticides to ensure production efficiency and conservation of soil, water, energy, and biological resources. Examples of alternative agricultural practices include use of crop rotation, animal and green manures, soil and water conserving tillage systems, such as no-till planting methods, integrated pest management, and use of genetically improved crops and animals. Consonant with sustainable agriculture, alternative agriculture focuses on those farming practices that go beyond traditional or conventional agriculture, though it does not exclude conventional practices that are consistent with the overall system.

Alternative farming. Production methods other than energy- and chemical-intensive one-crop (monoculture) farming. Alternatives include using animal and green manure rather than chemical fertilizers, integrated pest management instead of chemical pesticides, reduced tillage, crop rotation (especially with legumes to add nitrogen), alternative crops, or diversification of the farm enterprise.

Alternative fuels. Substitutes for traditional liquid, oil-derived motor vehicle fuels like gasoline and diesel. Includes methanol, ethanol, biodiesel, compressed natural gas, and others. The alternatives are promoted for pollution reduction properties and/or to reduce U.S. dependence on imported oil. Ethanol can be produced from grain, agricultural wastes, and excess crops.

Aluminium phosphide. A chemical which acts as a funmigant for grain. Available commercially as tablets; 4-5 days exposure allowed.

Alyssum. Genus of 80 species of plants, in the family Cruciferae, of which the Mediterranean perennial *A. saxatile* and its varieties are chiefly grown in gardens.

Amaryllis. Genus of the family Amaryllidaceae. The belladonna lily *A. belladonna* is a garden plant, native to southern Africa; it winters as a bulb and bears elegant pink flowers. The *Hippeastrum* species are sometimes loosely known as amaryllis.

Amel Corn. An inferior variety of wheat *(Triticum vulgare dicoc-cum)*, the larger spelt. Also called French Rice.

Amelioration. The process of enhancing the agricultural value of land by drainage, tillage, liming, manuring, etc.

Amendment. Any material like lime, gypsum, saw dust, sulphur or soil conditioner which is added to the soil to influence plant growth by improving the physical and biological properties of the soil.

Ammonia (NH_3). A pungent colourless gas used in the manufacture of fertilizers (e.g., Ammonium Nitrate, Sulphate of Ammonia). Anhydrous ammonia is a very concentrated nitrogenous fertil-izer (82% nitrogen) consisting of pure ammonia in liquid form under pressure and applied by injection into the soil, usually when moist, where it vaporises and combines with soil colloids. Aqueous ammonia, a by-product from gas works called gas liquor, comprising ammonia in solution, used to be commonly used as a fertilizer, applied to the soil surface.

Ammoniated mercury. A white poisonous powder which is soluble in acids. It finds use as antiseptic and parasiticide, especially in the form of ointment having about five per cent ammoniated mercury.

Ammoniated superphosphate. A fertilizer having 5 parts of ammo-nia to 100 parts of superphosphate.

Ammoniflcation. Means the decomposition of nitrogenous organic matter by micro-organisms, with the release of ammonia.

Ammonium fixation. The adsorption or absorption of ammonium ions by the mineral or organic fractions of the soil so that the ions become relatively insoluble in water and unexchangeable by cation exchange.

Ammonium nitrate. (NH_4NO_3). The commonest nitrogen fertilizer currently in use either as a straight fertilizer or in compound fertilizers. Now available as a pure salt, safe for agricultural use, but prior to 1965 it was only available as an unstable synthetic salt, liable to caking, and was usually blended with chalk or ground limestone, (see nitro-chalk).

Ammonium phosphate sulphate. A product obtained by treating a mixture of phosphoric acid and sulphuric acid with ammonia. Consists principally of a mixture of ammonium phosphate and ammonium sulphate. Used as fertilizer.

Ammonium phosphates. Mono and di-ammonium phosphates are occasionally used in the pure state as fertilizers but, more commonly, they are constituents of concentrated compound fertilizers. Both the phosphate and the nitrogen are soluble in water (see fertilizers, artificial fertilizers).

Ammonium sulphate. A white crystalline salt having about 20 to 21 per cent nitrogen, widely used as fertilizer. The nitrogen is not easily lost in drainage, because the ammonium ions replace calcium in the exchange complex and thus gets retained by soil particles. It is suitable for wetland crops like paddy and jute, also for other crops grown on variety of soils. It has acidic effect on the soil.

Amnion. A protective membranous sac containing amniotic fluid cushioning the developing embryo of an animal.

Amniota. Any reptile, bird or mammal, their embryos possess an amnion and an allantois.

Ampelopsis. Genus of the Vitaceae (grape) family. It is a deciduous climber with twining tendrils found in the Americas and Asia. Species include the pepper vine *Ampelopsis arborea* found in the USA.

Amphidiploid. An allopolyploid which is obtained by hybridization between two plant species or genera where the chromosome set contributed by each parent undergoes doubling and produces in the hybrid a chromosome number which is the sum of diploid numbers of the two parent forms.

Amphimixis. Normal sexual reproduction in which the morphologically dissimilar male and female gametes unite together for the formation of zygote.

Amphitheatre. The etymology is from amphi (both, or both sides + theatron (theatre). It means a circular theatre with seating on both sides. Garden amphitheatres are made with landform, planting or stonework.

AMS. Aggregate measure of support; Agricultural Marketing Service.

Amylopectin. A polysaccharide which is made up of branched chains having a large number of glucose units linked in the 1-4, 1-6 position. A constituent of most starches.

Amyloplast. Plastid capable of storing starch, usually as grains. It occurs in the endosperm of storage organs, tubers, roots and seeds

Amylopsin. An enzyme found in pancreatic juice which breaks down starch into sugars.

An dromon oecious. Condition in which both male and hermaphrodite flowers are borne on the same plant.

Anaerobes. Organisms capable of living in absence of free oxygen (gaseous or dissolved).

Anaerobic conditions. Environmental conditions in which free oxygen (gaseous or dissolved) is absent.

Analgesia. The term used for the absence of sensitivity to pain. This can be produced by medication given for the relief to pain and can also happen in some physical and emotional disorders.

Analogues. Organs of different plants with like function but of unlike origin.

Analytic seed sample. Seed sample which is obtained by dividing the mechanical seed sample in which the analysis for different tests is done conveniently.

Anaphylaxis. Refers to a state of increased sensitivity or hypersensitivity in an animal which follows the injection of an antigen as a protein.

Anaplasma marginal. Minute parasite which is responsize for anaplasmosis in the animals

Anatomy. The internal structure of plant.

Anchor root. A large root that holds a plant in a physical soil mass.

Androdioedous. Condition in which male and hermaphrodite flowers are borne on separate plants

Androecium. (1) Stamens of a flower, as a unit. (2) The collective term for stamens of a flower.

Androgenesis. Plant development from male gametophytes.

Anemophilous plants. Plants which are pollinated by the wind.

Anemophily (Anemophilous). Pollination by wind.

Aneurism. Means a spindle shaped, globular or cylindrical dilation of a blood vessel. It often has a heavy deposit of fibrin causing an interference with the circulation of the blood. Immature forms of one of the palisade worms is known to cause aneurism in horses.

Angel crops. Crops raised without chemical fertilisers, plant protection chemicals and growth stimulants

Angelica. (1) Angelica is a perennial herb sometimes called archangel, masterwort and ground ash. It is a member of the parsley family and grows in fields and damp locations. Plants are believed to have originated in Syria. Stems, leafstalks, seeds and roots are all used. The oil in the seeds is often distilled and used for food flavoring, medicines, and the flavoring behind the liquor "Galliano." The Lapps chew the roots as a tobacco substitute. Plants are 5 to 6 feet tall and resemble wild carrot or Queen Anne's Lace. The root is purplish. Plants thrive in a moderately cool location, with partial shade. Plants are propagated from seeds or old root division and spaced 18 inches between plants in 3 foot rows. (2) Any of a group of tall, perennial herbs with divided leaves and clusters of white or greenish flowers, belonging to the carrot family. Most are found in Europe and Asia. The roots and fruits have long been used in cooking and in medicine. (Genus *Angelica*, family Umbelliferae.) *Angelica archangelica* is a culinary herb, the stems of which are preserved in sugar and used for cake decoration. *A. sylvestris*, the species found in Britain, has wedge-shaped leaves and clusters of white, pale violet, or pinkish flowers. Angelica oil is used in perfume and liqueurs.

Angiosperm. (1) The higher flowering plants in which the flowers are more complicated and the seeds get enclosed in the fruits. They are of two types, (a) Dicotyledous-embryo of seed bearing two cotyledens (b) monocotyledous-embryo of seed bears only one cotyledon. (2) A member of the group of flowering vascular plants whose seeds are enclosed within a mature ovary (fruit) in contrast to the seeds of gymnosperms which are not enclosed in an ovary.

Angiospermy. Refers to the condition in which seeds are enclosed within an ovary—a characteristic of flowering plants (angjosperms) that distinguishes them from all other plants.

Anhydrase. An enzyme which is capable of accelerating the removal of water from a substance e.g., carbonic anhydrase removes water from carbonic acid.

Animalize. The chemical treatment of vegetable fibres, such as cotton or flax, so that the fibres will resemble animal fibres, like wool, in their behaviour toward wool dyes.

Anions. Anions are negatively charged ions such as sulfate SO_4- and phosphate PO_4-

Annual ring. (1) A distinct layer of growth as seen in the cross section of a woody stem. (2) The annual growth of secondary wood (xylem) in stems and roots of temperate woody plants, seen as a series of concentric rings in a stem cross-section due to the distinction in size between xylem elements formed in spring and autumn. Annual rings are used to

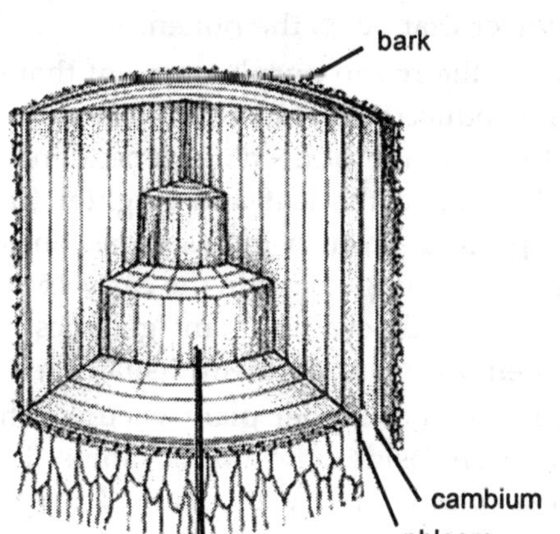

Fig.: Annual ring

estimate the approximate age of a tree. Growth is constant throughout the year in the tropics and no rings are seen. Annual rings may also be seen in cow horns.

Annual up timing (Crop timing). Adjusting planting and / or adopting cultural operations to facilitate harvesting of the crop at a specified time.

Annual. (1) A plant which completes its life cycle in one year, *i.e.*, grows from seed, flowers, fruits, and dies. (2) Any plant which completes its entire life cycles and dies within one year or less. (3) Plants whose life cycle lasts only one year, from seed to blooms to seed.

Annular. Ring-like; said of leaf scars that encircle the bud, or bundle scars that are circular with an opening in the center.

ANPR. Advance notice of proposed rulemaking.

Antacid or Antiacid. Medicine which is able to counteract acidity *e.g.*, sodium bicarbonate. Antiacids are used to relieve indiges-tion and other conditions due to hyper sensitivity. Some anti-acids possess laxative effect, when used too often.

Antagonistic medicine. Medicine which is able to counteract the action of another medicine.

Antecedent moisture. Refers to the degree of wetness of the soil at the beginning of a run off period frequently expressed as an index determined by summation of weighed daily rain-falls for a 10 to 20 day period preceding the runoff in question.

Antecedent soil water. Refers to the degree of wetness of the soil prior to irrigation or at the beginning of a runoff period. It is usually expressed as an index or as total inches of soil water.

Antemortem. Before slaughter. As used in the meat and poultry inspection program, the term refers to the examination that USDA meat inspectors are required to conduct of all live animals just before they are killed.

Anthemis. Genus of the Compositae (daisy) family, including mayweed *Anthemis cotula, A. tinctoria,* a source of yellow dye, and camomile *A. nobilis.*

Anther. (1) The pollen bearing part of a stamen, borne at the top of a filament or sessile. (2) The portion of the male flower that bears the pollen.

Anthocyanin. (1) Anthocyanin is the red to purple pigment that develops in leaves during the autumn. The pigment is produced after chlorophyll is destroyed due to environmental changes. Amounts of anthocyanin produced is dependent on stored sugar and fall sunlight. The more stored sugar in the leaf and brighter the autumn days, the more intense red. Anthocyanin production also increases as phosphates moves out of the leaves; the lack of phosphates in young leaves in the early spring due to cool weather can also lead to the production of anthocyanin. Plant genetics and heredity also have an effect on red colors. (2) Chemical plant pigment; usually red or blue

Anthocyans. Water soluble pigments of higher plant vacuoles, hav-ing the anthocyanidin flavonoid pigments and their glycosides, the anthocyanins. They are mainly responsible for the red, blue and purple colours of flowers, fruits, leaves, and buds.

Anthracnose. A fungus disease that causes spots and often death of foliage.

Anthrax bacterin. It tends to stimulate the treated animal to produce immune bodies i.e. active immunity. Its anthrax bacterin being sterile, is in itself incapable of producing disease in the treated animal, it is, therefore, safer than any of the anthraxspore vac-cines.

Anthrax vaccination. The term used for a preventive vaccine against a serious disease of livestock called anthrax.

Anthrax-Spore Vaccinations. These consist of living anthrax spores. They are used alone or in combination with anthrax serum. The spores in these vaccines become so weakened that they will not produce the disease in livestock under ordinary conditions, but some susceptible animals may die. For this reason these vac-cines should be used only on those premises where the disease has occurred earlier.

Antianthrax serum. It is responsible for increased resistance to anthrax in direct proportion to the quantity of serum which is given to the animal. It is of value both as a preventive and as a therapeutic agent. It should be used as a preventive only when immediate protection is needed because the immunity which it confers is of relatively short duration.

Anti-auxin. Substance which can inhibit growth by antagonising the activity of auxins.

Antibacterial agent. Agents which are able to kill or inhibit the growth of bacteria e.g., antiseptics, antibiotics and disinfectants.

Antibiotic feed supplement. A feeding material which is used chiefly for its antibiotic content *e.g.,* terramycin, auremycin, penicillin and steptomycin. Thus it is a single or a combination of antibiotics having growth-promoting properties. They give rise to a more rapid growth by altering the flora in the digestive tract of the animals.

Antibiotics. Chemical substances produced by microorganisms or synthetically that inhibit the growth of, or destroy, bacteria. Antibiotics are used at therapeutic levels to fight disease in humans and animals. Since the 1950s they have been used at subtherapeutic levels in animal feeds to enhance growth and prevent disease in livestock and poultry. Rules guiding the use of veterinary drugs and medicated animal feeds, including tolerance levels for drug residues in meats for human consumption, are promulgated by the Center for Veterinary Medicine of the Food and Drug Administration. The Food Safety and Inspection Service enforces the FDA rules through a sampling and testing program that is part of its overall meat and poultry inspection program. Antibacterial drugs obtained from living organisms, such as moulds, bacteria or other micro-organisms, which act by inhibiting the growth and multiplication of germs and other bacteria in the body. Examples are penicillin and the tetracylines. Antibiotics have no effect on illnesses caused by viruses.

Antiblackleg serum. A serum which is used for treating calves which are already affected with blackleg as well as for preventing it. It is produced by injecting a pure culture of blackleg organisms into the veins and later under the skin of horses and drawing the blood afterwards.

Anti-desiccants. Anti-desiccants, such as Wilt-Pruf, are used on broadleaved evergreens to reduce winter drying or desiccation. Most anti-desiccants contain a waxy formulation which prevents water loss from leaves. Winter thaws may breakdown the compounds, necessitating re-application. Avoid using anti-desiccants on the under side of leaves. Stomates may be blocked, limiting air exchange and killing leaves.

Antidotic. A chemical substance which is produced by microorgan-ism and has the capacity to inactivate the injurious effect of bacterial toxins.

Antigen. A proteinous substance which is capable of stimulating the production of neutralizing substance, when it is injected into a vertebrate.

Antigen-antibody reaction. Any reaction which takes place be-tween an antigen and its corresponding antibody it may involve agglutination, precipitation, complement-fixation etc., the sen-sitivity of allergic reactions of many individuals towards certain kinds of foods, pollen, dust, feathers, hair, drugs, etc., are all examples of antigen-antibody reactions.

Anti-infective. A remedy which is administered to overcome an infection in livestock *e.g.*, sulfona-mides, penicillin, etc.

Antimycin. An antibiotic functioning as a respiratory inhibitor in mitochondria, but generally not in bacteria.

Antioxidants. Substances which are antagonistic to oxidation process.

Antipruritic. An agent which is used for relieving itching.

Antipyretic. A remedy which is able to lower the body temperature.

Antirheumatic. Drug which is used to prevent or cure rhemuatism in animals *e.g.*, sodium salicylate.

Antiseptic dusting powder. Powder having boric acid, sodium perborate or iodoform induces healing of wounds and also prevents attacks of flies.

Antiseptics. These are chemical agents which are used to treat human and animal tissues, usually the skin, with the object of inactivating or killing those microorganisms which are causing an infection, *e.g.*, acriflavine bisphenols, ammonium compounds, etc.

Apetalous. Without petals. Ex. Grass flowers

Apex. (pl. apices) the top or tip of a structure, such as a leaf or stem.

Aphid(s). (1) Insect pests that look like small dots, enjoy most of the plants in your garden, suck their juices and spread viral diseases. Some gardeners get rid of them with insecticidal soap, some use organic methods such as beneficial predators and some of us just wring our hands. (2) A small green or white insect that sucks juices from plant parts. (3) Problem in almost all crops. Pierce leaves and suck out plants' juices.

Fig.: Apex

Aposematic coloration (warning coloration). The brilliant and striking patterns of colours which occur in insects especially those which are predator resistant, which results in a predator learning to avoid the species.

Appetitive behaviour. Any pattern of animal behaviour that results in the satisfaction of a specific need, *e.g.*, searching by an animal for food or a mate.

Apple blossom weevil. An insect *(Anthonomus pomorum)*, approxi-mately 1/2 cm long, laying its eggs in apple and pear flower buds, on which the larvae feed so that no fruit develops.

Apple sucker. A plant louse, *Psylla mali*, able to jump with its large hind legs. Its pale yellow, stump-winged offspring suck sap from opening apple flowers, causing fading and browning.

Applied insect control. Refers to the use of sprays, insecticides and other methods in insect control.

Appraised stumpage price (or appraised rate). On national forests, the Forest Service estimate of the market price for timber to be cut and removed. It cannot be less than the base rates. The appraised price is the advertised minimum for competitive bidding by purchasers.

Appressed. Flattened against the twig; not spreading.

Approach grafting. (1) A grafting method where rootstock is taken to the scion donor plant for grafting. Stock and scion are independent for their nourishment till the completion of the graft union (Synonym : Inarching). (2) Refers to a type of grafting wherein two inde-pendent, self-sustaining plants have been grafted together. The top of the stock plant has been removed above the graft and the base of the scion plant has been removed below the graft, after a union has taken place. It is a method of establishing a success-ful union between certain plants which are difficult to graft together otherwise. Approach grafting can be carried out at anytime of the year. The cut surface should be securely fastened together and covered with grafting wax.

Aquaculture. (1) The production of aquatic plants or animals in a controlled environment, such as ponds, raceways, tanks, or cages, for all or part of their life cycle. In the United States, baitfish, catfish, clams, crawfish, freshwater prawns, mussels, oysters, salmon, shrimp, tropical (or ornamental) fish, and trout account for most of the aquacultural growing production. Less widely established but growing species include alligator, hybrid striped bass, carp, eel, red fish, northern pike, sturgeon and tilapia. (2) The cultivation or rearing of aquatic plants or animals. (3) The technique of growing plants without using soil (hydroponics) either in circulated, aerated nutrient solutions or in inert media such as gravel, sand, peat, vermiculite or saw dust, to which a nutrient solution is added. (4) The raising of fish or other aquatic animals for their commercial value. (5) Hydroponic *horticulture*.

Aquatic plant. (1) Any plant that grows in or near water. (2) Plants that grow in, or live in or on water.

Aquifer. (1) A bed of rock which holds water and allows water to percolate through it. (Artesian Well). (2) A stratum of earth or permeable rock that stores significant quantities of water. (3) An underground geological formation, or group of formations, containing usable amounts of groundwater that can supply wells or springs for domestic, industrial, and *irrigation* uses. Removing more groundwater from an aquifer than is naturally replenished is called overdrafting, and can result in a dropping water table, increased pumping costs, land subsidence (which reduces the future recharge capacity), saltwater intrusion, reduced streamflows in interconnected ground- and surface-water systems, and exhaustion of groundwater reserves. Overdrafting groundwater occurs primarily in the Plains States and the West.

Aquifoliaceae. The *holly* family, which contains about 300 species of trees and shrubs with evergreen leathery leaves whose edges are often spiny. The flowers generally have their parts in fours. The male and female flowers are on separate plants, the females having red or yellow fruits, usually called berries, but technically drupes. *Ilex*, the holly, is the main genus.

Arabis. Genus of over 100 species of annual or perennial plants, of the cabbage family Cruciferae, chiefly of the northern hemisphere. *A. albida*, white flowering rock cress, and *A. caucasica* are popular rock garden plants.

Arable crops. Annual crops which require the land to be ploughed and cultivated (unless direct drilled) and seeds to be sown, as distinct from those which yield a product without such annual activities (*i.e.*, fruit, grass, etc.). The main arable crops grown in the India include the cereals, potatoes, rootcrops, and various vegetables.

Arable land. Land capable of being ploughed, that is, fit for tillage, as opposed to that only suited for pasture. The term generally means land ploughed at regular intervals.

Arable. Land which is suitable for the production of crops that need Tillage.

Arachnida. A class of Arthropods including spiders, ticks, mites, harvestmen and scorpions. They are mainly carnivorous, usu-ally with six pairs of limbs, four for walking and the first pair as jaws. They have no antennae, simple eyes and the head and thorax are fused together.

Arbor day. Arbor Day in Illinois is celebrated on the last Friday in April, per state statute of 1947. Technically, the day is "Arbor and Bird Day." Arbor Day is in honor of J. Sterling Morton, member of the Board of Agriculture in Nebraska, who founded the first Arbor Day program in 1872. Illinois' first Arbor Day was celebrated in 1887.

Arborcide. A type of herbicide chemical that kills trees.

Arboreal. Tree-like or pertaining to trees.

Arborescent. (1) Tree-like; defined arbitrarily as pertaining to a woody plant at least 20 feet high in maturity with a single stem and more or less definite crown. (2) Becoming tree like, nearly the size of a tree.

Arboretum. (1) Area set aside for growing and displaying different kinds of woody ornamental trees, shrubs and vines with proper labelling. (2) Organized collection of trees and other woody plants; a garden with a large collection of trees and shrubs cultivated for scientific or educational purposes.

Arboriculture. (1) Cultivation of trees and shrubs. (2) The science of the culture or care of trees. (3) The systematic culture and care of trees and shrubs. Are A unit of metric land measure, 100 sq. metres. (Hectare)

Arborist. A certified/licensed tree doctor specialist.

Archangel, yellow. Common name for the herbaceous perennial plant *Galeobdolon luteum*, a member of the Labiatae family with yellow flowers.

Archil or orchil. Violet dye which is obtained from several species of lichens, notably from *Roccella tinctoria*. As materials dyed solely with it fade readily in the sunlight, another dye is usually employed at first and archil is used to give added brilliance.

Arching. Curving gracefully.

Arctosis. These annuals and tender perennials are natives of South Africa and belong to the Daisy family, Compositae. They grow 1 to 2½ feet high, usually have fuzzy leaves and produce single-stemmed flowers, 2 inches across, in most colors except blue. The annual kinds are the most popular. When they are planted outside, they should be 12 inches apart. Flowering will begin in June and in favorable climates will continue into September, as long as the faded blooms are removed. The plants will fail if the nights are really hot. When they are grown in a greenhouse, care must be taken not to over water. They need plenty of sunlight and a night temperature between 45 and 50 degrees. Seeds can be sown in a greenhouse in the spring in fine, sandy, loamy soil. The temperature should be 50-55 degrees.

When the seedlings are an inch or so high, they're set 3 inches apart in flats filled with a porous soil. Water them well and place them in a cold frame until danger from frost has passed. They can then be planted outside. When they are used for decorating the greenhouse during late winter and spring, the seeds are sown in September and the new plants are potted separately in 3-inch pots and later in 5-inch pots and kept under glass.

Area yield options contract. A contract entitling the holder to receive a payment when the area yield is below (above) the put (call) option strike yield. The strike yield is the yield at which the holder of an option contract can exercise the option.

Arenaceous. Sandy, Arenaceous soil have a high proportion of sand particles. Sandstone is an arenaceous rock. (See Argillaceous)

Arenga. One of the palms that produce a type of sago and palm wine, found in all the islands of the Indian archipelago. The only species is *A. saccharifera*. The trunk is covered with coarse black fibres, which are made into rope. It yields a large amount of transparent sap, which soon becomes turbid and milky; when fit for drinking it is yellowish, astringent, and very intoxicating.

Areole. (1) A round or elongated often raised or depressed area on a cactus which is equivalent to a bud and from which spines, flowers, stems, or roots grow. (2) The area occupied by a group of spines or hairs on a cactus.

Argente. A breed of rabbit of French origin, with a silvery coat. There are several shade varieties, *e.g.*, Argente Blue, Argente Silver, etc.

Arid climate. A dry climate with an annual precipitation usually less than 10 inches. Not suitable for crop production without irrigation.

Arid fruit culture. The cultivation of edible fruits in arid region that are consumed either fresh or processed.

Arid. (1) Modified mesocarp covering the seed of nutmeg Also called mace. (2) An appendage or outer covering of a seed, growingfrom the hilum or funiculus. (3) A relatively dry climate in which annual precipitation is less than 10 inches, which generally is insufficient for crops to be grown without irrigation. Such areas usually are the focus of debate over federal water policies. (3) A term applied to land that is dry, parched and often devoid of vegetation. Also means deficient in rainfall-usually applied to a climate or region where rainfall is barely sufficient to support vegetation.

Aril. A fleshy appendage of the seed, usually a fleshy seedcoat.

Aristolochia. Genus mostly of shrubs, often climbing, in the family Aristolochiaceae; they are found in temperate and warm regions and have pungent aromatic roots and tubular flowers. *A. clematis*, birthwort, is cultivated as a medicinal plant, and is naturalized in Britain. *A. macrophylla* (synonym *sipho*), Dutchman's pipe, is a popular North American garden climber. This and related species have swollen tubular flowers with lurid purplish patches, and an odour which attracts flies to pollinate the plant.

Armed. With pines (barberry, Japanese quince) or prickles (brambles). Leaves that are pungent at the tip or around the margin (holly), are not included in this limited definition.

Armsby feeding standard. In this feed standard the feeding value has been expressed in net energy which is found out in heat unit. The net energy may be defined as the metabolizing energy minus the energy lost in the heat increment or so called work of digestion. It may also be known as the gross energy minus the energy lost in faeces, urine, combustible gases and in the work of digestion.

Aromatic plants. Plants which yield aromatic essential oils on steam distillation or solvent extraction e.g. lemongrass; patchouli.

Aromatic. (1) Chemical compound derived from the hydrocarbon, benzene (C6 H6), having a pungent odour. (2) Fragrantly scented, at least when broken or crushed.

Arrish. Stubble or a stubble field.

ARS. Agricultural Research Service.

Arsnic. A chemical element, steel-grey, brittle and crystalline. Found naturally in a number of minerals. Compounds are very poisonous and are used as weed-killers and pesticides. (Lead Arsenate)

Arterial drain. A watercourse which carries water draining from smaller ditches.

Arterial drainage. Drainage work on major rivers and water-courses which are under the control of Regional Water Authorities, Internal Drainage Boards or local Authorities. (Field Drainage)

Artesian well capacity. Refers to the take at which a well will yield water at the surface because of artesian pressure.

Artesian well. A well sunk into an aquifer sandwiched between impermeable strata in which the hydrostatic pressure normally forces a continuous flow of water up the well. The pressure is due to the well outlet being below the level of the water source. Normally occurs in basins (*e.g.*, the London Basin).

Arthritis. Refers to the inflammation of a joint. It results in an enlargement and impaired movement of the joint, causing ex-treme lameness, often accompanied by fever.

Arthropod. The Arthropoda from the most numerous phylum of the animal kingdom. Characterised by an external skeleton of hard-ened cuticle, divided into segments some of which bear jointed paired limbs (the word arthropod means jointed feet). Includes insects, spiders, millipedes, centipedes, crabs and shrimps, etc.

Artichoke. Either of two plants belonging to the sunflower family, parts of which are eaten as vegetables. The common or globe artichoke (Cynara scolymus) is a form of thistle native to the Mediterranean. It is tall, with purplish-blue flowers; the leaflike structures (bracts) around the unopened flower are eaten. The Jerusalem artichoke (Helianthus tuberosus), which has edible tubers, is a native of North America (its common name is a corruption of the Italian for sunflower, *girasole*). (Family Compositae.)

Articular membrane. A membrane consisting of the thin, enlarged base of the petiole, on which the leaf scar occurs.

Articular. A membrane consisting of the thin, enlarged base of the petiole, on which the leaf scar occurs.

Articulate. Having nodes or joints where separation may naturally occur.

Articulation. A joint in a stem or fruit, at which natural separation takes place.

Artificial brooding. Means handling of baby chicks without the aid of hen. It is carried out by using heated brooder. It has advan-tages over natural brooding. Brooding chicks are possible at any time of the year and in large numbers. It makes control of sanitary conditions, regulation of temperature and planned feed-ing possible.

Artificial long days. Interruption of dark period or extension of natural daylength to prevent flower bud initiation of short day plants.

Artificial parthenogenesis. Laboratory phenomena which involves the maturation of an ovum and its development into an individ-ual without involving fertilization; the essential stimuli are shaking, pricking, heating, and treating with dilute organic acids; the phenomenon can be initiated in frogs and many marine inverte-brates; in a very few instances animal produced by artificial parthenogenesis have been reared to adults.

Artificial regeneration. Renewal of forest crop by sowing, planting or other artificial means, also the crop so obtained.

Artificial seed. Encapsulated or coated somatic embryos used for planting like seeds. *Aseptic* Free of all micro-organisms

Artificial subsurface irrigation. The term used for the application of irrigation water below surface of soil having perforated or porous pipes which are laid underground below the root zone and water led into the pipes by suitable means. Method may make the soil saline or alkaline and the neighbouring land gets damaged due to seepage.

Artificial vagina. Arrangement for collecting semen particularly from bull. It consists of an outer cylinder of heavy rubber casing, glass, ebonite, or metal with a thin rubber inner-tube which gets attached to the outer cylinder in such a way that a watertight space gets created between the inner and outer walls. One end of the inner tube is kept open; the tube at the other end has been tapered to fit tightly over a graduated test tube in which the semen has been collected at the time of ejaculation. There is a small valve in the outer wall for introducing warm water into the space between the inner and outer surfaces. A temperature of 41°C has been found to be near optimum for most animal species. The inner tube is well lubricated.

Artificial vegetative reproduction. Refers to Vegetative Repro-duction.

Arugula. Arugula, also known as salad rocket, roquette, rocket salad, salad roquette, roquette salad and tira, is a leafy green, grown as an annual. Leaves are compound and have a spicy, peppery flavor. Arugula is a cool season vegetable best grown during the spring and fall. Plants mature within two to three months from seeds. Warm temperatures cause plants to bolt. Grow arugula as you would leaf lettuce.

Ascending. (1) Curving indirectly or obliquely upward. (2) Said of a bud that is between spreading and appressed. Curving indirectly or obliquely upward.

Ascomycetes. A class of fungi, known as Sac Fungi, which carry spores in a sac or ascus. They range in size from very small to quite large, the fruiting bodies of the latter often being rounded or tuberous, not toadstool-shaped. Many are parasites; many cause plant diseases; others 'ripen' cheeses.

Ascon. A type of sponge having a simple canal system in which openings in the body wall led directly to spongocoel, as in *Leucosolenia*.

Ascorbic acid. Vitamin C. An essential growth factor that is found in fruits (especially citrus) and vegetables.

Ascospore. (1) Spore which is produced because of sexual fusion subsequent to reduction division in Ascomycetes (2) Spore which is formed in an ascus and characteristic of Ascomycetes.

ASCS. Agriculture Stabilization and Conservation Service, predecessor to the Farm Service Agency (FSA).

Aseptic canning. A technique in which food is sterilised outside the can and then aseptically placed in previously sterilised cans which are subsequently sealed in an aseptic environment

Aseptic. Means free of microogranism which may cause contamination or infection.

Asexual propagation. (1) The multiplication of plants using a vegetative plant part. (2) Plant propagation not involving union of gametes. The propagation using parthenogenetic seed (as in mangosteen) is asexual. (3) Multiplying plants without use of seeds.

Asexual reproduction. Reproduction without Gametes. In plants it consists of either spore-formation or vegetative reproduction.

Ash. (1) Constituent of milk comprising various salt in solution, of which calcium and potassium phosphates and sodium chloride predominate, and also various vitamins (A, B, C, D and E). In normal milk the ash proportion remains fairly constant at about 0.75% (Milk Composition). A significant reduction may indi-cate added water. (2) A term used in the analysis of feeding stuffs for the residue left after through burning and the destruction of organic matter. Otherwise called mineral matter.

Aspartic acid. Refers to an amino acid which is having a chemical formula $HOOC.CH_2CH_2CH(NH_2)COOH$. It serves as a precur-sor of pyrimidines, which is needed for the synthesis of nucleo-tide, and is an intermediate in the urea cycle.

Aspergillaceae. Family of the Aspergillales *e.g.*, Aspergillus, Penicilium.

Aspergillales. An order of the Euascomycetes which are having a close ascocarp with the asci irregularly distributed in it.

Aspergillosis. A fungal disease of chicks resulting from inhalation of spores of *Aspergillus*

from damp or mouldy litter or feeding- stuffs, causing the growth of white nodules in the lungs, and resulting in breathing difficulties.

Asphodel. Either of two related Old World plants of the lily family. The white asphodel or king's spear (*Asphodelus albus*) is found in Italy and Greece, sometimes covering large areas, and providing grazing for sheep. The other asphodel is the yellow asphodel (*Asphodeline lutea*). (Genera *Asphodelus* and *Asphodeline,* family Liliaceae.) The ancient Greeks connected the beautiful plants of *A. lutea* with the dead, and they were supposed to grow in the mythological Elysian fields where heroes enjoyed new life after death.

Asphyxia. Suffocation or the suspension of animation because of suffocation.

Aspirator. It is seed processing machine in which seed separations are carried out by using air on the basis of differential terminal velocity of seed. A negative pressure or vacuum is produced within the machine by the fan aligned at the discharge point. Airrushing to fill in the vaccum produces a stream of air which is used to separate seeds,

Asporogenous yeast. Yeast which fails to produce ascospores. Ass. A small, usually grey, long-eared animal of the horse genus. Also known as donkey. Used as a draught animal and crossed with the horse to produce a mule.

Assart. To reclaim for agriculture by grubbing. Assarted land is that cleared of trees and bushes.

Assay. Means qualitative or quantitative determination of the components of a material, as an ore or a drug.

Assertive mating. Refers to the non-random sexual reproduction in a population; tendency of certain types of males to breed with certain types of females, resulting in parents that are mentally and physically more similar than would be expected by chance.

Assessment. Generally an automatic or mandatory deduction from a producer's marketing receipts used to fund activities that promote or otherwise support a particular farm product. Under certain agricultural marketing orders or commodity promotion programs, assessments may be applied against receipts to help pay for generic advertising or research. The term check-off is often used interchangeably with assessment. Federal deficit reduction marketing assessments have also been connected to certain commodity price support programs (dairy, peanuts, sugar, tobacco, and soybeans) to help reduce the federal budget deficit, which arguably is higher because of the programs.

Assimilate. Substance which is formed in plant during the manufacture of food.

Assimilation. Refers to the absorption and utilization of simple food-substances.

Assimilative capacity. The ability of a body of water to cleanse itself; its capacity to receive waste waters or toxic materials without deleterious effects and without damage to aquatic life or humans who consume the water.

Assimilatory roots. These are long, slender hanging roots which are green in colour. These roots are capable of carbon assimilation and manufacture carbohydrates *e.g.*, submerged roots of water chestnut.

Association. A major plant community in which more than one plant species is dominant, as in mixed deciduous woodland. Also applied in modern usage to any small unit of natural vegetation (Soil Association).

Aster. Any plant of a large group belonging to the same subfamily as the daisy. All asters have starlike flowers with yellow centres and outer rays (not petals) varying from blue and purple to white. Asters come in many sizes. Many are cultivated as garden flowers, including the Michaelmas daisy (*Aster nova-belgii*). (Genus *Aster*, family Compositae.) The China aster (*Callistephus chinensis*) belongs to a closely related genus; it was introduced to Europe and the USA from China in the early 18th century.

Asthenoblosis. Refers to the theoretical explanation of the phenomenon of diapause or arrested growth of an insect. It is assumed that there occurs an accumulation of waste products in the tissues and that these restrict normal development in much the same way as they may bring about muscular fatigue. During a resting stage these substances gradually disappear because excretion is continued while normal metabolism becomes slows.

Astingent lotions. Liquids having one or more astringent drugs. They often find use in the treatment of lameness in animals.

Astringency. A puckering taste to the mouth caused by tannins

Astringent. Drug which is able to contract tissues. It is locally applied to control bleeding relieve inflammation, etc.

Astroturf. A synthetic lawn, popularly used in developed countries in roof gardens, as well as in stadium (mainly in football and hockey).

Asynapsis. Refers to the failure of chromosomes to pair at meiosis or the absence of chiasmata formation.

At-foot. Suckling. Thus a ewe with lambs at foot has suckling lambs.

Athermanous. Substances which are able to absorb the radiant heat falling on them and, therefore, disallow it to pass through them are known as at hermanous substances e.g., water, wood, liquids, etc.

Athyrium. Genus of *ferns* in the family Aspleniaceae, of about 180 species, including the lady fern *A. filix-femina*, a native of damp places in northern temperate regions of Europe, Asia, and North America. It grows to a height of about 1 m/39 in, and is much cultivated as an ornamental.

Atmometer. An instrument which is used for measuring evaporation; also called an atmidometer or evaporimeter.

Atro. Botanical name prefix meaning dark or intensely colored.

Atrophic rhinitis. Nasal inflammation in young piglets with rapid uneven growth of the snout, causing deformity.

Atrophy. Dwarfing or reduction in size, often extreme, through disease affecting the whole plant or part.

Atrosanguinea. Dark or blood red or maroon.

Atrovirens. Dark or blackish green.

Attainment area. An area considered to have air quality as good as or better than the National Ambient Air Quality Standards as defined in the Clean Air Act. An area may be an attainment area for one pollutant and a non-attainment area for others.

Attar of roses. Perfume derived from the essential oil of roses (usually damask roses), obtained by crushing and distilling the petals of the flowers.

Attenuate. Showing a long gradual slender taper; usually applied to apices, but equally appropriate for bases of leaves, petals, etc.

Attenuation weakening. Refers to the reduction in virulence. Any procedure in which the pathogenicity of a given organism gets reduced or abolished.

Attractant(s). A chemical or agent that lures insects or other pests by stimulating their sense of smell. Attractants are a nontoxic technique for luring insects into traps and are heavily used in orchard crops. Though distinct from toxic baits, attractants are regulated as pesticides. Refers to the materials that are able to attract insects or other animals and make them to eat or contact poison bait or sprays, and consequently cause their death.

Auger. (1) A tool which is used for boring holes. A soil auger is used to bore into the soil and withdraw a small sample for observa-tion. (2) A mechanism which is involved for conveying granular materials (mainly grain) or slurry, comprising a power-driven helical screw in a closefitting tube.

Aureomycin. An antibiotic which is obtained from *Streptomycin aureofaciens.* It has been passed to be effective against both gram-negative and gram-positive micro-organisms, and certain viruses. It is administered intravenously and orally. It is also used in antibiotic feed supplement,

Aurescent. Becoming yellow or yellow part of the season.

Autecology. Refers to the ecology of an individual organism or texonomic group, as opposed to synecology which is the ecology of a community.

Autecology. The study of the interaction of a plant species with its environment.

Autoclave. An apparatus which uses stream under pressure for sterilization. The materials get sterilized by air-free saturated steam at temperature in excess of 100°C.

Autoecism. Refers to the capability of a parasitic fungus so as to complete its entire life cycle on the same host.

Autogamy. Self fertilisation or the fertilisation of a flower by its own pollen.

Autogenous vaccine. Vaccine prepared with germs which obtained from the patient's own blood and employed in the treatment of his disease.

Autograft. A type of graft which involves transplantation of tissue or an organ from one part of an individual to another part of the same individual.

Autoinfection. Means reinfection from bost's own parasites.

Autointoxication. Means self-poisoning which occurs due to the absorption of uneliminated toxins produced within the body, *e.g.*, from waste products.

Autolytic. Self-digestive. It refers to the process of self-digestion, which takes place in vegetable and animal tissues, particularly after they have ceased to be a normal part of the organism to which they belong; thus the tenderizing of beef during hanging is an example of autolytic process.

Automatic-planter. It is a simple bullockdrawn automatic seed-cum-fertilizer planter which is useful for metering correct quan-tity of seed and for placing fertilizer at proper depth. It is possible to attach four such planters to tool bar of tractor. It is also equipped with a pair of maker arms for aligning the rows.

Autonarcosis. Said of the state of being poisoned, rendered dorment, or arrested in growth, because of self-produced carbon dioxide.

Autophytes. Green plants which are able to prepare their own organic food from the raw inorganic materials absorbed from soil and the air.

Autopoiyploid. A polyploid organism in which all the chromosomes come from the same species. (Allopolyploid)

Autopolyploidy. Polyploidy with similar sets of chromosomes.

Autotroph. (1) A self sufficient organism that synthesizes its own food from inorganic molecules. (2) Any microorganism using only inorganic materials as a source of energy. Environment carbon dioxide forms the main or rather the sole source of carbon.

Autotrophic cells. Cells able to synthesize macromolecules from simple nutrient molecules like carbon dioxide, ammonia and water.

Autotrophic plants. The plants getting their foods exclusively from inorganic materials occurring in soil, air and water.

Autotrophic. Organisms able to manufacture their own food from inorganic materials using energy from an outside source. Most green plants are completely autotrophic, manufacturing organic foods from carbon dioxide, water, and mineral salts, using sunlight energy fixed by Chlorophyll. The process is known as photosynthesis. Some bacteria are autotrophic, using energy obtained by oxidising inorganic substances. Nitrifying Bacteria in the soil are examples.

Autumn crocus. Any of a group of late-flowering plants belonging to the lily family. The mauve meadow saffron (*Colchicum autumnale*) yields colchicine, which is used in treating gout and in plant breeding. (Genus *Colchicum*, family Liliaceae.) Colchicine causes plants to double the numbers of their chromosomes, forming polyploids)

Auxanometer. An instrument used to measure growth of plants.

Auxin. (1) A chemical that stimulates plant growth. (2) Plant growth regulator that functions more by stimulation of cell elongation than through an effect on cell division

Auxins. A group of plant growth-regulating substances or hormones, which are produced by growing tips of stems are roots. Artificial or synthetic auxins are also produced which are important in agriculture and horticulture, *e.g.*, inhibiting sprouting of potato tubers and thus lengthening the storage period, and preventing fruit drop in orchards. Some synthetic auxins exhibit differential toxicity, 2,4-D is toxic to dicotyledons, but not to monocotyledons, and is used successfully to control weeds in cereal crops and lawns.

Auxotroph. Strain of a microorganism which needs some nutrient not needed by the original strain from which it was derived. An auxotroph may be produced by the occurrence of one or more mutations in a prototroph.

Available nutrient. (1) That quantity of a nutrient element or compound in the soil that can be readily absorbed and assimilated by growing plants. (2) Chemical elements or compounds in the soil that can be readily absorbed and assimilated by growing plants ('available' should not be confused with 'exchangeable').

Available phosphoric acid. Refers to the sum of the water-soluble phosphoric acid (monocalcium phosphate) and citrate-soluble phosphoric acid (dicalcium phosphate) in the soil.

Available water supply. Refers to that amount of water which may be expressed either in terms of a rate of flow or as a volumetric quantity, and exists in a source of water supply like a stream, or reservoir, over and above the quantity necessary to supply valid and prior rights and demands.

Available water. The portion of water in the soil that can be readily absorbed by plant roots. Most soil scientists consider it to be the water held in the soil against a pressure of up to about 15 bars.

Average infiltration rate. Obtained by dividing the mass infiltration that takes place during the time rainfall intensity exceeds the infiltration capacity, by the time during which infiltration takes place at capacity rate.

Avidin. Protein material which can undergo combination with biotine making the vitamin to be unavailable to the body. Cooking renders avidin inactive.

Avitaminosis. Refers to vitamin deficiency.

Awn. (1) Refers to a small bristle at the tip of a flower or fruit, especially in the grasses, or of a leaf, the wan gets attached to the husk and encloses the grains. (2) A slender bristle at the end or on the back or edge of an organ. In grasses, the awn is usually a continuation of the mid nerve of the glumes or lemmas, rarely of the palea. (3) A bristle-like appendage.

Awner. The part of a threshing machine which cuts off the awns, consisting of knives rotating on a fast-moving spindle.

Axenic culture. (1) A culture without foreign or undesired life forms. (2) Refers to the growing of an organism of a single species of strain *e.g.*, bacterium in a medium free of other living organisms *i.e.*, a pure culture.

Axil. (1) Refers to the angle between the upper side of a leaf and the stem on which it is borne; the normal position for lateral buds. (2) Latin: the angle between a branch or leaf and the axis from which it originates. (3) The angle above a leaf; the point above a leaf scar. (4) The location on a stem between the upper surface of a leaf or leafstalk and the stem from which it is growing. (5) The angle between a petal and a stem; where buds occur above a leaf; the point above a leaf scar.

Axilary buds. (1) Buds that form in leaf axils. (2) A bud origination in an axil or point where a leaf stalk or branch forms an angle with the main stem. (3) Bud greating in the axil of a leaf.

Axillary. Growing from an axil. Thus an axillary bud grows from the axil between stem and leaf.

Axis. The main stem or central support of a plant.

Axle-tree. The beam on a cart or wagon connecting two opposite wheels.

Azalea. (1) Any of a group of deciduous flowering shrubs belonging to the heath family. Several species are native to Asia and North America, and many cultivated varieties have been derived from these. Azaleas are closely related to the mostly evergreen *rhododendrons.* (Genus *Rhododendron,* family Ericaceae.) Azaleas, particularly the Japanese varieties, make fine ornamental shrubs. Several species are highly poisonous. (2) Azaleas are found in subgenera Pentanthera and Tsutsusi of genus Rhododendron. These subgenera are commonly referred to as deciduous and evergreen, respectively. Azalea leaves are typically thin, flexible while leaves elsewhere in the genus tend to be leathery. Deciduous azaleas are so called because they drop their leaves in the fall or winter. Evergreen

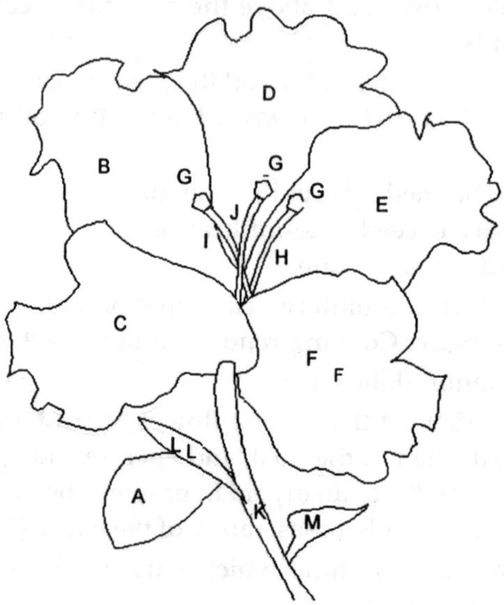

Fig.: Azalea

azaleas have two kinds of leaves: spring leaves emerge at or after time of flowering; summer leaves (usually smaller and somewhat thicker) emerge midsummer and remain on the plant through the winter. Another important difference is that the terminal buds of deciduous azaleas produce only flowers; those of evergreen azaleas produce new vegetative shoots as well as flowers.

Azoic. Means without life applicable to precambrian era or rocks.

Azonal soils. Refers to broad soil order in genetic classification of soils. It includes soils that have poorly developed profiles. Examples of azonal soils include sand dunes, loess, recent deposits of alluvium, and partially weathered bedrock.

Azotobacter. *Azotobacter chroocoaum.* It is a non-symbiotic nitro-gen fixing bacterium occurring in the soil. It is utilizing soluble carbohydrates; strongly aerobic, susceptible to phosphate deficiency.

B

B.H.C. Benzene hexachloride, a persistent organochlorine insecti-cide used mainly in sprays and dusts. Particularly used to control wireworms and other soil pests.

Babiana (baboon-root). (1) Dwaif half-hardy bulbs with brilliantly-coloured flowers of great beauty and pale-green Lairy" leaves. (2) This group consists of about 61, tender, deciduous plants that grow from corms. These natives of South Africa are commonly known as Baboon Flower and Baboon Root because baboons love to eat their corms. Baboon Flowers produce fans of short, slender, pleated, fuzzy leaves and spikes of colorful, fragrant flowers in late spring and early summer. Each corm produces numerous flower stalks and each stalk bears 10 or more flowers in white or shades of yellow, blue, and mauve. B. stricta is the most popular species and its lovely flowers may last as long as a month in the garden. Baboon Flowers can be grown in containers and placed in a greenhouse or a cool, sunny window. Baboon Flowers are hardy from zones 8 to 10. In areas where winter temperatures fall below freezing, these plants can be planted in a protected area with a southern exposure and mulched with a thick layer of leaves, salt hay, or similar material. They can be grown this way as far north as middle Atlantic States. In areas with wet summers, dig up the corms and store them in a warm, ventilated place until fall. The corms should be placed in a sunny or very lightly shaded location at the onset of cool, autumn weather. Plant them in well-drained soil consisting of two parts of loam and one part of leaf mold, with a small addition of well-decayed manure and sand. Plant them 5 to 9 inches deep and 2 inches apart. You can also plant up to 10 to 12 corms in a 6-inch pot. Place them deep in the container. Divide only when the plants become crowded. Water these plants regularly throughout bloom. Once the flowers have died and the leaves begin to fade, less water is applied; when the leaves have withered, the soil is kept dry until the fall.

Baccate. Pulpy, fleshy.

Bach (seed). The term used for the quantity of harvested crop kept into a container or a bin on a repetitive basis, specifically for treatment such as drying.

Bacillary white diarrhoea. An often fatal bacterial disease of young chicks caused by *Salmonella pullorum.* which causes serious losses between 4 and 14 days of age. It is inherited from parent stock which themselves have survived an outbreak. Carrier parents can be detected by a simple blood test. Known as B.W.D. for short. Also called pullorum disease.

Bacillus. A genus of spore-producing bacteria. Also a general term for rod shaped bacteria.

Bacitracin. An antibiotic which has been found to be effective against bacterial infections.

Back band. A strap, rope or chain passing over a cart saddle, and holding up the shafts of the cart.

Back board. A board at the back of a cart.

Back chain. A chain fastened across the back of a draught horse to the ridger rods on the shafts of a cart or wagon. Also called a ridger.

Back cross. A cross of a hybrid with one of its parents with the purpose of transferring a specific gene

Back end. The later part of a season; the late autumn.

Back fire. In forestry, a fire which is started intentionally ahead of and advancing fire for removing inflammable material by con-trolled burning and thus stop or control the main fire.

Back furrow. Refers to a raised ridge which is left at centre of the strip of land when ploughing is started from the centare to side.

Back hoe. A shovel mounted on the rear of a tractor, hydraulically operated to dig trenches or pits in soil.

Backcross breeding. Refers to the breeding method in which the backcrossing is done for several generations followed by subse-quent selection until a desired combination of traits from the recurrent parent and one or two desired traits from the non-recurrent parent enters the progeny.

Backcross. (1) A hybrid of two plants crossed once again back to one parent. (2) In plant breeding it refers to a cross of a hybrid *i.e.* F_1 with one of its parents. It is carried out to transfer a specific gene from an undesirable variety to another commercially desirable one lacking in that particular character. In genetics, it refers to a cross of hybrid with a homozygous, recessive parent and the aim is to test the gametic ratio of F_1.

Backwood. Remote, uncleared forest land.

Bacteria denitrificans, Denitrifying bacteria that are living side by side with the azatobacter and clostridium. They reduce nitrogen compounds in the soil and liberate nitrogen in the free state; they thrive under anaerobic conditions and may be active in soils as well as lower compact layers of manure pits.

Bacteria. Minute, single-celled organisms which cannot be seen by the naked eye. Most are rod-shaped from *c.* 0.0005-0.005 mm long (1/50 000-1/5000 in). They are commonly called 'genus' or microbes' and inhabit soil, water, air and other living organ-isms in millions. Some are responsible for diseases, *e.g.* Brucel-losis in cattle. Others are essential for maintaining soil fertility and aid in the decomposition or Organic Matter. Amongst the most important are those capable of Nitrogen Fixation in the soil and root nodules of certain plants.

Bacterial ooze. Discharge, exudate of sap or gum etc. from a plant because of bacterial disease.

Bacterial ring rot. An important disease of potatoes especially in Canada and U.S.A. caused by *Corynebacterium sepedonicum.* Symptoms vary—usually the first sign is wilting of the lower leaves, occasionally the top leaves, followed by leaves gradu-ally turning brown. Infected stems contain a creamy fluid. Infected tubers cut across at the stem end usually show a pale yellow vascular ring which later becomes rotten. The tuber flesh outside the ring typically easily separates from the inner flesh. Skin discoloration and cracking sometimes occur and soft rots subsequently develop, masking the ring rot symptoms. The bacteria is not soil borne, but is often spread from tuber to tuber in harvesting, dressing, storage and planting.

Bactericidal. Capable of killing bacteria.

Bactericide. A substance which is capable of killing Bacteria.

Bactericidin or bacteriolysin. Substance present in the sera of mammals which is capable of inhibiting the growth of bacteria. especially the gram positive ones, or of destroying them.

Bacteriochlorophyll. The main photosynthetic pigment of most photosynthetic bacteria.

Bacteriods. Irregular, enlarged forms of rodshaped bacteria which are especially of the rootnodule forming specis. These are ultimately absorbed by the cells of the rootnodule.

Bacteriology. Refers to the study of bacteria.

Bacteriolysin. Substance or agent which is able to cause rupture of bacteria.

Bacteriophage. A Virus that parasitises Bacteria. Also known as a phage.

Bacteriostat. An agent that only checks the growth of bacteria but does not kill them.

Badland. A type of land generally devoid of vegetation and broken as a result of serious erosion.

Bag closer. Machine which is used for sewing the filled seed bags.

Bag muck. An old term for Artificial Fertilizers, due to the fact that they are applied from a bag.

Bag. (1) A descriptive term for a cow with an enlarged udder prior to the birth of a calf, due to the development of milk vessels. (2) The udder of a cow.

Bagasse. Mill residues which are obtained from the cane-sugar industry. It consists of the crushed stalks from which the juice has been extracted. The word is also applicable to similar residues from other plants, such as sorghum, beet, or sisal, but it usually refers to sugarcane bagasse when unmoditied.

Bagger weigher. Small machines fixed beneath a bin which fill and weigh the seed bag simultaneously in a single operation.

Bagging hook. A sickle-like tool having a square tip and smooth cutting edge. It is also known as fagging-hook, bag-hook, badging-hook. Used for the cutting of com, pulse crops or grass. (See Sack)

Bail milking. The milking of cows in a portable shed or Bail, either moved from field to field frequently or kept in a relatively permanent and sheltered position. Sometimes the cattle are kept in temporary or permanent yards during the winter and milked in an adjacent bail.

Bailiff. A person employed to look after or manage a farm on behalf of the owner, receiving either a salary or share of the profits.

Bakers. Potatoes sold for human consumption, which are of Wart-Potato standard, and which will not pass through a 65 mm horizontal mesh when manipulated, but will pass through a 90 mm mesh. (Mids)

Balance plough. An old type of one-way plough with the right and left-and left-hand bodies attached to separate beams and handles, the beams being connected together at an angle, and having a common axle and pair of wheels. While one plough operated the other was carried more or less vertically in front of it. At the end of a furrow the working beam was raised and the other lowered. Also called a tipping plough.

Balanced daily ration. A ration in which the nutrient ingredients are correctly proportioned for the needs of the type of stock eating it. (Maintenance Ration, Protein Equivalent, Starch Equivalent, Nutritive Ration) for 24 hours,

Balanced feed. Feed containing all the known required nutrients in proper amount and proportion. It is prepared by the recommen-dations of recognised authorities in the field of animal nutrition. The species for which it is intended and conditions of growth and development should be indicated.

Balanced fertilizer. Soil additive having suitable proportions of each necessary mineral element to develop a plant or a crop.

Balata. A product analogous to "gutta percha" obtained from the bark *of Mimusops globosa.*

Bale loader. A device which is attached to the front loader arms of a tractor to lift or clamp groups of Bales for loading onto a trailer or stack, or into a barn. Commonly employed in conjunc-tion with a bale sledge drawn behind a baler. Also called Bale Slave.

Bale sledge. A sledge towed behind the delivery point of a ram Baler, on which the operator stacks bales (usually 6 or 8), although many are now unmanned. These stacks are removed from the sledge periodically. This process allows bales to be grouped rather than scattered indiscriminately about a field by the baler, and facilitates loading and transport to the bam.

Bale. (1) A compressed package of hay or straw for easy handling and storage, tied usually with wire, sisal or plastic twine. Bales may be square, rectangular, round or flat depending on the type and normal operating density of the Baler used. (2) A portable field milking shed. Also called Bail or Hosier bale.

Baler. A machine that picks up Hay or Straw from a Swath or Windrow, compressing and typing it into Bales. The density to which the hales are compressed can be adjusted in accordance to the crop condition or other requirements. The most popular balers produce rectangular bales between 45 and 90cm long, but normally 35 X 45 cm in cross section, and weighing 9-18 kg. Balers producing large 500 kg bales are becoming increasingly popular when large quantities of straw need handling.

Ball planting. Refers to a method of planting young plants in which plants are removed from the ground with a compact ball or lump of earth round their roots for providing nourishment and help their growth in a new location.

Balled and Burlapped (B&B). (1) A field plant whose root has been balled and burlapped for transplanting. (2) Any plant dug with a soil ball and tied up in burlap.

Balling. (1) Measuring of sugar percentage in sample syrup at fixed temperature. Also called brix. (2) The aggregation of macerated tea leaves into lumps or balls during rolling or fermentation.

Ballistic. Fruits with explosive dehiscence and discharge of seeds.

Balsam. (1) A prelty half-landy annual for sunny beds or borders with rich soil or for the green-house. (2) Any of various garden plants belonging to the balsam family. They are usually annuals with spurred red or white flowers and pods that burst and scatter their seeds when ripe. (Genus *Impatiens*, family Balsaminaceae.) In medicine and perfumery, balsam refers to various oily or gummy aromatic plant resins, such as balsam of Peru from the Central American tree *Myroxylon pereirae*.

Balustrade. A baluster is a short pillar with a curved outline and a balustrade is a barrier made with pillars of this type and topped with a coping or rail. The word comes from the Greek word (balustion) for a pomegranate flower which resembles the shape of a baluster.

Bambusa (Bamhoo). There are three great classes of bamboos. Must species do will in the open, specially if the soil is of moist, deep, tight, loamy nature and has some heat or lead-mould in it.

Banana flour. Also known as banana meal. Meal prepared from fully mature unripe bananas.

Band application. The spreading of chemicals over, or next to, each row of plants in a field, as opposed to broadcast application.

Band placement. Refers to the fertilizer placement in bands on one side or both the sides of the row, about two inches below the seed and two inches to one side of the seed or plant grown.

Band seeding. The term used for placing the seeds in rows directly above, hut not in contact with a band of fertilizer.

Banding. (1) The tying of bands of grease round fruit trees to trap insects. (2) The fastening of corn in Sheaves.

Banding-in. The tying together, at breast height, of the several Bines arising from a hop plant.

Bandster. A person who binds sheaves of corn after the reapers (obsolete).

Bank for Cooperatives (BC). Lending institution within the Farm Credit System that provides credit to agricultural cooperatives and rural utility cooperatives nationwide. Nationally chartered CoBank Agricultural Bank (Denver) has the authority to finance U.S. agricultural exports and to provide international banking services to farmer-owned cooperatives.

Barbed bristles. With terminal or lateral spine-like hooks that are bent sharply backward.

Bare fallow. Land, particularly heavy land, left fallow for a year during which it is ploughed several times so that the sun and wind dry the soil during the summer months, desiccating and killing perennial weeds. Success depends on summer rainfall. Bare following is used less frequently nowadays due to the availability of herbicides.

Bare land holding. Farm land lacking the basic buildings necessary for agricultural production.

Bare root (Bare root plant). (1) Rooted plant without soil, opposite to balled and burlapped. (2) A plant dug as bare root for transplanting. (3) Said of plants with the soil removed from their roots; not shipped in pots or with a full root ball.

Barfored's test. Biochemical test which is used for testing monosac-charides. Barfored's reagent is an aqueous solution of copper acetate and acetic acid. It is added to a solution or suspension of the suspected monosaccharide. The positive result is a red precipitate; the negative result is the absence of any precipitate. In this lest there occurs the reduction of cupric accetate to red insoluble cuprous oxide.

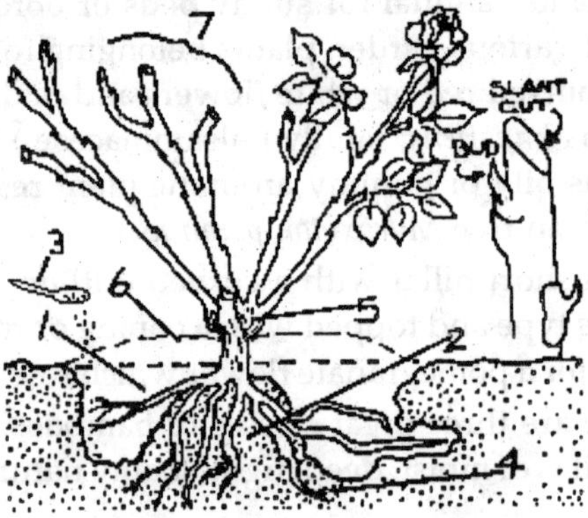

Fig.: Bare root

Bargaining association. A farmer cooperative intended primarily to influence farm prices or other terms of trade between the members and the buyers of the commodities they produce.

Bari. Tea plantation in which trees are allowed to grow without pruning to produce seed.

Barium. Carbonate.White, water insoluble powder which is used as rodenticide.

Bark inversion. A method of grafting for dwarfing a plant in which a complete ring of a bark is removed from the trunk, inverted and regrafted into place.

Bark miner. An insect the larva of which spends most of its develop-mental period in the outer corky bark.

Bark ringing. The removal of a ring of bark from the trunk of a plant to regulate shoot growth and flowering

Bark. (1) In medicine and horticulture, it refers to all the tissues outside the cambium and so includes the phloem. (2) Refers to the cork and other dead tissues to the outside of the functioning phellogen. (3) Bark is a collection of dead phloem cells, usually protecting the phloem, cambium and xylem vessels. Bark naturally sheds from trees, though some trees shed faster than others. Sycamore, birch, and beech trees have thin bark, though this doesn't necessarily make them more susceptible to injury. (4) The rough outer part of the cortex; loosely used for the entire cortex. (5) The tough exterior covering of a woody stem or root outside the cambium. A method of grafting in which dormant scion is inserted between the bark and wood of a rootstock.

Barley loose smut. Fungal disease of barley which is caused by *Ustilago nuda*. It is charaterized by smutted heads where grains are replaced by black powdery mass of spores, leaving the rachis later naked.

Barley water. (1) Fruit beverage containing at least twenty five per cent fruit juice or pulp, thirty per cent total soluble solids and 0.25 per cent barley starch. (2) Squash of fruit in which part of added water is replaced by an extract of dehusked and polished pearly barley.

Barley. (1) A cereal probably the first to be cultivated by man, grown for its grain which is used either as stock feed or for beer making. The bearded seeds in the ear are arranged in 2, 4, or 6 rows. (2) Barley was first cultivated in Egypt about 5000 B.C. The raw grain is enclosed in a tough husk. Pearl barley is produced in a revolving drum where the cereal is ground until the hull and germ are removed from the grains, leaving starchy balls or 'pearls'.

Barleycorn. The grain of barley from which malt is made (personi-fied as John Barleycorn). Also a term fora single grain of barley.

Barn machinery. A general term for implements or machines which are not used in the fields, *e.g.* Hammer Mill, Grain Drier, weighing machines, chaff cutters, etc.

Barn. A building in which grain, hay and implements, etc, are stored.

Barn-cured. Cured within a barn, as contrasted to sun-or air-cured.

Barnevelder. A heavy breed of fowl, with black and white varieties, although similar in colour to a partridge. It has a single comb, yellow legs, and the hens produce dark brown eggs.

Barnyard. The area adjoining a barn. The farmyard.

Barrage sexual. Repulsion between sexually incompatible hyphae. Incompatible strains will grow away from each other.

Barren. (a) (Of plants) not producing fruit or seed (b) (Of land) bare of vegetation, incapable of producing useful crops; arid. (c) (Of animals) incapable of producing offspring, due to infertility or sterility.

Barrier. Any type of obstruction-physical, chemical or biological that does not allow the migration of animals, or gradual exten-sion of their territories.

Barrows and gilts. A barrow is a young castrated male hog; a gilt is a young female hog. Both are raised for pork. Market news reports of prices paid for "barrows and gilts" are of keen interest to producers and packers alike.

Basal area. Refers to the area of the cross-section of a stem. It is usually 6 ft. above union of a tree at breast height. When it is applied to a crop ("Crap basal area"). The sum of basal areas of all the stems or the total basal area per unit of area. It may be expressed in true or in quarter girth units.

Basal doso. Manures or fertilizers which are applied to the soil before the crop is sown or planted or transplanted.

Fig.: Barn

Basal rosette. Ground level ring of leaves (single or multiple layers) around the plant's central stem where it joins the roots.

Basal. Pertaining to the extremity of an organ by which it is attached to its support; leaves located at the base of the plant only.

Base (or contract) acreage. A farm's crop-specific acreage of wheat, feed grains, upland cotton, or rice eligible to enroll in commodity programs under previous legislation, and subsequently eligible for production flexibility contracts under the FAIR Act of 1996. Base acreage equaled land planted for harvest to the crop, plus any land enrolled in Acreage Reduction Programs, plus land considered planted to the crop in zero, 50/85-92 or under permitted normal flex or optional flex acreage during a specified period of time. A farmer's crop acreage base was reduced by the portion of land placed in the Conservation Reserve Program (CRP), but increased by CRP base acreage leaving the CRP.

Base. The bottom of a structure, such as a leaf.

Base exchange capacity. Amount of base which is held on a clay under specified conditions e.g., of pH.

Base exchange Process in which soil absorbs certain positively charged ions from the soil solution and releases other cations in equivalent quantities.

Base period price. The average price for an item in a specified time period used as a base for an index — such as 1910-14, 1957-59, 1967, 1977, or 1982. Time series of data are often deflated to a base period price. Such deflated time series are referred to as constant dollar values (versus nominal dollar values).

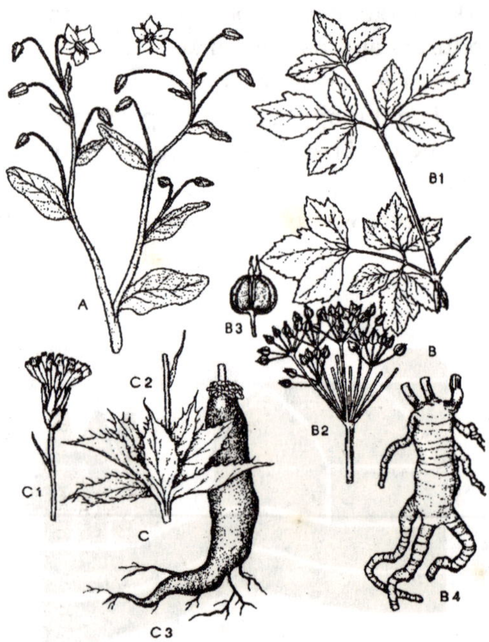

Fig.: Basal rosette.

Base saturation percentage. The extent to which the Adsorption Complex of a soil is saturated with exchangeable Cations other than Hydrogen and Aluminium. It is expressed as a percentage of the total Cation Exchange Capacity.

Base saturation. Extent to which a material is saturated with ex-changeable cations other than hydrogen, expressed as a percent-age of the cation exchange capacity.

Base status (High or low). Quantity of nutrient bases available in the soil.

Basic cleaning. Means cleaning of the seed by separating material smaller as well as larger than the good seed, general size grading, and cleaning. For example, cleaning by Air Screen Cleaner.

Basic commodities. Six agricultural crops (corn, cotton, peanuts, rice, tobacco, and wheat) declared by permanent law as requiring federal price support.

Basic Formula Price (BFP). Calculated monthly by USDA, the BFP is the base price for all milk regulated by federal milk marketing orders. Currently, the BFP is based on the preceding month's average price of Grade B milk paid by processors in Minnesota and Wisconsin, adjusted by current-month changes in the value of certain manufactured dairy products.

Basic slag. A by-product of steel manufacture which is used as a Fertilizer. It is a fine, powdery, grey-black, and rich in Phos-phate and lime.

Basidiomycetes. (1) A large group of Fungi, having most of the familiar types like mushrooms, toadstools, bracket and jelly fungi and puff balls, and also parasitic forms such as Rusts and smuts. Spores are produced by club-shaped cells called basidia. (2) Large class of fungi, with perhaps 12,000 species, which includes mushrooms, toadstools, bracket fungi, puffballs, smuts, and rusts. The thallus, the body of the plant, is a mass of branching threads. The reproductive spores are borne on a club-shaped organ called a basidium (plural basidia), and are called basidiospores. There are usually four on each basidium. The basidia are grouped into conspicuous fruiting bodies, from which they are dispersed by the wind. Mushrooms and brackets are fruiting bodies.

Basidiospore. Sexually-derived spores which are produced following the union of two nuclei on a specialized clublike structure called a basidium.

Basil or sweet basil. Plant with aromatic leaves, belonging to the mint family. A native of the tropics, it is cultivated in Europe as a herb and used to flavour food. Its small white flowers appear on spikes. (Genus *Ocimum basilicum*, family Labiatae.)

Basil. (1) An aromatic herb which is used in cooking. (2) A sheepskin, which is roughly tanned and undressed.

Basing point. A geographical site used to establish fixed rates and/or prices for federal milk marketing orders. Generally, rates or prices increase according to the distance from the basing point. The FAIR Act of 1996 authorizes USDA to consider the use of multiple basing points for pricing milk under federal milk orders.

Basipetal. (1) Development in succession towards the base, oldest at the apex and younger at the base. (2) Transport of substances away from the apex, within a plant.

Basket brooder. Brooder which is made with an ordinary domeshaped basket, lowerhalf of which gets lined with guuny sack or cement mixture wash and upper half left open for allowing passage to smoke hurricane lantern source of heat, sleeved off by 1/2 inch wire mesh; floor has been covered with straw or saw dust.

Bast phloem. The inner bark of a tree which iş used in long strips for basket-and mat-making, and for tying plants to stakes.

Bastard fallow. Land left fallow for half the summer, between the harvesting of one crop (usually silage or hay) and the sowing of the next crop, during which it is ploughed to kill perennial weeds by desiccation. The procedure remains the same as for a bare-fallow except that it docs not waste a year. Success is very dependent on favourable weather conditions. Also known as half allow, pin fallow short fallow or summer fallow.

Bastard trenching. A method of deep digging in which the soil layers get replaced in their natural sequence, as opposed to ordinary trenching in which the sequence of soil layers taken with each spit get reversed when replaced.

Bat. A pole which is used in hop gardens for supporting wire-work usually Sweet Chestnut or Larch.

Batch drying. A method of drying Bales of Hay in batches in a bam, usually using heated air blown through a ventilated floor. Batches of about four layers have been dried at a time until moisture content gets reduced below 20%, then removed for storage, and the next batch introduced. This method has been land to be less popular nowadays than storage drying and tunnel drying,

Batesian mimicry. Refers to the resemblance of a harmless animal to a poisonous or damagorous one providing protection to the former because predators will tend to avoid both. The clearwing both resembling a bee is a well-known example.

Batfowling. Catching birds at night by showing a light and beating the bushes.

Bating. One of the steps which are used in the tanning of leather. It is following dehacring with lime. In this the skins are soaked in a bath containing enzymes and ammonium salts. The aim is to remove residual lime, hair, skin, and glands.

Batology. The study of brambles.

Batt. (1) The long wooden handle of a Scythe. (2) A faggot of stout sticks.

Baule unit. Baule postulated that the unit of fertilizer of any other growth factor, may be taken as that amount which is necessary to produce a yield that is 50 per cent of the maximum possible. According to this concept one Baule unit of a growth factor is equivalent to one Baule of any ohter growth factor in terms of growth-promoting ability. The values of the Baule unit in pounds per acre of N, P2O5 and K2O have been 223.45 and 76 respectively based on Mistcherlich's work.

Baulk. (1) An unploughed strip of land left to mark the boundary between fields. (2) A ridge created by ploughing and then left unploughed. (3) A beam of a timber. (4) A ridge (balk) in which potatoes are planted.

Baulking plough. A type of plough bearing one or more double mould boards, each throwing two furrows simultaneously, one to each side.

Bay. (1) The space between two columns in a barn, or any recess in a farm building. (2) A stall in a stable. (3) The combined cry of hounds hunting an animals, or the last stand of an animal cornered by hounds. (4) A description of a horse, reddish brown to chestnut in colour. (5) Any of various species of *laurel* tree. The aromatic evergreen leaves are used for flavouring in cookery. There is also a golden-leaved variety. (Genus *Laurus*, family Lauraceae.)

B-cells. Beta cells of islets of Langerhans.

B-complex. Group of accessory food factors (vitamins), which include thiamine or vitamin E1; riboflavin B2; pantothenic acid B3; niacin or P.P. factor : pyridoxin B6; biotin (H); inositol; cholinc; para-amino benzoic acid and folic acid etc.

Beaked. Ending in a point, especially on fruits.

Beam. (1) A large straight piece of iron or timber which is usually square in cross-section and is forming one of the main structural members of a building, normally supporting rafters. (2) The main shaft of a plough, wooden or metal, to which are fixed the Mouldboard, Coulters and wheels, and the handles of horse drawn ploughs. This part transmits the power of the animal to the plough. (3) The main trunk of a stag's horn.

Beamage. An allowance on an animal carcase for loss of weight by evaporation.

Bean(s). Various leguminous plants, and their seeds borne in pods, distinguished from vetch, clover and peas by having a square, hollow steam. They are tall, erred plants, varying in length according to variety, and develop a strong tap root bearing nodules containing bacteria capable of nitrogen Fixation. When ploughed-in, valuable nitrogen and organic Matter is returned to the soil. The seeds or beans are rich in Protein.

Bean straw. The Haulm of bean plants.

Bearberry. Any of a group of evergreen trailing shrubs belonging to the heath family, found in high and rocky places. Most bearberries are North American but *Arctostaphylos uva-ursi* is also found in Asia and Europe in northern mountainous regions. It has small pink flowers in spring, followed by red berries that are dry but edible. (Genus *Arctostaphylos*, family Ericaceae.)

Beard. (1) The hairy tuft on the lower part of a goat's jaw, or on a turkey's neck (also called Brush). (2) A term for an Awn or threadlike spike, as on the ears of barley. (3) Patch of hairs on a flower or other part

Bearing habit. The relative position of flowering and fruiting on the shoots of a plant (See Fruiting habit).

Beaters. (1) Steel bars on the drum of a threshing machine of combine harvester which dislodge the grain out of the cars of cereals as they rotate. Also known as beater bars. (2) Those who rouse game in shooting or hunting.

Beaumont period. A period of 48 hours during which temperatures do not fall below 10°C (50°F) and relative humidity remains above 75%. Outbreaks of Potato Blight may be expected within 3 weeks of such a period.

Bed planting. Planting on elevated level beds, which are separated by narrow furrows.

Bed. (1) Narrow flat-topped ridge on which crops are grown with a narrow furrow on each side to facilitate irrigation and drainage. (2) An area in which seedlings or sprouts are grown before transplanting. (3) A garden plot. (Seed Bed). (4) A layer or stratum of rock. 3. A sleeping place for an animal. (Bedding)

Bedded set. A young hop plant, having rooted in a nursery bed from a cutting.

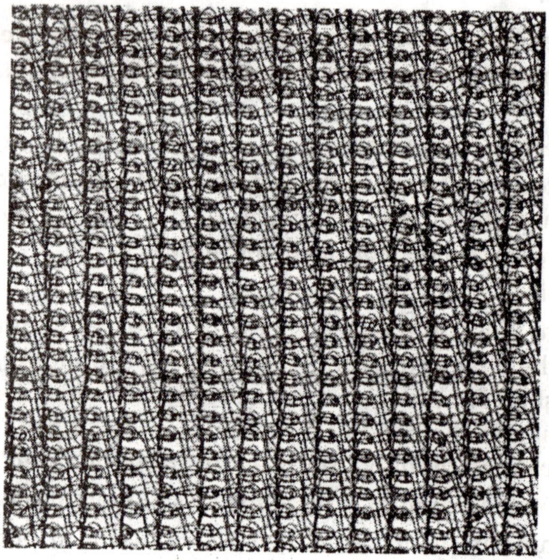

Fig.: Bed planting

Bedding land. The ploughing, grading or otherwise elevating the surface of fields into a series of parallel beds or 'lands' with shallow surface drains separating them.

Bedding plant. (1) A plant used in displays of colourful plants. Usually the plants are annual or biennial and start their life in conservatories. (2) Plants that are raised by commercial growers and sold in the early stages of growth for use in outdoor gardens.

Bedding soil. Arranging the surface of fields by ploughing and grading into a series of elevated beds separated by shallow ditches for drainage.

Bedding. Litter for farm animals to sleep on, usually straw, shavings, sawdust, etc.

Bedrock. The solid rock beneath the soil.

Bee beard. (1) The pollen of flowers collected by bees and fed, together with honey, to the larvae. (2) The local name of several plants yielding nectar.

Bee brush. Brush or whisk broom often used to brush off bees from a honey-comb before it is taken away for extraction.

Bee candy. Artificial food for bees which is made from sugar and substances like honey.

Bee escape. A device which allows bees to go through a self-closing exit in a beehive.

Bee farming. The term used for the keeping of bees for honey and beeswax they produce, and for their pollinating of planars.

Bee fold. An enclosure for beehives.

Bee hole. Commonly applied to the hole which is bored in teak by *Xyleutes ceramicus*, (the beehole borer).

Bee parturage. Also bee forage. Plants yielding necter and pollen, and within flying distance of a hive.

Bee. A furry insect of the Order Hymenoptera. There are solitary and social bees. The latter live in colonies. Each colony has one queen bee together with large numbers of female 'workers' and a fewer male 'drones'. The most highly socialised is the honey-bee kept in hives by beekeepers for commercial Honey produc-tion. Bumblebees have much smaller colonies. A growing practice is the hiring and placing of Beehives in Orchards to promote pollination and fruit setting.

Beech. The common beech of the northern temperate regions belongs to the genus *Fagus*, of which there are 10 deciduous species. In the southern hemisphere there are 36 species of the genus *Nothofagus*, both deciduous and evergreen.

Beehive. A box in which bees are kept to produce honey. Hives contain two separate chambers, a lower brood chamber in which the queen lays and rears her brood, above which is the honey chamber or supers containing frames in which the bees store surplus honey which is periodically removed. (See Fig. on p. 80).

Beeswax. Wax secreted by bees and used to construct the cells of the honeycomb. Used in polishes and cosmetics.

Beet curly top virus. Occurs in chile. Causes severe stunting and chlorosis. Results in reduced yields. Spread by leafhoppers.

Beet cyst nematode. A cyst-forming, plant-parasitic eelworn or rounderworm *(Heterodera schactil)* which attacks beet.

Beet leafhoppers. Spread beet curly top virus in chile.

Beet. (1) A plant *(Beta spp.)* of the goose foot family with a succulent root, used as a food and as a source of sugar. (2) A sheaf of harvested flax.

Fig.: Beehive

Beetle engraver. Certain members of family Scolytidae, whose bark tunnels form characteristic patterns having one or more egg galleries issuing from central nuptial chamber.

Beetle longhorn. It is a member of the family Ceramycidae. Its larva bores in the wood of cambium mainly of injured, fallen, dying or dead trees.

Beetle power-post. A member of the families Bostrychidae or lycti-dae, It bores mainly in sapwood of seasoned timber and leavidg a fine flour-like dust.

Beetle. (1) An insect of the Order Coleoptera, in which the forewings are reduced to hard and homy protective covers for the hind-wings and body. (2) A heavy wooden hammer with a long handle used for driving wedges and posts, or crushing paving stones, etc.

Begonia. Any of a group of tropical and subtropical plants. They have fleshy and succulent leaves, and some have large, brilliant flowers. There are numerous species in the tropics, especially in South America and India. (Genus *Begonia,* family Begoniaceae.)

Belgain hare. A breed of rabbit with the appearance of a hare, having an abnormally long head, ears and legs. Rich red or chestnut in colour with dark, wavy ticking.

Belgian blue. A breed of cattle found in Belgium, South Netherlands and Luxembourg.

Bell. (1) To bellow or roar. The cry of a stag at rutting time. (2) The catkin containing the female flowers of the hop. Also to be "in flower.

Bellflower. General name for many plants with bell-shaped flowers. The *harebell* (*Campanula rotundifolia*) is a wild bellflower. The Canterbury bell (*C. medium*) is the garden variety, originally from southern Europe. (Genus *Campanula,* family Campanulaceae.) The clustered bellflower (*C. glomerata*) is characteristic of chalk grassland, and is found in Europe and northern Asia. Erect and downy, it has tight clusters of violet flowers in late summer.

Below-cost timber sale. A timber sale from national forest lands in which the expected federal revenues are less than the estimated federal expenses to sell the timber.

Belt transect. A narrow strip of land taken as a sample of an area in studying the vegetation.

Belted galloway. A variety of Galloway cattle with a white belt round the body. Also called Beltie.

Benazolin. A translocated herbicide, used only in mixtures, for controlling broad-leaved weeds.

Bench cross-section. Refers to the cross-section for contour benches. It is an important phase in planning for irrigation. Width of strip between ridges is kept such that it fits the farm

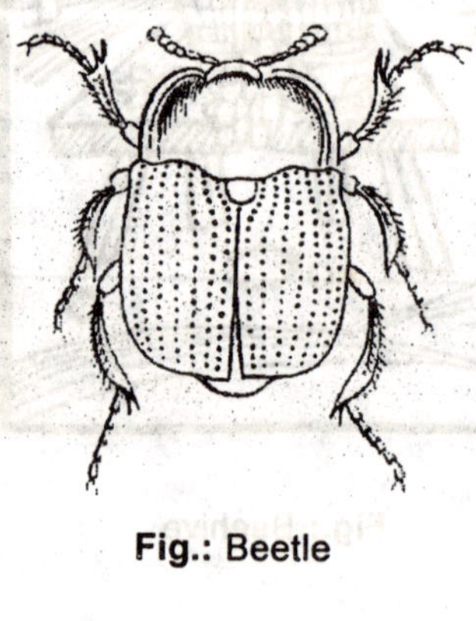

Fig.: Beetle

equipment to be used 'height of the bend is kept sufficient to contour both normal irrigation stream and storm runoff.

Bench grafting (root grafting). Grafting where roots are used as the rootstock

Bench terrace. It is the commonest type of terracing which is followed in India. It is constructed to make sloping land culti-vable or stable. It has a series of platforms or mostly level benches which are cut into the slope in a step sides often with retaining walls supported by rock or vegetation. Vertical drop 2 to 6 ft. depending on slope and soil conditions. Small shoulder bund (1 ft. height) is also constructed along outer edge of terrace.

Benedict solution. An alkaline copper sulfate solution, blue in colour. When a few drops of a reducing suger solution are added after heating the copper solution, the cupric copper gets reduced to the cuprous state and a change to a reddish colour is observed. It finds use as a test for the presence of sugars in solution.

Benedict's test. A biochemical test which used for testing sugars. The Benedict's solution is prepared by mixing an alkaline copper sulphate solution with sodium citrate. The positive result for relatively large quantities of sugar and the negative results have been the same as those of Fehling's test.

Beneficial cultivations. Those cultivations which enhance the con-dition of the land, or are good for crops.

Beneficial predators. Insect allies that kill off pests like aphids. Why do beneficial bugs stick around? Well, for one reason, your garden is such a good source of food.

Benomyl. A systemic fungicide.

Bent. A general terms for short flowering stems of grasses projecting above the turf and old dried grass-stalcks.

Benzene hexachloride. Also called BHC. It is an effective and economic insecticide available as dusting powder, water dis-persible powder, emulsifiable concentrate or solution and as fumigant in highly volatile solvents. It has 5 isomers of which gamma has been most important due to insecticidal value; insects killed by contact or by stomach of fumigant action, can be used in mixture with other insecticides and fungicides, except those having strongly alkaline materials like Bordeaux mixture, lime or lime sulphur.

Benzoin. Resin (thick liquid that hardens in the air) obtained by making cuts in the bark of the tree *Styrax benzoin*, which grows in the East Indies. Benzoin is used in cosmetics, perfumes, and incense.

Bergamot. Small evergreen tree belonging to the rue family. A fragrant citrus-scented essence is obtained from the rind of its fruit and used as a perfume and food flavouring, for example in Earl Grey tea. The sole source of supply is southern Calabria, Italy, but the name comes from the town of Bergamo, in Lombardy. (*Citrus bergamia*, family Rutaceae.)

Beri-beri. An illness by an acute deficiency of vitamin B (thiamine).

Bermudagrass. Bermudagrass is a warm season grass thriving when temperatures are between 65 and 95 degrees F. Bermuda grass, like zoysia, will turn brown in mid October and not green until mid-May. It's leaf blade is narrower than zoysia. Advantages and disadvantages are similar to Zoysia. Bermudagrass may suffer winter injury in central Illinois. *Seeding Rate:* Bluegrass should be seeded at the following rates: New turf area:

1 to 3 pounds per 1000 square feet. Overseeding: 1 to 3 pounds per 1000 square feet (the same as new seeding).

Berry. (1) A fleshy, indehiscent, pulpy, multi-seeded fruit resulting from a single pistil. (2) A simple fruit with one or more carpels and having the entire pericarp fleshy. (3) A synonym of small fruits or soft fruits.

Best Management Practices (BMP's). A conservation practice or combination of practices designed to maintain agricultural productivity while reducing point- and nonpoint-source water pollution. State water quality agencies (or their designees) determine BMPs to fit local conditions and to make the most efficient use of natural resources and purchased inputs. The term has been defined as generally recognized farmer management practices designed to reduce or prevent contamination of ground water and surface water, erosion, and runoff from cropland, including the use of conservation tillage, no-till, ridge plating, strip tillage, contour farming, strip cropping, irrigation water management, judicious fertilizer application, slow-release fertilizers, soil and tissue testing, and vegetative buffer strips.

Betaine. A colourless, tasteless, crystalline substance present in sugar beet, producing a strong fishy taste on decomposition.

BFP (dairy science). Basic formula price, determined by the U.S. Department of Agriculture, that serves as a basis for pricing milk in the United States. The BFP is calculated monthly based on previous BFP and the price of cheese.

B-horizon. Subsurface layer in which certain leached substances (*e.g.* iron) gel deposited.

Bi-. (1) Prefix meaning twice or doubly. (2) Twice or doubly.

BICO Report. The Foreign Horticulture Service's report of U.S. agricultural export and import data on Bulk, Intermediate, and Consumer-Oriented (BICO). In addition, the data base includes forest products and edible fish and seafood products. These trade data are further classified among 46 separate product groups. Data are available in both calendar and fiscal year format and for 16 world regions and 35 individual country markets. The BICO data can be accessed at.

Bicolor. A plant or flower in two colors.

Biennial plants. (1) A plant that grows, flowers, produces seeds or fruit, and dies in two years. Some herbacous flowers and vegetables are biennial. Most biennial plants produce foliage the first year and bloom the second year. (2) A plant that flowers and hears fruit only in its second year, then dies. Examples include carrot and beetroot. (3) A plant that usually only lives two years, normally producing flowers and seed the second year; a plant that grows foliage the first year and flowers the second year.

Biennial. (1) Ordinarily applied to plants that live only two seasons; during the first season only leaves and stems are produced above ground, while the flowers and seeds are borne the second summer. Here used in a special sense in separating the biennial canes of the raspberries and blackberries from the stems of other woody plants. In these species, the canes themselves are biennial from underground perennial stems. (2) Germinate from seed during the growing season and often produce an over-wintering storage root or bulb the first summer. Quite often they maintain a rosette growth habit the first season, meaning that all the leaves are basal. They flower and develop seeds the second summer, followed by death. In the garden setting, we grow many biennials as annuals (e.g., carrots, onions, and beets) because we are more interested in the root than the bloom. Some biennial flowers may be grown as short-lived perennials (e.g., hollyhocks). (3) Plants that flower in their second year and then die, such as Sweet Williams and hollyhocks. (4) Plants that live only two years and die are called biennials. Plants usually grow vegetatively (producing leaves) the first year and flowering the second. Common biennials include:

- Cabbage
- Canterbury bells
- Foxglove (Digitalis)
- Hollyhocks
- Leeks (Allium)
- Lupines
- Money Plant (Lunaria)
- Parsley
- Sweet Williams (Dianthus)

Bifarious. Arranged in two opposite rows.

Bifida/bifidum. Divided into two parts.

Bifoliate. Having two leaves or leaflets.

Bifurcate. Forked, as some Y-shaped hairs.

Big bud. A swelling of current buds due to a gall-mite.

Bigeneric hybrid. Species formed by crossing two different genera.

Bilateral cordon. Training system as in grapes, where the trunk is divided into two branches extending horizontally on a supporting wire.

Bile. A bitter, yellowish green secretion of the liver, important in the digestion of fats, and stored in the gall bladder. Also called gall.

Biliary pigments. Pigments formed in the liver from the haemoglobin of old red cells that are destroyed there.

Bill hook. A hatchet with a curved point, used in hedge trimming, etc.

Bin sampler. Large size trier constructed on the principles of big triers for drawing samples from bins.

Bin size. Ideal compost sizes range from 3x3x3 to 5x5x5. Smaller piles will not generate the heat necessary for plant material to decompose. Larger piles are harder to manage and may not decompose uniformly.

Bin. (1) A large receptacle for storing feedingstuffs, com, fruit, etc. (2) A canvas container on a collapsible wooden frame used initially for storing hops when picked. Also called a crib.

Binder. A machine (full name: reaper-binder) which cuts and hinds corn in sheaves. Powered by p.t.o., although older and heavier machines obtained power from a large hull wheel. It produces long, unbroken straw, used in thatching.

Bink. A wasps' or bees' nest.

Binomial system of nomenclature. The assignment of names to organisms using two Latin words, the first denoting the genus and the second denoting the species e.g. *Mangifera indica.*

Bioaccumulation. The absorption and concentration of toxic chemicals, heavy metals, and certain pesticides in plants and animals. Toxicity can be expressed in several ways: lead that is ingested by calves can bioaccumulate in their bones, interfering with calcium absorption and bone development; stored chemicals may be released to the blood stream at a later time, for example, during gestation or weight loss; and chemicals may concentrate to lethal levels at upper ends of the food chain. Bioconcentration is a synonym for biaccumulation.

Bio-aesthetic planning. Planting of trees and plants for beautifying roads, parks, public and private places.

Bioassay. (1) Determination of the biological activity as well as relative strength of a substance (hormone, drug etc.) by comparing its effect on a test organism with that of a standard preparation. (2) Refers to the evaluation of the effect of a substance on an organism under controlled conditions e.g., the bio-assay of fungicides.

Biochemical Oxygen Demand (BOD). A measure of the amount of oxygen consumed by natural, biological processes that break down organic matter, such as those that take place when manure or sawdust is put in water. High levels of oxygen-demanding wastes in waters deplete dissolved oxygen (DO) thereby endangering aquatic life. Sometimes referred to as "biological oxygen demand. Chemical oxygen demand (COD) is a measure of the oxygen consumed when organic matter is broken down chemically rather than biologically. COD can be determined much more quickly than BOD and more accurately reflects the amount of organic matter in a water sample. BOD is a standard measure of water quality.

Biochemical. Deals with the study of chemical substances occuring in living organisms and the reactions and methods for identifying these substances.

Biochemistry. Deals with the study of chemical processes and sub-stances occurring in living cells.

Biodiesel. Biodiesel is registered with the Environmental Protection Agency as a pure fuel or as a fuel additive and is a legal fuel in commerce. It is typically produced through the reaction of a vegetable oil or animal fat with methanol in the presence of a catalyst to yield glycerin and biodiesel (chemically called methyl esters). It is an alternative fuel that can be used by itself or blended with petroleum diesel for use in diesel engines. Its use can result in substantial reduction of unburned hydrocarbons, carbon monoxide, and particulate matter. Provisions of the Energy Conservation Reauthorization Act of 1998 (ECRA, P.L. 105-388) amended the Energy Policy Act of 1992 (EPACT, P.L. 102-486) to allow that the use of biodiesel added to conventional diesel at blends of 20% and higher would produce credits to offset up to 50% each year of alternative fuel vehicle acquisition requirements. Farmers and processors anticipate that increased use of biodiesel will strengthen the market for soybean oil.

Biodiversity (or biological diversity). In general, the variety and variation among plants, animals, and microorganisms, and among their ecosystems. It has 3 levels: ecosystem diversity, species diversity, and genetic (within species) diversity. The concept of maintaining biodiversity holds that civilization should preserve the greatest possible number of existing species so that a highly diverse genetic pool, which can be tapped for useful and beneficial characteristics, will be available into the future. Genetic diversity provides resources for genetic resistance to pests and diseases. In agriculture, biodiversity is a production system characterized by the presence of multiple plant and/or animal species, as contrasted with the genetic specialization of monoculture.

Bioengineering. Deals with the manufacture and use of artificial replacements for body organs that have been lost or are mal-functioning, like artificial limbs, heart pacemakers.

Biogas. A mixture of methane (70%) and carbon dioxide produced as an alternative source of energy by the anaerobic digestion of waster material—mainly animal slurries, hut also crop residues and industrial wastes. Research continues to perfect the tech-nique and includes the use of digested slurries as animal feeding-stuffs.

Biological clock. Physiological mechanism in plants and animals which makes them to have a knowledge of time and to take account of it, even if the normal indicator of day and night luminosity are lacking.

Biological control. Controlling plants, diseases, and animal pests using natural enemies; or inhibiting the reproduction of pests by methods that result in the laying of infertile eggs, etc. The practice of using beneficial natural organisms to attack and control harmful plant and animal pests and weeds is called biological control, or biocontrol. This can include introducing predators, parasites, and disease organisms, or releasing sterilized individuals. Biocontrol methods may be an alternative or complement to chemical pest control methods. Biocontrol is part of the Animal and Plant Health Inspection Service program to control several economically important pests of food and fiber crops; it also is researched and used by other USDA agencies that promote integrated pest management.

Biological Interchange. The term used for the interchange of ele-ments between organic and inorganic states in a soil or other substrate through the agency of biological activity. It occurs due to the biological decomposition of organic compounds and the liberation of inorganic materials on the one hand (mineralira-tion), and the utilization of inorganic materials in synthesis of microbial tissue on the other (immobilization). Both processes commonly take place continually in normal soils.

Biological mineralization. The conversion of an element occurring in organic compounds to the inorganic form because of biologi-cal decomposition.

Biological monitoring. Using living organisms to test the quality of either effluent to be discharged into receiving waters, or waters downstream from a discharge.

Biological oxygen demand. See biochemical oxygen demand.

Biological weathering. Physical and chemical weatherings which are assisted by biological agencies.

Biologics. Immunization vaccines, bacterins, antigens, and antitoxins and other preparations made from living organisms and their products, intended for use in diagnosing, immunizing, or treating humans or animals, or in related research. The Animal and Plant Health Inspection Service has responsibility for approving some animal biologics. See Veterinary biologics.

Biomagnification (or biological magnification). The increase in the concentration of bioaccumulated toxic chemicals in organisms higher on the food chain due to preferential storage of the toxic chemical in edible body parts. For example, chlorinated pesticides concentrate in the fat and skin of fish in contaminated lakes and streams and are biomagnified when those fish are eaten by larger fish, and perhaps eventually by mammals or birds of prey.

Biomass. (1) The generic term for any living matter that can be converted into usable energy through biological or chemical processes. It encompasses feedstocks such as agricultural crops and their residues, animal wastes, wood, wood residues and grasses, and municipal wastes. Refers to the total weight of living matter in a population. It is usually expressed in terms of dry weight per unit area. (2) Substance of living tissue, often a measure of weight or volume. (3) The combined dry or wet weight of all the organisms in a habitat.

Biometrics. The science dealing with the application of statistical procedures for the study of biological problems

Biopesticide. A pesticide that is biological in origin (i.e., viruses, bacteria, pheromones, natural plant compounds) in contrast to synthetic chemicals. Transgenic Bt cotton and corn are biopesticides because Bacillus thuringiensis (Bt) is a naturally occurring soil bacterium that has been genetically engineered into the plants.

Biopsy. Diagnostic examination of a piece of tissue which has been removed from the living organism.

Bio-security. "Bio-security" refers to the policies, and measures taken, for protecting a nation's food supply and agricultural resources from both accidental contamination and deliberate attacks of "bio-terrorism." Now viewed as an emerging threat, bio-terrorism might include such acts as introducing pests intended to kill U.S. food crops; spreading a virulent disease among animal production facilities; and poisoning water, food, and blood supplies. The federal government is now increasing its efforts to improve bio-security because of the recognition that the United States is currently vulnerable, both from a civil and military standpoint.

Biosynthesis. (1) Biological synthesis, building or forming of biochemical compounds in a living organism. (2) Refers to the coining together of chemical building units to form new materials in living plant or animal.

Biosystematics. The science of classification of organisms.

Biota. Fauna and flora of a region.

Biotic. Pertaining to life.

Biotin. It refers to one of the ten recognised vitamins of the B complex. It is widely distributed in foods.

Biotype. Physiological race or form within a species which is morphologically identical, but is different in genetic, physiological, biochemical or pathogenic characteristics.

Bipinnate. Twice compound with the leaflets along each side of a common axis. The leaflets are further divided into pinnules.

Bird baths. A garden adornment consisting of a bowl shaped container fixed over a pillar, storing water for birds.

Bird chilli (Tabasco chilli). Perennial *chilli (Capsicum frutescens),with* small sized pungent fruits.

Bird minding. Scaring birds away from field and orchards by creat-ing noise, *e.g.*, shouting, firing guns, etc. Also called bird scaring or bird starving.

Bird scarers. Device for making loud noises at intervals which scares away the birds.

Bisexual. (1) Both sexes present and functional in the same flower. (2) Said of a flower with both stamens and pistils.

Bite. Grazing. Early bite is grazing in the early spring. Late bite is grazing at the end of growing season.

Biternate. Twice ternate; structure basically ternate, but whose primary divisions are again each ternate.

Bitter pit. A disease of apples, characterised by brown spots and depressions.

Biuret reaction. It is a specific test for the detection of peptide linkage. It involves adding of alkali and a few drops (3-4 drops) of a dilute solution of copper sulphate (about 0.02%) to the solution. This produces the development of a bluish to pink colour depending on the type of protein present.

Biuret test. *See* Biuret Reaction.

Black heart. A type of winter injury in which the inner wood gets darkened while the cambium and the bark remains alive *e.g.*, nursery trees are susceptible to this type of injury.

Black land. Heather moors, dark in appearance. Also a term for the dark, humus-rich, fenland soils.

Black mustard. A cruciferous plant *(Brassica nigra)* grown for its seeds to produce table mustard. Seeds also contain up to 22% oil, extractable under pressure.

Black rust. A fungal disease *(Puccinia graminis)* of wheat, oats, barley, rye and several grasses. Forms reddish brown (becoming black) spots or lines on stems and leaf sheaths. Its alternative host is common barberry *(Berberis vulgaris).*

Black soil. Occurs in low lying areas having 55 to 65% clay content, good supply of lime and the pH is 7-8. Nitrogen content has been medium, potash is fairly well supplied P2O5 is low. It is quite fertile and suitable for wheat, cotton, chillics and jowar.

Blackcurrant. A shrub *(Ribes nigrum)* grown for its small black berry.

Blackleg. (1) A bacterial disease *(Erwina atroseptica)* of potatoes which causes blackening and rotting of the stem base. Leaves lose colour, turn brown, and stems die. Brown rot infects the tuber. Principally a seed borne disease. (2) A fungal disease *(Phoma batae* and *Pythium spp.)* of sugar beet and mangold seedlings, which become black at ground level, and threadlike, then wilt and die. 3. A bacterial disease *(Closlridium spp.)* of sheep (also called gas gangrene) and cattle (also called blackquar-teror felon). Insheep *C. chauwaei* infect ewes via wounds, often at lambing, dipping or shearing, and lambs at castration or docking. It causes hot painful swellings in muscles which darken and may be gassy, and death within hours. In cattle, bacteria from soil infect minute wounds, particularly in calves, producing symptoms similar to sheep.

Blackout system. A means of covering plants with black polythene or cloth to shorten the photoperiod to promote flowering of short day plants.

Blade. (1) The flat or expanded part of a leaf. Cereals are 'in blade' when flat, the thin, long leaf is formed but before the corn ears have developed. (2) The thin cutting edge of a knife, axe, scythe, etc. (3) The expanded portion of a leaf, the lamina.

Blae. Hardened clay or carbonaceous shale, often blackish or dark bluish, and spread on arable land to improve soil texture.

Blanching. (1) Heating of a fruit or vegetable product in boiling water or steam for a brief period to inactivate enzymes before processing. (2) The technique of covering plant parts with soil or other materials to exclude light. (3) Providing conditions under which growth gets forced to take place in darkness, resulting in tissues or organs with little or no green colouring matter.

Blaney-criddle formula. It states that the amount of water con-sumptively used by crops during their growing season gets closely related with mean monthly temperature and daylight hours.

Blemish. Includes green patches, brown spot and patches, damage which occurs due to pest and disease, breakage in handling, sponginess and black spots, as in tobacco.

Blend. (1) To cross-broad plants or animals to obtain an offspring having characteristics of each (2) To mix two or more ingredi-ents to get a standardized product.

Blending inheritance. The inheritance of characters so that the off-spring, and successive generations become intermediate be-tween the original parents. This is because the character has been controlled by several genes.

Blending. In grain marketing, the combining of two different qualities of grain in order to change the total value of both lots. For example, it is common to blend grains of differing moisture or different foreign material content to achieve the requirements of a contract order.

Blet. Incipient internal decay in fruit.

Blight. A plant disease caused by fungal parasites and various insects, Aphids, Mildew, Rust, Smut, etc. Usually a blight will attack a whole crop and sometimes a particular crop throughout a region. Nowadays the term is usually restricted to Potato Blight.

Blind cultivation. Cultivating with a harrow weeder, rotary hoe, or other implement to kill weeds before a seeded or planted crop has come up.

Blind eyes. Shoots which do not produce flowers.

Blind shoot. *A* shoot in which the apical bud aborts, resulting in no further growth

Blind. A plant which fails to flower or produce fruit.

B-line. (1) The term used for the fertile counterpart of the A-linc. The B-line is not having fertility-restoring genes and is used as the male parent to maintain the A-linc *i.e.* A-line *x* B-line repro-duces the A-line (2) Maintainer of the A-line.

Bloat. The swelling of a cow's rumen due to gas from fermentation of green food, particularly lush grass containing white clover, or from frosted or muldy food, and obstructions in the gullet. It causes respiratory distress and, in acute cases, death. Also called hoven, rumen tympany or blast.

Block. A main territorial division of a forest, which is generally bounded by natural features and bearing a local proper name.

Bloom. (1) The state of flowering. (2) Blossom. (3) Delicate waxy or powdery substance on the surface of leaves, berries etc. (4) A waxy coating found on stems, leaves, flowers and fruits, usually of a grayish cast and easily removed. (5) Grains, short rods, or crusts of wax on the surface of some leaves and fruits.

Blossom honey. Honey produced wholly or mainly from the nectar of blossoms.

Blossom wilt. A fungal disease *(Sclerotinia laxa)* of apples, pears, plums and cherries, causing fruit to rot and blossom and young leaves to wilt and die.

Blotch mine. Dark patch on a leaf which is caused by a minute insect larva mining or burrowing between the upper and lower epider-mis. Various insects do this, including the larvae of small moths like the Lilac leaf-miner *(Gracilaria syringella)* and of Diptera such as the Celeyfly *(Acidia heraclei)*. Blotch Mines are only one of various patterns which are produced by leal-mining insects.

Blotch. Disease which shows large and irregular spots on stems, shoots and leaves.

Blow fly. (1) A general name for a number of species of fly which deposit their eggs in flesh and carcases. Also called flesh fly. (2) The Bluebottle, a two-winged fly with metallic blue wings. The larvae feed on animal matter and dung.

Blow. (1) To develop bloom or blossom. (2) The condition of a cow with its stomach swollen by gases, mainly methane. (3) A term for soil blown by the wind.

Blown out land. An area from which all or almost all the soil has been removed by wind erosion. Usually barren, and unfit for crop production.

Blowout (erosion). An excavation in areas of loose soil, usually sand which is produced by wind action.

Blue ointment. Mild mercurial ointment having a bluish colour. It is used for the destruction of external parasites particularly for the extermination of lice.

Bluing. The objectionable development of blue pigment in flower petals usually after harvest.

Blush. Pink or reddish tint to a plant part

Board foot. A measure for lumber, equal to a 1-inch thick board that is 1 foot long and 1 foot wide in nominal dimensions (a 2x4, for example, is less than 2 inches thick and 4 inches wide, but a 1-foot long 2x4 is still counted as 2/3 of a board foot); typically reported in thousands of board feet (mbf). Also used to estimate the volume of lumber that can be produced from logs and standing trees.

Boarding. The practice of tilting a plough towards the ploughed land to increase the pressure on the Mouldboards.

BOD. Biochemical oxygen demand; biological oxygen demand.

Bodge. An old measure for oats, approximately a gallon, or half a peck.

Body. The operating part of a plough comprising a coulter, share and Mouldboard. Most modern ploughs have several bodies.

Bog soils. An intrazonal group of soils having a muck of peaty surface underlain by peat, developed under swamp or marsh types of vegetation. It is mostly in humid or subhumid climate.

Bog spavin. Accumulation of synovial fluid in and around the hock joints of horses.

Bog. In general terms it refers to a spongy, usually peaty, wet area of marshy land. The term is applied more strictly to wet, very acid Peat, characterised by *Sphagnum* moss, found particularly in areas of heavy rainfall. Distinct from a fen which has an alkaline to slightly acid peaty soil.

Bogey. A low heavy cart used in the past for carrying Haycocks.

Boiler. A pullet or Hen which is sold for the table at the end of the laying period.

Bole. (1) The basal region of the stem or trunk of a tree. (2) The stem of a tree.

Boll weevil. An insect pest of cotton that is the subject of an Animal and Plant Health Inspection Service eradication program cooperatively funded and managed by cotton producers. The larva of a noctuid moth *Heliothis armigera.* It feeds on cotton bolls and other seeds.

Boll. (1) An old measure of capacity for grain. In northern England varying from 2 to 6 Bushels in Scotland usually 6 bushels. (2) An old measure of weight for grain equal to 140 Ibs (63.5 kg).

Bolt, carriage type. Commonly used bolt in wood. It is having a round head and a square or splined neck which grows into the wood to keep the bolt from turning.

Bolt. (1) A bundle of willow shoots or reeds about 3 ft in circumference. (2) To break away, as when a horse suddenly dashes off. 3. A sieve for separating Bran from flour. (Bolting Cloth).

Bolter. (1) A term used of biennial plants, which normally take two years to flower and set seed, but which get out of hand and produce seed instead in the first year, thus reducing the harvest yield. Examples of crops which sometimes bolt include cabbage and sugar beet. (2) A Bolting Cloth.

Bolting. (1) Formation of elongated stem or seed stalk. It usually takes place during the second season of the growth in biennial plants. (2) Process by which plants produce flower stalks and flowers. (3) Vegetables quickly flowering rather than producing the food crop. Usually caused by late planting and too warm temperatures.

Bone ash. Produced by burning bones with free access to air and contains 15 to 16.5 percent phosphorus. It is used as a feed supplement.

Bone chewing. Animals on Iced which are low in phosphorus commonly chew bones, wood dried or decayed animal carcasses as an instinctive reaction to the dietary deficiency. Affected animals get unthrifty, and milk production reduces to a minimum. The remedy involves feeding of phosphorus rich feeds such as oil cakes.

Bone feedstuff. The term used for a common supplementary source of phosphorus for animals. It is a natural source of calcium phosphate which is provided by bones in various forms. They supply phosphorus and calcium in animal diet.

Bone flour. Finely ground bones, containing phosphate, used as Fertilizer and also as a Feedingstulf.

Bone marrow. Yellowish tissue that occupies a bone cavity.

Bone meal. Bones, coarsely ground, after removing most of the rat. Used as a feedingstulf of fertilizer. (Meat-and-Bone-Meal).

Bone seeker. Compound or ion that migrates in vivo preferentially into bone.

Bonsai style. Training plants in bonsai-like fashion.

Bonsai. (1) A living woody plant dwarfed by careful pruning and container culture. (2) Japanese art of growing miniature trees and shrubs in shallow containers by extreme dwarfing treatments.

Bonus commodities. From the agricultural perspective, these are commodities donated to domestic feeding programs that USDA acquires for unexpected surplus removal reasons or because Commodity Credit Corporation holdings are not needed for other purposes, or are in danger of waste or spoilage. For example, if meat prices fall, USDA may buy beef and donate it to the National School Lunch Program, or if the CCC is holding an excess of cornmeal that is in danger of spoiling, it might donate this to the lunch program. From the food program perspective, these commodities are those donated in addition to the commodities that must be provided under mandatory requirements in food program statutes.

Bonus crop. An additional crop of economic importance taken in intercropping / mixed cropping systems without any extra inputs.

Boot. (1) Upper leaf sheath of a grass. (2) Stage at which the inflores-cence expends the boot.

Borage. Borage is an annual plant, growing between 90 cm/3 ft and 120 cm/4 ft high. It has hairy leaves and bright blue, star-shaped flowers. It originated in E Europe, but is now found throughout most of Europe and N America.

Borcic acid. Boric acid, white powder used in medicine as very weak antiseptic.

Bordeaux mixture. (1) A mixture of copper sulphate (bluestone or blue vitriol). Lime, and water, used as a fungicidal spray on crops. (2) Bordeaux mixture was originally used in France in the 1860s to control grape diseases. The standard formula for bordeaux mixture is four pounds of copper sulfate, four pounds of hydrated lime and 50 gallons of water. Mix four pounds of the lime in four gallons of water. Do the same for the copper sulfate. Strain the lime mixture through cheesecloth, add to 42 gallons of water, and then add the sulfate mixture. Use immediately. Small amounts can be made by mixing four ounces of hydrated lime in 2 gallons of water. Mix four ounces of copper sulfate in 1 gallon of water. Pour the copper sulfate mixture into the lime mixture. Bordeaux mixture can cause damage to plants if used improperly. Damage or injury results more in humid weather and when the mixture doesn't dry quickly. Bordeaux mixture will leave a bluish-white deposit on the plant.

Border irrigation. Also known border strip irrigation. It refers to system of irrigation in which the field has been divided into border strips by using low ridges; water turned into upper end of each border strip moves down the slope in a thin sheet. It is an efficient, rapid and reltively easy method of irrigation. It is useful for all close-growing crops, some row crops and orchards where topography and soils are suitable.

Border plants. Plants located at the perimeter of landscapes or experimental plots.

Border rows. In hybrid seed plot, recommended number of rows of the male parental line grown on all the sides of the field.

Border strip. Grassed thickly vegetated strip at the edge of a field, along outlet channels, or at ends of rows to check or prevent erosion.

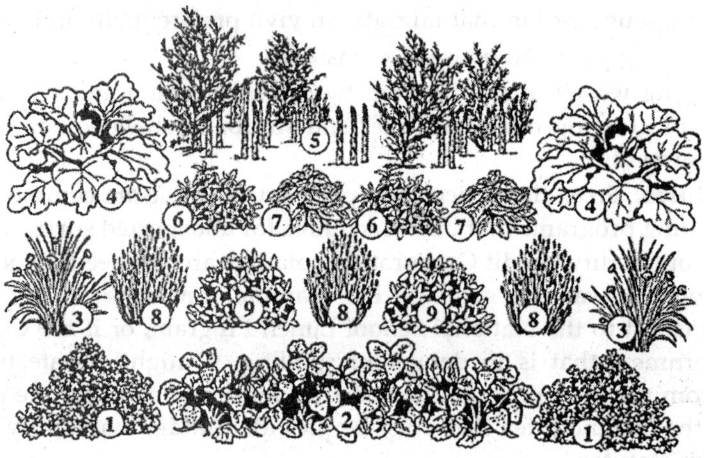

Fig.: Border plants

Border. Strip of land, which varies in width and length edged on both sides by a low ridge which guides the flow of water within that strip.

Boron (B). (1) A chemical element present in borax and boric acid, and essential to crops, epecially roots, in trace quantities only. Excess lime can induce deficiency, particularly in light soils, and result in deficiency diseases (e.g. Brown Heart, Hearts Rot). (2) Boron affects the absorption of other nutrients. It also affects the germination of the pollen tube once the pollen grains land on the stigma. Without germination, the seeds are not fertilized and fruits abort..

Signs of Deficiency: Multiple buds will develop with a lack of boron. Leaves will be small and heart rots and corkiness can develop.

Bosco. An Italian word, usually applied to a wood of evergreen oak (Quercus ilex) with a mysterious air. The Sacro Bosso at Bomarzo is, literally, a sacred wood - inspired by the ancient idea of making a Sacred Grove in association with an Egyptian temple.

Botanic garden. The idea of making a collection of plants is found in Egyptian and Mesopotamian gardens. In classical Greece and Rome, such plants were used for medicinal purposes. This practice was continued in the monasteries of medieval Europe and flourished anew with the scientific concerns of the renaissance. The world now has a large number of botanic gardens most of which are used for the scientific study of plants. (2) A garden used for growing and displaying ornamental plants for educational and research purposes

Botanical name. (1) Scientific name assigned to a plant species. (2) The Latin or scientific name of a plant, usually composed of two words, the genus and the species.

Botanical pesticides. (1) Pesticides whose active ingredients are plant-produced chemicals such as nicotine, rotenone, or strychnine. Also called plant-derived pesticides. Being "natural" pesticides, as distinct from synthetic ones, they are typically acceptable to organic farmers. (2) Pesticides obtained from plants.

Bottle garden. The art of making miniature garden by growing delicate plants inside bottles

Fig.: Bottle garden

Bottom growth. Those grasses and clovers in a pasture growing close to the ground, as opposed to the taller plants, or top growth.

Bottom heat. The undersurface heat provided in the soil by electric cables or hot water pipes.

Bottom land. Alluvial deposits such as a floodplain.

Bottom. Usually synonymous with "vessel" or "ship." A ship of American registry may be referred to as a "U.S. bottom," whereas if registry is other than U.S., the ship, in U.S. usage, may be called a "foreign bottom."

Botulism. Food poisoning which occurs due to the toxin of Clostridium botulinum. Fungal food poisoning lakes place in man and animals Fungal toxins include, aflatoxins, phallotoxins and muscarine.

Bougie. Cylindrical instrument which is used for insertion. It is used in the treatment of strictures.

Bound water. Water held by the cell against freezing.

Bovin malignant catarrh. Highly infections discase which is charaterized by fever, catarrhal inflammation of the oral, nasopharyn-geal and conjuctival mucous membranes, general lymphoid hyperplasia and frequent involvement of the central nervous system.

Bovine growth hormone. A proteinous substance produced and released by the pituitary gland which is located at the base of the brain.

Bovine ketosis. A disease of cattle caused by an excessive drain on blood sugar for the production of Lactose, which impairs energy metabolism and results in the accumulation of toxic levels of ketones in the blood, milk and urine. Symptoms include loss of appetite, sweet-smelling breath or milk (due to the presence of acetone and acetoacetic acid), falling milk yields, loss of weight, lethargy, and nervous behaviour. Most animals recover after treatment with glucose, glucose-forming substances or glu-coneogenetic hormones. Also called acetonaemia or acidosis.

Bovine leucosis. A cancerous disease, Leukaemia, of cattle. It may remain begin throughout life, or become malignant usually in cows 4 to 8 years old, resulting in fatality.

Bovine Somatotropin (bST). Also called bovine growth hormone, bST is a naturally occurring protein that has been genetically engineered as a synthetic compound (now manufactured in large quantities and commercially available to farmers) that causes cows to increase the efficiency of milk production per unit of feed consumed. Its use has caused public controversy, and some states require retail dairy product labels to identify the use of synthetic bST.

Bovine tuberculosis. A highly contagious disease of cattle that causes severe economic losses, especially in dairy herds. The Animal and Plant Health Inspection Service quarantines infected herds and works with producers to eradicate the disease. The target date for total eradication has been pushed back from 1998 to after the year 2000. A chronic contagious bacterial disease *Myco-bàcterin tuberculosis.*

Bovista. Genus of fungi belonging to the class Basidiomycetes. There are about 35 species, all found in temperate regions. They are small puffballs, up to 8 cm/3 in in diameter, and differing from *Lycoperdon* species mainly in having a mass of sterile, threadlike tubes or fibres among the spores. *B. plumbea* and *B. nigrescens* are found on heaths and pastures in Britain. They have no stem, but are held to the ground by a thread; when this is broken by the wind, the puffball is blown around, releasing its spores as it goes.

Box ridging. A system in which box shaped ridge is formed around the base of the plant to control erosion and aid irrigation.

Boysenberry. Boysenberries, like dewberries and logan berries are a trailing type of blackberries. Plants grow horizontally, compared to the erect habit of most blackberries. To grow, most plants are tied to supports or wires to limit their trailing habit. Fruits are sweeter, ripen sooner and are in looser clusters. Boysenberries are not winter hardy and need protection from cold temperatures. Plants should be untied from supports and lain horizontally after the ground starts freezing. Cover canes with straw or loose hay at least a foot or more. Prune boysenberries in a similar fashion to blackberries. Boysenberries, like most trailing blackberries, are difficult to grow in Illinois. Choose a protected location and expect a crop every five to ten years. Be proud if you succeed more often.

Bracing. Method of grafting done to support branches which are in danger of breaking off.

Bracken. A tall fern *(Peteridium aquiliunu,* also called Brake Fern. It is common on hillsides, heaths, moors and in woods, where burning or overgrazing allows it to establish itself extensively excluding heather or grasses.

Bract. (1) A modified leaf, often small and associated with leaves or inflorescences, but sometimes large and showy. (2) A modified leaf of an inflorescence. Several bracts form an involucre. (3) A modified, often very small leaf / leaf-like structure subtending a flower or inflorescence. It may appear to be the part of the flower. (4) Leaf-like or petal-like structure, which may appear with or below flowers, or have a bright flower-like appearance. (5) The reduced leaves of the inflorescence and upper part of a shoot.

Bracteate. With bracts. Bracted.

Brahman. Indian or Zebu cattle characterised by loose skin on the throat and dewlap, with well developed seat pores, a muscular hump over the neck and shoulders, and large drooping ears. Commonly steel-grey in colour, but variations from black to white are found, and some red strains. They are long-lived cattle, tolernt of high temperatures, and resistant to insects and various tropical diseases.

Braird. The first sprouting shoots of com or other crop. Also called breer or brere.

Brake harrow. A heavy harrow for breaking up clods.

Brake. (1) A harrow. (2) A frameworkd in which a restless horse is confined during shoeing. 3. A light carriage for breaking a horse to harness. 4. Brushwood or thicket. 5. An implement for crashing or braking flax or hemp. Also cattle a flax brake. 6. A scissor-like tool for stripping willow bark for basket-making.

Braky. Full of braken or brake, or rough and thorny.

Bramble. A fruit which is a member of the Rubus genus.

Bramley. A well known variety of apple.

Bran. Thus husks of ground com separated from the flour by bolting. A palatable feedingstuff with a useful protein and fibre content, especially for cattle and poultry. It is often fed to sows prior to farrowing and, as bran mash, is given to sick animals with no appetite. Also used in making brown bread.

Branch. One of the coarser divisions of a trunk or main stem; loosely, any division of the stem.

Branchlet. A smaller division of a branch.

Brand. (1) A mark burned onto an animal's hide for identification purposes by means of a hot branding iron. (2) General term for Blights or fungal diseases of grain crops.

Brash. (1) Loose disintegrated rock and rubble, or a soil containing many stone drippings or rock fragments. (2) Pruned branches of conifers.

Brashing. The pruning of the lower branches of young conifers between 15 and 20 years old up to a height of about 5 fit to allow better access into a plantation.

Brassica. The generic name for cabbage, and its related plants including cauliflower, broccoli, kale, savoy, Brussels sprouts,, and for turnip and swede.

Braxy pasture. Grassland infected with *Clostridium septique.* (Braxy).

Bray's nutrient mobility concept. According to this, as the mobil-ity of a nutrient in the soil diseases, the amount of that nutrient required in the soil to produce a maximum yield (the soil nutrient requirement) increases from a variable net value. It is deter-mined by the magnitude of the yield and the optimum percent-age composition of the crop, to an amount whose value tend to be a constant. According to Bray :

$$\log (A-Y)=\log A-C1b-CX$$

Where Y represents the yield obtained when X units of a plant nutrient are added to a soil having b units of the same immobile but available from a nutrient. A represents the maximum yield obtained under the given con-ditions. C1 represents a constant representing the efficiency of b for yields. C represents the efficiency factor for X which is the added fertilizer form of the nutrient b.

Breadcorn. Corn, from the flour of which bread is made.

Break. (1) To tame or train a horse to wear a saddle or for draught work. (2) A change of crop in a rotation programme. Thus an arable break could be corn grown for several years in a field following usually a root crop and to be succeeded by perhaps a ley. 3. To cut up an animal's body. (3) New lateral shoot, often developing after the removal of apical dominance by pinching

Breaker. A device at the end of a hose to permit application of water at low velocity.

Breaking ground. Ploughing uncultivated or fallow land.

Breast plough. An ancient but simple plough, spade-like, with a Song shaft and cross bar which was pressed against the chest. Used for paring turf.

Breast. The Mouldboard of a plough.

Breed count. Microscopic method of counting bacteria in a dried, stained film of milk.

Breed in, breed out. To introduce or remove a characteristic from an animal breed or plant variety by continuous breeding of those individuals lacking or having the characteristic, respectively, until it becomes fixed in or is lost from the breed or variety. (In Breeding, Character).

Breed. (1) To reproduce (both animals and plants). (2) To promote the reproduction of animals and plants, often under control, in order to select certain characteristics for transmission to offspring. (Bloodline, Cross-Breeding, Plant Breeding). 3. A strain, race, variety, stock or kind of plant or animal. Mostly the result of a continuous cycle of hybridisation followed by inbreeding of an isolated pocket of plants or animals.

Breeder seed. The seed raised and controlled by the originating or sponsoring institution or breeder.

Breeding crate. An apparatus designed to take the weight of bull which is too heavy for the heifer or cow with which is to be mated. Also called a service pen or service crate.

Breeding stock. Farm animals selected for producing offspring, as opposed to being fattened for slaughter, etc., in order to maintain or increase the size and quality of a herd of flock.

Brewers' grains. The residue of barley after being used in brewing beer, consisting mainly of protein and fibre, used in either wet or dried as an animal feedingstuff. Also called Draff.

Bridge grafting. (1) Refers to a form of repair graft ing for plants whose root system is damaged but bark of the trunk is injured. Trees of some species, like the cherry, and pecan can heal over exten-sively injured areas by the development of callus tissue. (2) The repair grafting done when the root system of the tree is injured.

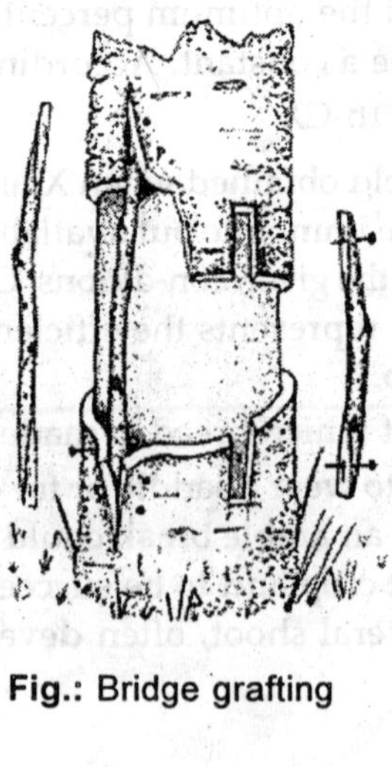

Fig.: Bridge grafting

Brim. (1) A sow on heat. (2) To put a boar to a sow for mating.

Brining. Treating vegetables with solution of salt in water (brine) to improve flavour and to serve as a heat transfer medium for facilitating processing.

Brittle Hoofs. Referto an abnormally dry state of the horn. The hoofs tend to become almost of the consistency of stone, chip and crack easily, cause contracted heels, and lead to difficulties in shoeing. Long-continued dryness or stabling on dry, hard floors has been conductive to the trouble.

Brix saccharometer. An instrument which is used for measuring the degree and gives the direct reading of the total solids dissolved in the juice. It has been found to vary from 22 to *225* depending on the variety.

Brix. (1) Refers to the percentage of total solids in sugarcane juice. It is read off from brixometer; range from 18 to 22; 15 in juice of young and immature cane. (2) The percentage of total soluble solids in fruit juices, plant saps etc., measured using Brix hydrometer.

Brixometer. Special hydrometer which is calibrated to read off directly the quantity of total solids in sugarcane juice.

Broadcast application. The spreading of pesticides or fertilizers over an entire area (see band application).

Broadcast fertilizer distributor. A machine which broadcasts fertilizer. Various distributive mechanisms exist including spin-ning discs, revolving fingers, rollers, brushes and chains etc.

Broadcast. Means sow or scatter seed on the surface of the land by hand or by machinery.

Broadcaster. A machine for sowing seeds, mainly grass and clover, consisting usually of a hopper supplying seed to one or more revolving brushes by which it is scattered (Fiddle).

Broadcasting. The process of scattering agricultural inputs, such as seed, fertiliser and manure on the surface of the soil by hand or by implements for uniform distribution over the entire field.

Broad-elliptic. Wider than elliptic.

Broadleaf. (1) A term applied to non-coniferous trees, almost cotermi-nous with deciduous. Leaves are usually broad, flat and then, as opposed to linear needles, and veins are networked as opposed to parallel. Broadleaf trees or hardwoods which are slower growing than Conifers and produce compact hard wood, take longer to provide a profit when grown commercially. They are suited best to lower hill slopes and deep fertile lowland soils. (2) A term applied to dicotyledonous plants, usually weeds, e.g., dock, Fot Hen, e c.

Brocket. (1) A stag in its second year with its first horns, unbranched and dagger-like. (2) Mottled.

Broken furrow. A type of furrow slice turned by a digger plough, rough and broken, as opposed to the smooth continuous furow left by a Lea Plough.

Broken work. Ploughing which leaves the furrows broken. Also called broken rib work. (Broken Furrow, Whole work).

Brome grass. A large genus *(Bromus)* of long-awned grasses strongly resembling oats, but more flowery, mostly unimportant as fod-der and generally regarded as weeds in Britain and becoming increasingly troublesome.

Bronchitis. Refers to the inflammation of the mucous membrane of the bronchial lubes.

Bronchopneumonia. Refers to the inflammation of the bronchi and of the lungs.

Bronzing. A metallic bronze or coppery color, especially of foliage after a winter.

Brood cell. Cell in which the larvae of bees are reared.

Brood chamber. Part of the beehive in which the brood is reared.

Brood comb. A wooden frame enclosing a wax sheet from which bees build the cells of the Honeycomb, placed in the brood chamber of a hive. (Beehive)

Brood frame. Wooden frame having the brood comb.

Brooding. Refers to the rearing of chicks after hatching till maturity; natural and artificial methods followed.

Brown budding. budding method commonly adopted in rubber using buds taken from bud wood of one year growth and stock plants often months or more growth.

Brown forest soils. An intrazonal group of soils having very dark brown surface horizons which have relatively rich humus (mull) grading through lighter colour soil into the parent material and characterised by a slightly acidic reaction, little or no illuviation of iron and alumina, and moderately high content of calcium in the soil colloids. Such soils have been developed under decidu-ous forest in temperate humid regions from parent material relatively rich in bases.

Brown head. A bloom that is excessively open.

Brown heart. A deficiency disease of swedes due to lack of boron, causing browning or mottling of the root making it unpalatable.

Brown Oil of Vitriol (BOV). Sulphuric acid, formerly used as a weed-killer, particularly against broadleaves such as charlock.

Brown rot. A fungal disease of apples, pears and plums, which turns the ripe fruit brown and rotten. They become mummified, hanging on the trees.

Browning. The process of browning of cut surfaces of fruits and plant parts due to polyphenol oxidation

Browse. (1) To feed and nibble on young shoots, buds and twigs of plants. (2) The tender shoots of shrubs and trees which cattle may feed on.

Brucclla abortus vaccine. It is obtained from a strain of *Brucella abortus* whose virulence has been reduced. It is widely used for the vaccination of calves between 4 and 8 months of age and is highly effective in preventing brucellosis.

Brush. (1) Vascular strands remaining on the pedicel of grapes after the berry is detached. (2) Commonly refers to undesirable shrubs and small trees. Loppings and trimmings from trees, shrubs and hedges. Also to lop or trim such plants, or to cut down weeds using a Brushing Hook.

Brushing hook. A sharp sickle-shaped tool which is used to trim hedges and clear undergrowth.

Brushwood. Low, rough, scrubby, undergrowth or thicket.

Brutting. A method of summer pruning used mainly by nut growers, but also applied to old

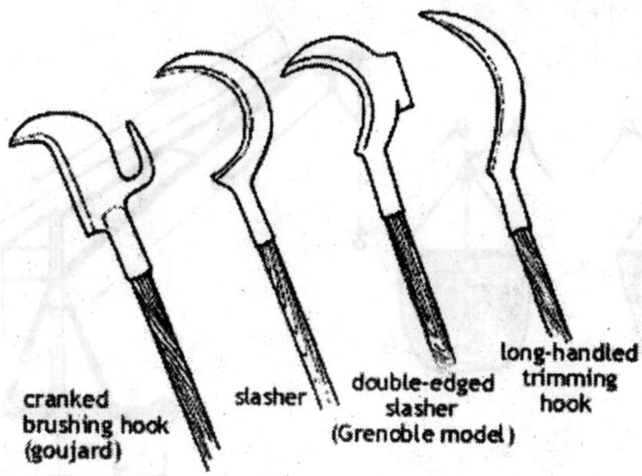

Fig.: Types of Brushing hooks.

fruit trees, in which side shoots of the bushes are snapped off near to their base by hand, and the ends left hanging, preventing secondary growth.

BST. Bovine somatotropin, commonly referred to as growth hormone. Produced naturally by the cow, stimulates metabolic functions related to growth and milk production.

Bt. Bacillus thuringiensis is a naturally occurring soil bacterium commonly known as Bt. It is a biological pesticide (biopesticide) used as a spray or dust and also in several genetically engineered plants. The plants have a gene from Bt inserted into their own genetic material. This new gene produces a natural protein that kills insects after the protein is ingested. The toxins are specific to a small subset of insects. Cotton has been genetically altered to control the tobacco budworm, bollworm and pink bollworm. Potatoes have been altered to control the Colorado potato beetle. A new hybrid of Bt corn, altered to be resistant to the European corn borer, became available for the 1997 planting season. Bt degrades rapidly to non-toxic compounds. It is not known to present any human or animal hazards. However, recent reports suggest that it may harm certain beneficial insects, such as monarch butterflies. Pest resistance management (PRM) plans are required by the Environmental Protection Agency as part of the registration.

B-trials. The term used for a series of microplot trials for ascertaining the value of a new variety over the area of its probable greatest usefulness.

Bucket elevator. An equipment which is used to lift the seeds to the top of bins/machines for cleaning, sorting etc.

Bucket-feeding. Feeding a young animal with milk and gruel from a bucket, rather than allowing it to suckle its mother.

Buckhound. A small staghound used for hunting bucks.

Buckrake. A simple implement with close-set long lines, usually rearmounted on a tractor hydraulic lift linkage, used for collect-ing and transporting cut grass and Green Crops for Silage making. Also used for other purposes such as transporting straw and hay hales.

Buckwheat. Two frost-sensitive cultibated plant species, common buckwheat (*Fagopyrum Esculentum*) and Tartarian buckwheat (*F. tartarcum*), often classed as Cereals, but actually

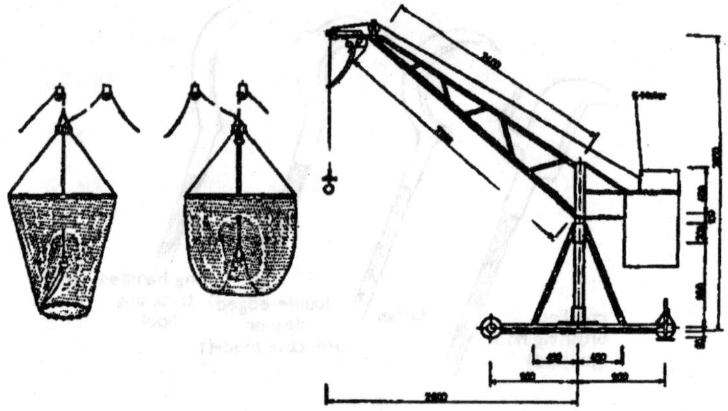

Fig.: Bucket elevator.

of the dock family. Grown mainly in patches, providing grain for peasants, and sometimes as a forage crop. Common buckwheat has white or pink flowers and dark brown seeds, sharply triangular in cross-section. Also called brank.

Bud. A structure of compact embryonic tissues, frequently enclosed in scales and becoming obvious in winter during plant dormancy.

Bud break. The resumption of growth of resting buds.

Bud eye. A potential growth point that is used for budding.

Bud mutation. (1) Mutation in bud, causing variation in the arising branch, flower or fruit (Synonym: Bud sport, Bud variation). (2) Mutation in the bud which brings about variation in it and produces a branch, flower or fruit unlike rest of the plant.

Bud rubbing. The removal of buds from spurs or shoots by pushing or rubbing with hand.

Bud scale. (1) A modified leaf or stipule (there may be one, a few, or many) protective of the embryonic tissue of the bud. (2) Protective scale-like leaves which cover the shoot apex, flower and embryonic leaves.

Bud sport (Bud mutation). (1) A plant or portion of a plant arising by spontaneous mutation in the apical meristem of a bud. (2) Mutation in bud, causing variation in the arising branch, flower or fruit (Synonym: Bud mutation, Bud variation)

Bud stick. shoot or twig used as a source of bud for budding.

Bud take. The success of budding operation as observed by the presence of green, healthy bud which is in the process of bud union.

Bud union. (1) A swollen and distinct node when a bud has been grafted. (2) The term used for the place of union of the bud with the plant-stock in a budded plant.

Bud wood. (1) Plant material which is used in budding of fruit and other trees. It is obtained from healthy vigorous and high yielding trees or branches which generally bear fruit of good quality, taken from mature wood of current year's growth, and from only round twigs, especially in the case of citrus species, the best being those from basal portions of young shoots. (2) Shoot from which bud is selected for budding.

Bud. (1) Undeveloped shoot or flower, often enclosed by reduced or specialised leaves called bud scales. (2) A vegetative outgrowth from yeast. (3) A structure of embryonic tissues, which will become a leaf, a flower, or both, or a new shoot. Especially the stage in which a growing point spends the winter or a dry season. May be naked or enclosed in bud scales. (4) The rudimentary or resting end or branch of a stem; usually referring to the stage in which the growing tips pass the winter or dry season; also applied to undeveloped flowers or flower clusters. A dormant growth point that is usually covered by scales.

Budded stump. Planting material comprised of the stock plus a dormant or recently activated bud, the stock being cut off a few centimeters above the bud

Budding and grafting knives. Knives which find use in budding and grafting operation in the asexual methods of plant propagation.

Budding. (1) A form of grafting using a single bud. (2) Production of a shoot from an underground stem. (3) The new growth in the spring. (4) The process by which yeasts produce vegetative outgrowths

Buddleia. Any of a group of ornamental shrubs or trees with spikes of fragrant flowers. The purple or white flower heads of the butterfly bush (*Buddleia davidii*) attract large numbers of butterflies. (Genus *Buddleia,* family Buddleiaceae.)

Budsports. Branch showing changes in one or more inheritable characters that could be perpetuated by asexual means.

Budstick. Shoot whose buds have been removed for preparation.

Buffer soil compounds. Clay, organic matter, and such compounds as carbonates and phosphates which make the soil to resist appreciable change in pH value.

Buffer strips. Slender areas of permanent vegetation, often planted along the edge or the contour of a field, usually to slow the flow of water or the velocity of wind, in order to capture sediment and other materials before they leave the farm and become pollutants. Types of buffers include filter strips, field borders, grassed waterways, field windbreaks, shelterbelts, contour grass strips, and riparian buffers.

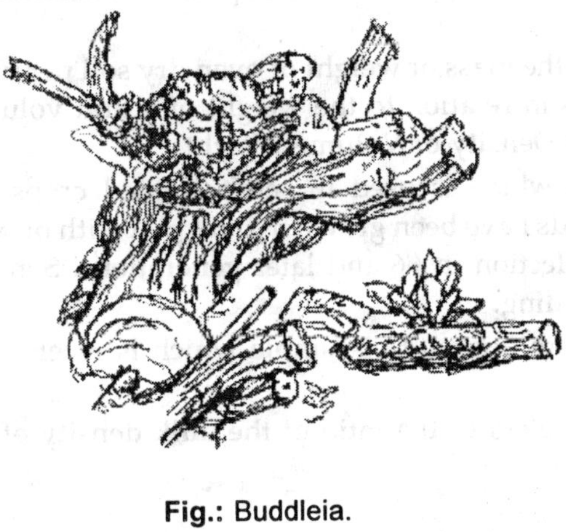

Fig.: Buddleia.

Bug spittle. Nymph of a sucking insect of the family Cercopidae. It commonly lives surrounded by a white, frothy mass resembling spittle.

Bug. Term strictly used only for members of the order Homoptera.

Bulb planter. Tool used to dig single holes for *bulbs.* This cylindrical, hand held tool minimizes disturbance to nearby *tree* or *shrub roots.* Best used to *naturalize* bulbs in *lawns.*

Bulb syrinage. It consists of a soft rubber bulb with a hard rubber pipe to be introduced into body openings *e.g.*, into the vagina.

Bulb. (1) A modified underground stem comprised of shortened central axis surrounded by fleshy scale-like leaves. (2) A specialized underground plant organ consisting of a greatly reduced stem (basal plate) surrounded by fleshy, modified leaves called scales. (3) The thickened underground storage organ of the group of perennials which includes daffodils and tulips. A modified underground stem comprised of shortened central axis surrounded by fleshy scale-like leaves.

Bulbar paralysis. Refers to the paralysis of the prolongation of the spinal cord into the brain.

Bulbel (bulbils, bulblets). (1) Small or miniature bulbs arising from the parent bulb. (2) Aerial bulblets. *Bulb cutting:* Cutting of bulb used for propagation.

Bulbil. (1) Small bulbs arising around the parent bulb. (2) Small bulbs arising in the leaf axils.

Bulblet. Small bulbs arising in the leaf axils.

Bulgur. Wheat that has been parboiled, dried, and partially debranned for later use in cracked or whole grain form.

Bulh dips. Chemical compounds in which bulbs are immersed to provide protection against pests and diseases.

Bulk bin. Large container which is used in handling and strong fresh fruits and vegetables. Most bulk bins have a capacity of 10 to 20 bushels.

Bulk commodities. Generally, high volume, low value unprocessed agricultural commodities, which are treated as though they are homogeneous (fungible) in nature prior to processing. Grains, oilseeds, and cotton are considered bulk commodities. Contrasting categories are high value commodities, semiprocessed and processed commodities, and consumer ready commodities.

Bulk density. Refers to the mass or weight of oven dry soil per unit bulk volume, including air space. This mass in relation to the weight of a unit volume of water was formerly termed as Apparent Density or Volume Weight.

Bulk method. Method which is used in self-polinated crops in which the segregating generations of hybrids have been grown in bulk plot with or without mass selection, and with single plant selection in F6 and later generations. Synonyms, bulk breeding and bulk method of breeding.

Bulk seed sample. Entire quantity of sample which is received in a seed laboratory for testing.

Bulk specific gravity. Refers to the ratio of the bulk density of a soil to the mass of unit volume of water.

Bulky feedingstuff. A feedingstuff having little nutritive value rela-tive to its volume, e.g., hay, straw or chaff in contrast to concentrates.

Bull head. (1) The corm of a banana plant which has flowered. (2) A condition when the short petals or excess number of petals give a blunt, broad appearance to a flower.

Bull. The inflorescence of banana during its ascent up the pseudostem.

Bulldog. (1) A deformed calf with a bulldog-like appearance due to its short legs and jaws and swollen abdomen. (2) A split ring which grips a bull's nostrils through which a rope can be passed for holding or leading.

Bulket. A dose of mineral, e.g., cobalt or magnesium, introduced to an animal by a special 'gun', providing it with a long-lasting supply, preventing Deficiency Diseases.

Bullimong. A mixture of oats, peas and mixed grains shown as a Forage Crop. The custom was confined to certain limited areas and has probably now ceased.

Bulling heifer. A maiden Heifer which has reached the right size and age for mating.

Bullock. A castrated bull. A bullock calf is a castrated male calf.

Bull-wheel. A large wheel which is providing drive to the machinery of a binder or reaper.

Bunchgrass. Grass that grows as distinct plants that get larger over time. Also called *clump-forming grass.* Contrast with *spreading grass* (also called *sod-forming grass*), which expands using running roots that create new plants from existing ones.

Bund former. An implement which is used for making bunds or ridges by collecting the soil. Bunds are needed to hold water in the soil, thereby conserve moisture and prevent run-off.

Bund. Low ridge of soil which is built along the field contours.

Bundle scar. Small dots or lines on the surface of the leaf scar marking the point of original departure of the vascular conducting strands into the leaf. Also called bundle trace.

Bunt. A fungus disease *(Tilletia caries)*, mainly of wheat but also of rye, filling the ears with a mass of black, fish-smelling, greasy spores. Bunted grains burst when threshed, the spores contami-nating and discolouring healthy grain, so that it is useless as seed or for milling. Also called stinking smut.

Bur, Burr. (1) The prickly seed-case or fruiting head of certain plants, particularly burdock, which adhere to animal fur and clothing. (2) A knob at the base of a deer's horn. (3) A knotting growth on a tree, leaving marks in timber. (4) The catkin or cone of the hop. (5) Arenaceous rock from which millistones (burr-stones) are made. Also one of the corrugations of a millstone. (6) Any rough or prickly seed envelope.

Burdock. Any of several bushy herbs characterized by hairy leaves and ripe fruit enclosed in burs with strong hooks. (Genus *Arctium*, family Compositae.)

Burgundy mixture. A misture of copper sulphate and sodium car-bonate (washing soda) used as a fungicide.

Burl. A knot or woody growth of very irregular grain.

Burley tobacco. The main type of air-cured tobacco; a cigarette tobacco that together with flue-cured tobacco account for more than 90% of total U.S. production. Burley tobacco production is limited by national marketing quotas and eligible for nonrecourse price support loans. Its production centers in Kentucky.

Burning. The browning of leaves caused by excess fertiliser

Burnt lime. A form of Lime with the highest neutralising value. Mainly Calcium Oxide and various residues produced by burn-ing lumps of Chalk or Limestone, and usually kibbed or ground before use. It readily absorbs water to form Calcium Hydroxide and cannot be stored for very long. Caustic and irritant, it is also used to destroy carcases of animals which have died from infectious diseases. Also called quicklime, lump lime and shell lime.

Bush drain. A type of field Drain which is constructed by packing bushes into a trench which is then refilled.

Bush fruit trees. A fruit tree, the lowest branches of which are below 75 cm (30 in) above the ground.

Bush fruit. Fruit which is grown on bushes as opposed to trees or Cane, e.g. blackcurrants, gooseberries.

Bush harrow. An old type of light harrow having a barried wooden frame with hushes or branches woven through it, used for covering grass seeds. Developed from the primitive harrow with a log tied on top of a bush.

Bush. (1) A low woody plant, without a distinct trunk, but with a number of branches at or near the ground level. (2) Term which is used to signify a region covered with forest or scrub not yet cleared for cultivation.

Bush-and-Bog. Refers to heavy cut-away disktillage implement used on rough or brushy pasture land.

Bushel. (1) A dry volume measure of varying weight for grain, fruit, etc., equal to four pecks or eight gallons (2150.42 cubic inches). A bushel of wheat, soybeans, and white potatoes each weighs 60 pounds. A bushel of corn, rye, grain sorghum, and flaxseed each weighs 56 pounds. A bushel of barley, buckwheat, and apples each weighs 48 pounds. (2) A dry measure by which grain was once generally computed, and also used for fruit, containing 1.28 cubic feet or 8 gallons. Many farmers still Drill seed in bushels per acre or refer to grain in terms of Ibs per bushel when discussing its quality, but, following metrication, grain traders now use kg per hectolitre. Fruit is now usually picked into bulk bins rather than bushel boxes. Also called sieve.

Business incubator. A facility that supports the development and operation of a number of small start-up businesses. Tenants of the facility share a number of support services including computers, support staff, telecommunications equipment, and janitorial services. Occupants also may receive technical assistance, business planning, legal, financial, and marketing advice.

Butcher system. A design, once popular, now rare, for hop garden wire-work, with three horizontal wires attaching to the poles the lowest at about 15 cm (6 in.) and the middle (breast wire) at about 120 cm (4 ft.).

Butt. (1) A tree trunk. (2) The base of a leaf stalk. (3) One of the short furrows ploughed where a field runs to an angle or at an awkward corner.

Butterfat. The fatty substances contained in milk, mainly glycerides of palmitic and oleic acids, present as minute fat globules, the size of which vary according to breed, and

diminish from time of calving. The largest and present in Jersey milk, churn easily and are best for butter-making. The smallest are present in Ayrshire and Freisian milk and are best for cheese. The butterfat content of milk is determined by the GerberTest and nowadays also by a number of semi-automatic instruments (e.g. Foss Milkotester) which are themselves calibrated using the Gerber Test.

Butterfly garden. Planting to attract, feed and support Lepidoptera.

Button center. A double rose flower with unexpanded central petaloid parts.

Button. (1) In coconut, female flower which is a small spherical body about 1.3 cm in diameter with resemblance to a small nut. (2) The young emerging fruit of a mushroom.

Buttress. Outgrowth of the trunk and roots at or above the surface of the ground to give the plant additional support

Bycatch. Fish or other sea creatures caught unintentionally during commercial fishing operations.

C

C Plant. A plant in which the first product of carbon dioxide fixation has been the 3-carbon compound phosphoglyceric acid. Ex-amples of this group include most crop plants except the grasses.

C_3 plants. Plants showing decreased carbohydrate production due to photorespiration and in which three-carbon sugars are the first stable intermediates in the Calvin Benson cycle of photosynthesis

C-4 Pathway. A cycle reported in a few tropical and temperate plants, It is characterized by using higher light intensities for fixing CO, than does the Calvin cycle.

C_4 Plant. (1) A plant in which the first product of CO2 fixation has been the 4-carbon compound oxaloacctic acid. Examples of C4 plants include many tropical grasses, corn, sugarcane and some weed species. (2) Plants that can photosynthesize at a faster rate than C3 plants due to special leaf anatomy and a unique biochemical pathway which begins with a stable, four-carbon sugar intermediate

Caatinga forests. Forests which are found in Brazil. In these forests the leaves fall in the dry season.

Cabbage lettuce. A type of lettuce which resembles a cabbage both in shape and its possession of a distinct heat.

Cabbage root fly. A fly *(Erioischia brassicae)*, which resembles the housefly. It lays white eggs in the soil surface near the steins of seeding Brasricas. The maggots, which hatch within a week, feed on the young roots and steins, causing plant water loss and frequently death.

Cabbage. (1) The piece of banana pseudostem discarded from the head or large corm. (2) The terminal bud of coconut. (3) *Brassica oleracea* var. *capitata,* the vegetable coming. (4) Abiennial cruciferous plant which is grown both as Fodder and for the table. It is characterised by a short steam with a very large head comprising tightly overlapping leaves. Cabbages are commonly fed to cattle, usually after milking, and pigs.

Cabotage. Trade or transport in coastal waters between ports within the same country. U.S. "cabotage" legislation—notably the so-called Jones Act — is designed to support the maritime industry.

Cacti. Nearly all its species may be grown in a moderately-heated green house.

Cade. A lamb or colt reared on the bottle. Also called Cosset Lamb.

Caecum. Refers to the blind, sac-like end of the large intestine which is connected to the small intestine. Also called blind gut.

Caffeine. The purine base alkaloid in tea and coffee, acting as a stimulant.

Cage wheel. Wheel or an attachment to a wheel having spaced cross bars to improve the reaction of the tractor in a wet field. It has been generally used in paddy fields.

Cairns group. An informal association of 15 agricultural exporting countries formed in 1986 at Cairns, Australia. During the Uruguay Round this group sought the removal of trade barriers and substantial reductions in subsidies affecting agricultural trade. These goals were in response to depressed commodity prices and reduced export earnings stemming from subsidy wars between the U.S. and the EU. The members account for a significant portion of the world's agricultural exports. The group includes major food exporters from both developed and developing countries: Argentina, Australia, Brazil, Canada, Chile, Colombia, Fiji, Indonesia, Malaysia, New Zealand, Paraguay, the Philippines, South Africa Thailand, and Uruguay. The Cairns Group was a strong coalition in the Uruguay Round of multilateral trade negotiations.

Calabash. Tropical South American evergreen tree with gourds (fruits) 50 cm/20 in across, whose dried skins are used as water containers. The Old World tropical-vine bottle gourd (*Lagenaria siceraria*, of the gourd family Cucurbitaceae) is sometimes also called a calabash, and it produces equally large gourds. (*Crescentia cujete*, family Bignoniaceae.)

Calcareous soil. An alkaline soil which is having sufficient calcium and magnesium carbonate to bring about visible effervescence when treated with hydrochloric acid.

Calcification. Refers to the process of the deposition of insoluble lime salts, especially calcium carbonate or phosphate in tissues (2) Refers to the process of accumulation of a layer of lime in the soil profiles in a dry climate, as it is happening in the soils of Rajasthan.

Calcified nodules. The nodules which are caused by nodular disease. They occur in knotty guts.

Calcium carbonate. ($CaCO_3$). A white insoluble solid which is found naturally as chalk, limestone, marble and calcite.

Calcium cyanamide. (1) ($CaCN_2$). Nitrolime. An artificial fertilizer which is powdery or granular, blue-black in colour, supplying lime and nitrogen to crops. It is converted by soil water to ammonia. (2) A very effective fumigant to control, insects and rodents.

Calcium hydroxide. $Ca(OH)_2$. Slaked lime or hydrated lime, It is in a finely divided condition by horticulturists to correct soil acidity.

Calcium levulinate. White powder which is soluble in water and therefore used increasingly for the preparation of sterile solu-tion.

Calcium nitrogen. A quick-acting fertilizer with about 48% calcium and 16% nitrogen.

Calcium oxide. (CaO). Quicklime. A lumpy white powder. It is caustic and irritant and is used to destroy carcases of animals which have died from infectious diseases. Also called burnt lime, and lump lime.

Calcium. (Ca) A soft white chemical element which is essential to life, and considered to be an important constituent of bones and teeth, and of the several compound forms of lime used in agriculture (Calcium Carbonate, Calcium Hydroxide and Cal-cium Oxide).

Calcium-phosphorus ratio. Normal ratio of calcium to phosphorus in feed ratios has been between 2:1 and 1:1. The ratio 2:1 is generally preferred. Vegetable protein concentrates and poor in calcium and phosphorus.

Calender of operations. Refers to graphic or tabular presentation which shows the kind of farm operations to be carried out during the season and specified time limits within which work has to be carried out.

Calendula. Hardy annuals which thrive in beds or borders and is almost any soil.

Calf crop. The number or percentage of calves produced in a herd within a given year relative to the number of cows and heifers exposed to breeding.

Caliche. More or less cemented desposit of calcium carbonate often mixed with magnesium carbonate at various depths. It is the characteristic of many of the semiarid and arid soils of the Southwest.

Calli clone. Plants regenerated from callus.

Callus tissue wound tissue cambium. A thin layer of meristematic cells between the bark and wood that gives rise to new *phloem* and *xylem* cells.

Callus. (1) Refers to the protective tissue which grows, as a swelling, over wounds in woody plants. (2) The cartilaginous growth, having collagen, which first grows to heal a bone fracture, later becoming bone tissue. (3) Any thickned tissue which is often hardened or leathery. (4) A hard protuberance, or the new tissues formed in response to a wound. (5) A thickened tissue that develops at the base of stem cuttings; new tissues formed in response to a wound. (6) Mass of parenchymatous cells, developing on or around a wound or from an explant.

Caloric energy of feed. May be defined as the inherent power of feeds for supplying nutrients which are essential to the repro-duction, maintenance, growth and milk production of animals.

Calorie. The quantity of heat required to raise the temperature of 1g of water by 1"C. A kilo-calorie (K.cal) or Large Calorie (Cal-orei) is 1000 calories.

Calvatia. Genus of fungi of the order Lycoperdales, in the class Basidiomycetes. They are puffballs; the pestle puffball *C. excipuliformis* has a rounded head with a tough outer wall on a long, thick stalk. It is common in woods, grows to 15 cm/6 in high and 10 cm/4 in diameter, and is edible when young.

Calve, Calve Down. To give birth to a call.

Calyx. (1) Sepals and often greenish bracts underneath a flower; the outer floral envelope. The outer set of perianth segments, usually green in color and smaller than the inner set. (2) The outer set of perianth segments or floral envelope of a flower, usually green in color and smaller than the inner set.

CAM plants. Desert plants in which Crassulacean Acid Metabolism (CAM) system is prevalent where the CO2 is fixed at night e.g. *Bryophyllum calycinum* and cactus plants

Cambium. (1) Cambium is the layer of actively dividing cells beneath the bark. Cambium cells that divide toward the trunk or center of the plant and form xylem tissues. Those that divide toward the outside produce phloem cells. The cambium layer is seldom more than 10 cells thick, usually averaging 5 cells. (2) A thin ring of tissue within the stem, branch, and trunk that continually forms nutrient and water-conducting vessels. (3) A lateral meristem in vascular plants which produces increases in diameter of stems and roots through the production of xylem, phloem and parenchyma cells. (4) A layer of actively dividing cells in plants which produces xylem, or wood, to its inside and phloem, or bast, to its outside in the process of secondary thickening. It is located just beneath the bark of trees and shrubs. (5) The thin membrane located just beneath the bark of a plant.

Camellia. Any oriental evergreen shrub with roselike flowers belonging to the tea family. Many species, including *Camellia japonica* and *C. reticulata*, have been introduced into Europe, the USA, and Australia; they are widely cultivated as ornamental shrubs. (Genus *Camellia*, family Theaceae.)

Campagna. This is the Italian word for open country. The 'Roman campagna' is the countryside around Rome - which was painted by landscape artists (eg Claude and Poussin) and helped to give form to English gardens in the eighteenth century.

Campanulata. Latin name for a plant having campanulate or bell-like flowers.

Campylobacteriosis. A diarrheal disease often caused by the type of bacteria known as *Campylobacteria jejuni* (*C. jejuni*) associated with poultry, raw milk, and water. There are an estimated 2.5 million cases annually in the United States with 200 to 730 deaths. Campylobacteriosis has been linked to Guillain-Barre syndrome (a disease which paralyzes limbs and breathing muscles) as well as Epstein-Barr, Cytomegalovirus, and other viruses. USDA has estimated that this disease costs the United States between $1.2 to $1.4 billion annually in medical costs, productivity losses, and residential care.

Canal System. A system which comprises of ail the canals. It constitutes a complete irrigation system.

Canal. Artifical watercourse which is usually cut.

Cancer eye. Malignant tumour which primarily attacks the eye of cattle and may be caused by irritation of the eyes by dust, sand, insects or strong rays of the sun. Its treatment involves the total removal of the cancerous tissue togheter with the eyeball by an experienced veterinarian.

Candied fruit. (1) Fruit impregnate with sugar, drained and dried is known as candied fruit. It is not sticky and is plump, tender and exceedingly sweet with high flavour. (2) A fruit / vegetable impregnated with cane sugar or glucose syrup, and subsequently drained free of syrup and dried

Cane sugar. A type of storage sugar (Sucrose) present in sugar cane, sugar beet, ripe fruits and tree sap (*e.g.*, maple sugar). It is broken down by intestinal enzymes into glucose and fructose which can then be absorbed by the bloodstream to provide energy.

Cane. (1) A main stem of a small fruit plant or vining crop. (2) In grapes the term applies to previous season's growth or new shoots after they lose their leaves. (3) Stem from a thin multi-stem plant; a long woody pliable stem rising from the ground. (4) The stem of reaspberry blackberry and other similar plants, and of the larger grasses, such as sugarcane or bamboo. (5) A long woody pliable stem rising from the ground. The long shoots of blackberry, grape, etc.

Canescent. Ash-colored, with gray hairs.

Canker. (1) A general term which is used for a number of fungal diseases of tree (often due to *Nectria sp.)*, especially those affecting fruit trees, in which growing tissues are destroyed and cortical tissues malformed, forming cancerous growths and open wounds. (2) A fungs disease which causes softening of the horn of a horse's hoof, inflammation and the production of a soft chessy material. (3) A colloquial term for ear mange in dogs and for the pussy conditions which can develop if not treated. (4) A dead area caused by a fungal disease. (5) Bacterial or fungal disease causing bark lesions

Canning. (1) Refers to the process of preserving food by the application of heat high enough for destroying essentially all microorgan-isms present together with sealing the food in air-tight sterilized canes so as to disallow recontamination and to preserve the food as early as possible in the condition in which it could be served. (2) A method of preserving food in a fresh condition by processing and sealing in airtight containers.

Cannula. A tube having a sharp-pointed plunger (a Tractor) which has been inserted to puncture an animal's rumen. The cannula is left in position when the trocar has been withdrawn to allow gas to escape.

Canopy class. Usually refers to a classification in accordance to completeness of canopy, *see* Canopy Density.

Canopy density. Refers to the relative completeness of the canopy. It is usually expressed as a decimal coefficient, taking closed canopy as unity. The following classification of canopy density is in vogue : *closed if* the density is 1.0 : *dense* if the density is between 0.75 and 1.0; *thin* if the density is between 0.5 and 0.75 and *open* if the density is less than 0.5.

Canopy. (1) Refers to branches, leave etc. which are formed by woody plants at some distance above the ground. (2) The overhead cover or horizontal projection of a plant formed by its leaves, branches etc. (3) The uppermost layer in a forest, formed collectively by tree crowns.

Cant. (1) An area of woodland clear-felled at one time. Often the unit of sale (by auction) for sweet chestnut *Castanea saliva)*, commonly about an acre in extent. (2) An area which is designed to a ploughman to plough in a ploughing competition.

Canterbury hoe. A type of hoe having three long, flat, sharp progns. Used break up clods.

Capability class, soil or land. A rating which indicates the general suitability of a soil or land for agricultural use.

Capacity, field moisture. Amount of water which is expressed in percentage of dry weight, that a soil is able to retain against the pull of gravity *i.e.* after saturation and allowing excess water to drain away.

Capillary action. Refers to the phenomenon of a liquid rising up a narrow tube, or the formation of drops, bubbles and films, due to high inter-molecular attraction within the liquid. Important in soil as the natural forces of capillary attraction cause water to rise through the minute 'capillary space' between soil particles and through soil pores.

Capillary fringe. Refers to the water in the layer of soil or subsoil into which ground-water is entering due to capillary rise. The ef-fective height of the capillary fringe is dependent on the pore sizes of the soil.

Capillary migration. Refers to the movement of water through a rock or soil material for the water. It takes place in the zone of aeration.

Capillary potential. The term used for the measure of the attractive forces with which water is held by soil.

Capillary water. Refers to the water which is held in the soil in pores, and capillary spaces, and as a thin film surrounding soil par-ticles, held under tension against the pull of gravity. Capillary water moves from water to drier parts of the soil under hydro-static gradient. It is actually the 'permanent' soil solution having dissolved substances such as applied plant nutrients.

Capillary zone. Zone in which soil water is held by capillary forces. The interstices may be completely filled (Saturation zone) or partly filled (aeration zone). Pressures within this zone have been less than atmospheric.

Capillary. Hair-like; very slender.

Capillary-fring belt. Refers to that portion of the aeration zone which is situated between the intermediate belt and water table.

Capitata. (1) Latin name for a plant with a head-like or large clustered feature. (2) Headlike, in a dense rounded cluster.

Capped blossom. Apple blossom in which the petals fail to open and from a brown 'cap' over the buds, as a result of the Apple Blossom Weevil eating the heart of the blossom.

Capped hock. Refers to the firm swelling which appears on the point of the hock of livestock. This blemish may be of the size of an apple or so small that it escapes notic. It occurs because of constant irritation, such as may be produced by nibbing or kicking the walls of the stable. The condition is not considered serious because it is not accompanied by lameness.

Capping. The term used for the formation of a crust on a soil during a hot, dry spell, usually after heavy rain which makes the surface soil particles to slake down, blocking up the soil pores.

Caprifig. Refers to the race of fig that does not form edible fruit but gives food for the wasps which pollinate the figs.

Capsicum. Any of a group of pepper plants belonging to the nightshade family, native to Central and South America. The different species produce green to red fruits that vary in size. The small ones are used whole to give the hot flavour of chilli, or ground to produce cayenne or red pepper; the large pointed or squarish pods, known as sweet peppers or pimientos (green, red, or yellow peppers), are mild-flavoured and used as a vegetable. (Genus *Capsicum*, family Solanaceae.)

Capsid bug. Refers to one of a large family of greenish or brownish plant Bugs, slender and elongate, about 6 mm (1/4 in.) long. The common Green Capsid is a pest of fruit and vegetables. The Apple Capsid which preys on the stem tips, fruit and foliage of apple trees may be now largely controlled by D.D.T. and other insecticides.

Capsid. Protein coat of a virus particle, which consists of capsomeres.

Capsule. (1) A dry fruit derived from a compound pistil, and opening in one of a number of ways. (2) A simple, dry, dehiscent fruit, formed of two or more united carpels. (3) A type of hard, dry, dehiscent fruit which is opening in various ways to release the seeds, *e.g.*, poppy. (4) A gelatinous envelope which is founder around certain kinds of bacteria. (5) A soluble gelatinous case having a dose of medicine, admini-stered orally to an animal.

4. The fibrous membrane which envelopes the various organs of the body, *e.g.* liver, spleen. (6) A thin metal envelope in an incubator, having a liquid (*e.g.*, ether) which expands and contracts, maintaining an even tem-perature regime in the incubator.

Captan. A fungicide which is used in seed dressings for peas and vegetables. Also used effectively against apple scab.

Captive supply. Products that manufacturers or processors own or contract to purchase for future delivery so as to have a predictable source of raw materials for their plants. In agriculture, the term often is used, for example, to refer to the cattle that beef packers own or contract to purchase 2 weeks or more before slaughter. Examples of such contracts include an exclusive agreement with an individual feedlot in which the price is based on market prices at time of slaughter; or a contract in which the price is specified in advance or is based on some other formula. At issue is the effect that captive supplies have on prices paid to cattle producers.

Car ivore. Any flesh-eating mammal which has teeth and claws *e.g.*, dogs, cats, bears etc.

Carbamates. A group of alkaloid insecticides which are selective and to some extent systemic in their action, and are non- persistent in the environment, *e.g.*, carbaryl, methiocarb.

Carbohydrates. A large group of compounds having carbon, hydro-gen and oxygen only, of the general formula $Cx(H_2O)y$. They are produced by plant Photosynthesis and are essential to me-tabolism in all living organisms. Carbohydrates form the largest part of the food of animals which have only small amounts of glycogcn and sugars in their bodies. The complex polysaccharides in plants are broken down by animals into simpler sugars by digestion before they can be absorbed and used. The energy stored in carbohydrates is re-leased, to power living processes and provide heat, by 'burning' with oxygen during cellular respiration.

Carbolineum. Wood preservative which is prepared from coaltar. It consists of anthracene oil and is an effective disinfectant and insecticide. It is widely used for the preventation or control of mites and ticks infecting stables, poultry houses, barns etc.

Carbon cycle. The circulation of carbon atoms between living organisms and the atmosphere. Carbon dioxide is assimilated during Photosynthesis by plants to produce carbohydrates and other complex carbon compounds. Animals feed directly on plants or on herbivores to obtain these compounds which are broken down during both animal and plant respiration to release carbon dioxide back to the atmosphere. Carbon dioxide is also released from the decaying remains of plants and animals by the action of bacteria and fungi.

Carbon dioxide. (CO_2). A colourless, odourless, gas occurring in the atmosphere. It is utilised by plants in Photosynthesis to produce Carbohydrates. A by-product of animal and plant Respiration and combustion (*e.g.*, straw burning).

Carbon nitrogen ratio (C.N ratio). The ratio of carbonaceous materials to nitrogen in plant or soil.

Carbon sequestration. Retention of carbon in ways that prevent or delay its emission to the atmosphere as carbon dioxide. This may help mitigate climate change by reducing the amount in the atmosphere. Silvicultural practices that encourage rapid, long term tree growth are an example. Crop residue retention practices designed to prevent erosion and improve the productivity of soil, such as conservation tillage, also retain larger amounts of carbon compared to many traditional cultivation practices.

Carbon. (C). A chemical element. The principal constituent of all organic compounds and essential to life.

Carbon-nitrogen ratio. (1) Refers to the ratio of the weight of total organic carbon to the weight of total nitrogen in a soil or in an organic material. The C:N ratio of wheat straw has been nearly 70:1 where as that of soil is 10:1. When undecomposed straw with a high C:N ratio is applied to the soil, its C:N ratio gets reduced through bacterial decomposition. To speed up bacterial decomposition, nitrogenous fertilizer has been added to the soil at the time of turning under straw.

(2) The ratio of the weight of organic carbon to the weight of total nitrogen (mineral plus organic forms) in soil or organic material.

Carboxydismutase. Enzyme which is responsible for the fixation of inorganic CO_2 into organic compounds in the dark reactions of the C3 photosynthesis cycle; also known as ribulose diphosphate carboxylase.

Carcinogen. Any substance that produces or promotes cancer. This is a key consideration in evaluating the safety of pesticides and other chemicals.

Cardinal temperatures. Minimum, optimum and maximum temperatures which are affecting plant growth. The minimum temperature is defined as that point at which growth begins; the optimum temperature is that point at which growth is best; and the maximum temperature is defined as that point at which growth stops.

Carminative. Drug relieving flatulence and colic by stimulating the movements of stomach and intestine.

Carnivorous plants. Plants which are able to capture small insects or animals and feed upon the nitrogenous compounds derived from them *e.g.* pitcher plant, bladderwort etc.

Carol bean. An alkaloid-containing poisonous plant.

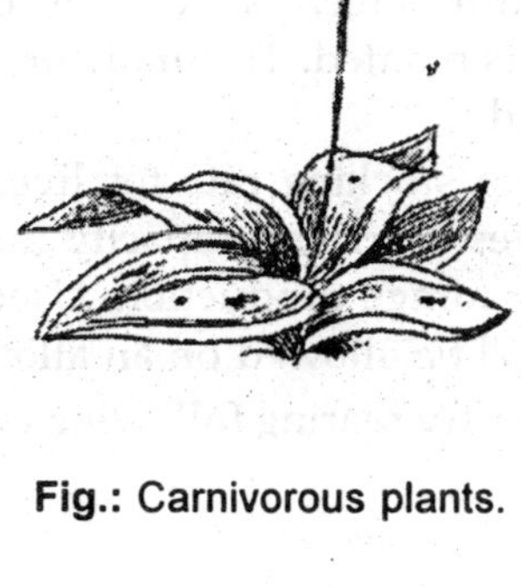

Fig.: Carnivorous plants.

Carotene. Carotene is yellow pigment found in many yellow or orange fruits and vegetables including carrots, squash, corn, bananas and cantaloupes. The yellow in buttercups and daffodils is caused by the carotene pigment. Carotenes are present in lesser amounts than the other yellow pigment, xanthophyll. Like xanthophyll, carotenes are present throughout the growing season, but masked by the chlorophyll (green) pigment.

Carotenes. A group of reddish yellow pigments. They are unsatu-rated hydrocarbons (c40H56), occurring in carrots and butter, providing their distinctive colours. Carotenes are widely distrib-uted in plants, acting as photosynthetic pigments (Photosynthe-sis) in cells lacking chlorophyll, and are converted into vitamin A by animals.

Carotenoid pigments. Orange or yellow pigments which are occurring in cells are crystals, usually in chromoplasts or in plastids.

Carotenoids. A group name for lipochromes or carotenelike plant pigments (Yellow compounds), which through food, get depos-ited in animal tissues.

Carotine. An orange pigment occurring in certain plastids; precursor of vitamin A.

Carpel. A single pistil in a female flower part containing several pistils (2) One of the innermost floral structures which bears or encloses the ovules.

Carpet bedding. (1) Covering the beds in gardens with dense, low growing herbaceous plants. (2) The nineteenth century practice of using bedding plants to create carpet-like patterns. The idea derives from the older ideas of knot gardens and parterres. At the end of the nineteenth century 'carpet bedding' became a term of abuse for annual displays of plants, but it remains popular in show gardens and public parks.

Carr. Boggy ground or land reclaimed from such areas by drainage. Also a fen wood, common in East Anglia, dominated by alder.

Carrecting strip. An irregular strip or area of land which is laying between contour strips.

Carrot fly. A small pestilent fly *(Psila rosae).* green-black in colour, attacking carrots, cellery, parsley and parsnips. Eggs are laid in the soil near the plants and the slender white larvae burrow into the roots near the growing tips. Growth is retarded and foliage loses its freshness. With serious infestation roots may fork or show distorted growth, and in some cases crops may fail.

Carrot. A root crop bred from the wild carrot *(Daucus carota)* with a long, tapering, reddish or yellowish root, which is sweet and edible. Grown mainly for human consumption, often on con-tract, for canning or freezing. Surplus or unmarketable roots are used as stook feed, mainly for cattle.

Carry over-soil moisture. Moisture which is stored in crop root zone between cropping seasons or before the crop gets planted. This moisture contributes to the water needs of crops before they are irrigated.

Carrying capacity. The maximum stocking rate for livestock possible without damaging vegetation or related resources. Carrying capacity may vary from year to year on the same area, due to fluctuating forage production. Used by the government in decisions about how many livestock will be allowed on an allotment on public lands.

Carry-on-period. The stage in poultry rearing following brooding and before chicks are able to survive without assistance.

Carryover. (1) Existing supplies of farm commodity not used at the end of a marketing year, and remaining to be carried over into the next year. Marketing years generally start at the beginning of a new harvest for a commodity, and extend to the same time in the following year. (2) The supply of a farm commodity not yet used at the end of a marketing year and carried over into the next marketing year. An excessively large carryover is typically described as a surplus condition that causes prices to fall. When the carryover falls below normal, there may be concerns of a shortage contributing to price escalation.

Cartel. An alliance or arrangement among industrial or commercial enterprises or nations aimed at limiting competition or exercising monopoly power in a market.

Carvopsis. (1) An indehiscent fruit with one seed which is completely fused to the inner surface of the pericarp. (2) The fruit of members of the grass family. (3) The fruit of members of the grass family; not basically distinct from an achene.

Cash grain farm. A farm on which corn, grain sorghum, small grains, soybeans, or field beans and peas account for at least 50 percent of value of products sold

Cash in lieu of commodities. Refers to cash provided to food program operators (e.g., elderly nutrition programs, child care food programs, and some school food programs) in lieu of mandated commodity assistance. Recipients may use the cash to buy whatever foods they need to operate their meal service programs.

Casting (ploughing). If a plough works round a strip of unploughed land it is termed to be casting.

Castor oil. An oil extracted from castor bean. It is a pale yellow liquid and is widely used as a mild purgative. It is sometimes effective for the removal of small wart *e.g.*, those on udders of cows.

Casual. (1) An introduced plant which has not become established, although sometimes found in places where it is not cultivated. (2) Plant occurring in a plant community of which it is not a regular inhabitant.

Catabolism. Metabolic process in which complex materials are converted into simple compounds e.g., respiration

Catalpa. The catalpa tree is widely cultivated in the southern United States as an ornamental shade tree. Growing to a height of 12 m/40 ft, it has spreading branches and large heart-shaped leaves. The white tubular flowers, tinged with purple, are succeeded by long, beanlike pods.

Catapult mechanism. Refers to the method of seed dispersal which depends on the jerking of a long stalk swaying in the wind.

Catastrophic Crop Insurance (CAT). A component of the federal crop insurance program, authorized by the Federal Crop Insurance Reform Act of 1994, that compensates farmers for crop yield losses exceeding 50% of their average historical yield at a payment rate of 60% of the projected season average market price. CAT coverage requires that a farmer realize a yield loss of more than 50% and only makes payments on losses exceeding the 50-percent threshold. Producers pay no premium for CAT coverage, but except for cases of financial hardship must pay an administrative fee of $50 per crop, up to a maximum of $200 per county and $600 in total (across all counties) for CAT protection. Under the

Reform Act of 1994 producers were required to obtain coverage at the CAT (or higher) level for crops of economic significance (accounting for 10% or more of their farm's crop production value) in order to be eligible for various other USDA program benefits. The FAIR Act of 1996 relaxed this requirement. A producer has the ability to purchase additional insurance coverage beyond CAT coverage, but must pay a premium, partially subsidized by the government, for that additional coverage.

Catch crop. (1) A quick growing crop incidentally planted and harvested between two major crops in consecutive seasons. (2) A contingency crop grown to replace a major crop which has failed

Catch wire. A wire that serves as an attachment for developing grape shoots or other plants with tendrils

Catechu. Name used for two drugs both of which are astringents obtained from trees. They have been compatible with zinc, calcium or iron compounds. Catechus are sometimes used in the treatment diarrhoea.

Catena. Group of soils always occurring in association with one another in some characteristic pattern related to topographical factors.

Caterpillar. The larva of a butterfly (e.g. Cabbage White) or moth, often causing damage to crops by feeding on foliage.

Cath crop. (1) Quick-growing crop which could be planted and harvested between two regular crops in consecutive seasons, or between patches of regular crops in the same season (2) Crop which is planted after another has failed or when the season has been too late for the usual crops.

Cathartic. A medicine acting by stimulating the movements of the intestines or by liquifying the intestinal contents. Mild cathartics like sulphur, mineral oil etc. are called laxatives while more powerful cathartics like castor oil, epsom salt and also are purgatives.

Cation Exchange Capacity (C.E.C.). (1) The total amount of exchange-able cations that a soil can absorb, expressed in milliequivalents per 100 g of soil or of other adsorbing material such as clay. It is a measure of the potential of a soil to hold nutrient cations for plant absorption. Clayey or organic soils usually have a large C.E.C., whilst the opposite applies to sandy or weakly organic soils. Agriculturally productive soils require application of fertilizers rich in cations and the C.E.C. is a guide to the quantity and frequency of application. Also called Total Exchange Capacity, Base Exchange Capacity, and Cation Adsorption Capacity. (2) The clay and organic components of the soil have a negative charge. As a result of these charges, positively charged ions (cations) such as hydrogen H+, potassium K+, ammonium NH+4, calcium Ca2+, magnesium Mg2+, aluminum Al3+, etc. may be held at the surface of the clay or organic particles and exchanged with other ions in the solution or with ions at the plant root's surface. This ability of a soil to hold cations is termed its *cation exchange capacity* (CEC). Since many cations are plant nutrients the cation exchange capacity is a measure of the soil's ability to hold such nutrients. Expressed in milliequivalents per 100 grams or per gram of soil (or other exchangers such as clay).

Catkin. (1) A single-sex spike with no petals; a slender, spikelike, drooping flower cluster. A spike-like inflorescence comprised of scaly bracts subtending unisexual flowers. (2) A spike-like inflorescence comprised of scaly bracts subtending unisexual flowers; often somewhat flexuous and pendulous but not necessarily so. Ex.- the flowers of willows or poplars.

Caudiform. A plant with a fleshy swollen base that is not a true bulb.

Caulescent. Having an evident leaf-bearing stem above ground.

Cauliflorous. (1) Bearing inflorescence on the trunk or woody branches e.g. cocoa; jack. (2) Flowering on the trunk or on specialized spurs from it or from the larger branches.

Cauline. Arising from or inserted on the stem.

Cave biology. Study of cave-dwelling organisms (troglobionts).

Cavings. The small pieces of straw and leaves broken up by a Threshing Machine and delivered from it after shaking out the grain and chaff.

Cell wall. The limiting layer of a plant cell enclosing the protoplasm, comprising the plasma membrane and cellulose laid down as a crystal lattice.

Cellar gardening. Using a basement to grow plants with artificial light.

Cellulose. A fibrous polysaccharide, the fundamental constituent of the Cell Walls in plants, (Carbohydrates). Cellulose is a common material in plant *cell walls* and was first noted as such in 1838. It occurs naturally in almost pure form only in *cotton* fiber; in combination with *lignin* and any *hemicellulose,* it is found in all plant material. Cellulose is the most abundant form of living terrestrial biomass *(R.L. Crawford 1981).* Cellulose, especially *cotton linters,* is used in the manufacture of *nitrocellulose,* historically used in *smokeless gunpowder.* Some animals, particularly *ruminants* and *termites,* can digest cellulose with the help of *symbiotic* micro-organisms - see *methanogen.* Cellulose is processed to make *cellophane* and *rayon,* and more recently *Modal,* a textile derived from beechwood cellulose. Cellulose is also used within the laboratory as a solid-state substrate for *thin layer chromotography.* Cellulose is the major constituent of *paper.*

Cemented soil. Soil in which the grains or aggregates adhere firmly and are bound together by some material which serves as a cementing agent such as colloidal clay iron silica, or alumina hydrates, lime etc.

Census of agriculture. A comprehensive set of quantitative information on the agricultural sector of the U.S. economy, broken down to the state and county levels (i.e., number of farms, land in farms, crop acreage and production, livestock numbers and production, production expenses, farm facilities and equipment, farm tenure, value of farm products sold, farm size, type of farm, among other data). The Census, conducted every 5 years, was the responsibility of the Commerce Department's Bureau of the Census. However, the FY1997 USDA appropriations act (P.L. 104-180, August 6, 1996) transferred funding for the Census of Agriculture to USDA's National Agricultural Statistics Service (NASS). NASS released the results of the 1997 Census in 1998.

Center for veterinary medicine. An agency within the Food and Drug Administration that is responsible for assuring that all animal drugs, feeds (including pet foods), and veterinary devices are safe for animals, are properly labeled, and produce no human health hazards when used in food-producing animals.

Center pivot irrigation. A self-propelled irrigation system in which a single pipeline supported on towers rotates around a central point. These systems are typically about one-quarter mile long and serve 128 to 132 acre circular fields.

Center plants. Provide compact, upright growth to fill in the crown of the container.

Central leader. (1) System of training a fruit tree in which it is trained to form a trunk extending from the soil to the top of the tree. (2) In this system a plant or a tree has been trained to form a trunk which extends from the surface of the soil to the top of the tree.

Central seed laboratory. Under Sub-section (i) of Section 3 and of the Indian Seeds Act. 1966; the Seed Testing Laboratory at IARI New Delhi has been recognised as the Cental Seed Testing Laboratory.

Centre of diversity. An area where the genetic and specific diversity is very high.

Centrifugal divider. An instrument which is used for dividing a sample in the Seed Testing Laboratory. It uses centrifugal force to mix and scatter seeds over the dividing surface.

Cereal forage. Cereal crop which has been harvested when imma-ture for either hay, silage, green feed or as pasturage, *e.g.*, Bajri fodder.

Cereal substitutes. Substances used by animal feed compounders to provide the carbohydrate content of the feed instead of traditional cereals. The common substitutes include manioc, maize gluten, rice bran, milling offals, grain screenings, citrus and beet pulp, and brewing and distillery waste.

Cereal(s). Cultivated members of the grass family whose seeds or grain are used to provide flour for breadmaking or as animal feed. The grain is rich in Starch but also contain valuable Proteins and Vitamins. The main cereals grown in India are Barley, Maize, Oats, Rye and Wheat.

Cernuous. Drooping or nodding. Having branches or flower heads that bend downward.

Certification (seed). Refers to the system of maintaining the genetic purity and quality of seeds. The crops offered for certificating are grown as per requirement of seed certification standards which have been established by a seed certification agency. The crop has been inspected several times to ensure purity and quality of seeds.

Certification sample (Seed). Refers to a smaple of seeds which has been drawn by a certification agency or by a duty authorised representative of the agency. The sample would be for the purpose of determining if the lot of seed from which the sample was collected is having statisfactory germination and purity to get tagged and sold as certified seed.

Certification tag. Lobel of specified design which indicates the certificate issued by the seed certification agency.

Certified plants. Plants produced under strict guidelines and inspections controlled by a regulatory agency to ensure trueness to type and freedom from damage, insects, diseases etc.

Certified seed producer. Person growing or distributing certified seed as per procedures and standards of the certification agency.

Certified seed sample. Samples that have to be submitted by a seed certification officer or inspector for escertaining the quality to be tagged and sold as certified.

Certified seed. (1) Seed which is grown from certified seed stocks and that which is meeting the standards set up by a certifying agency for germination, presence of weed or other seed and impurities, freedom from disease -carrying germs. (2) The progeny of breeder, foundation or registered seeds, so handled as to maintain satisfactory genetic purity and identity.

Certified/certification. Seed or stock verified to be true or lacking viruses.

Cetidiology. Division of biology which deals with the study of galls on plants caused by insects, mites and fungi.

Chafers. Various beetles, the grubs of which are pests in old grass-lands, usually those which are in slightly dry situations.

Chaff cutter. Machine used for chopping fodder.

Chaff. In general terms any worthless matter. The husks of corn separated from the grain during. Threshing or Winnowing Also short lengths of cut hay or straw. Also known as dowse or chop.

Chain harness. A type of harness used to draw ploughs or other im-plements having no shafts. Also known as sling gear.

Chain harrow. A type of Harrow, lacking a rigid frame, comprising a series of flexible chain links bearing spikes or knifetines, generally double-ended and reversible. This arrangement per-mits the harrow to follow uneven ground very closely.

Chain. A measure of length equivalent to 22 yds (20.1 m) or 100 links.

Chaldron. A unit of dry measure formerly used in England, equal to 4 quarters or about 32 bushels for grain and 36 bushels for coal.

Chalk. (1) White limestone product or soil. A soft white or greyish, very porous type of limestone, very rich in calcium carbonate. It is spread on the land to correct soil acidity and calcium deficiency, either as rough chalk or after crushing (ground chalk) when it is more effective. It is some-times klin dried, finely ground, and sold as ground carbonate of lime.

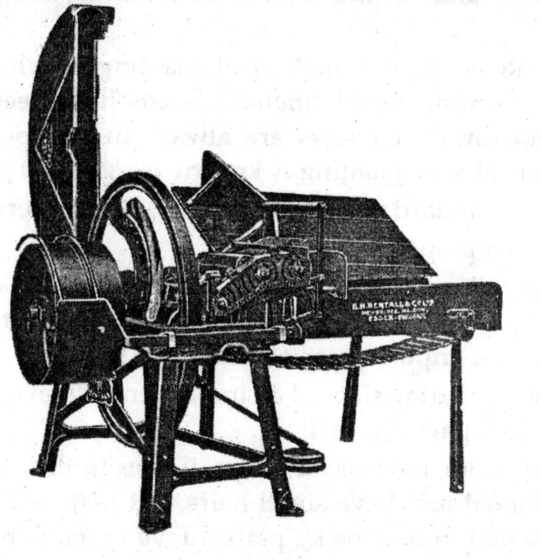

Fig.: Chaff cutter

Chambered. Said of pith when divided into small, empty compartments separated by transverse partitions; with cavities separated by plates or disks.

Channel terrace. Refers to the type of contour terrace for soil conservation, wide and relatively shallow channel excavated at suitable intervals on a falling contour with a suitable longitudi-nal gradients excavated soil deposited as a wide, low ridge along the lower ridge of the channel. It is recommended for relatively impervious soil and in heavy rainfall tracts.

Channeled. Grooved lengthwise.

Channelization. Engineering watercourses by straightening, widening, or deepening them so water will move faster. While improving drainage, this process can interfere with waste assimilation capacity, disturb fish and wildlife habitats, and aggravate flooding in other areas.

Charcoal. A soil additive to increase moisture intake.

Chartreuse. Variable green-yellow in coloration, often considered ineffective or detrimental if referring to autumn foliage color; lack of uniform color.

Check dam. Small low fixed dam which has been constructed of brush, logs, timber, loose rock masonry or concrete, in an eroded channel for reducing the slope of the water flowing therein during high stages, and also the resulting velocity, thereby preventing excessive scour and erosion and inducing deposition.

Check irrigation. Method of irrigation in which the water has been applied to a field or orchard which is divided into a series of checks by earth ridges, the water flowing from one check to another along the slope.

Check method of irrigation. Method of irrigation by which water has been run into relatively levelled plots which are surrounded by small ridges. It is the most common method in India and is suited to irrigation of grain and fodder crops in heavy soils where water is absorbed very slowly and so must stand for relatively long time to assure adequate penetration.

Check row planting. Refers to the method of planting in which row to row and plant distance has been uniform. In this method, seeds have been planted precisely along straight parallel fur-rows. The rows are always in two perpendicular directions. A machine used for check row planting is known as *check row planter.*

Check. A row or plot of standard variety included in field experimentation for comparison

Checking. The tilting of a plough to the land.

Check-off program. Usually, a reference to the generic research and commodity promotion programs for farm products that are financed by assessments applied to sales of those products by producers, importers, or others in the industry.

Cheek. One of two corresponding sides of an implement. Often used for the side of a plough facing the unploughed part of a field.

Cheilanthes. Genus of about 180 species of small *ferns* in the family Polypodiaceae. Their fronds are often curved and have small hairs that help to conserve water, since these ferns are usually found in dry, rocky places. Five species are native to Europe, and C. *pteridioides* will grow in Britain under sheltered conditions.

Chemical analysis. The determination of the composition of a substance using chemicals to break it down into simpler substances. The analysis may be qualitative, showing only the constituents present, or quantitative, showing the actual amounts present. Used, for example, to determine the Fertilizer require-ment of soils, or the compositional quality of milk.

Chemical mutagen. A chemical capable of causing genetic mutation.

Chemical score. Refers to a method which is used for determining the protein quality of feed in which the content of each essential amino acid has been determined and expressed as the percent-age of standard, the lowest percentage is considered to be the score.

Chemical thinning. Thinning of blossoms by the use of chemical substances.

Chemical weed control. The application of herbicides to crops on hare land to kill weeds.

Chemigation. The application of a pesticide and/or fertilizer through any irrigation system. This delivery technique raises some concern that it may increase pollution.

Chemosterilant. A chemical that controls pests by preventing reproduction, thereby causing the population to collapse. This contrasts with chemicals that directly kill pests.

Chemotroph. An organism (bacterial) that uses chemical energy for the synthesis of food.

Chemotropism. A bending or turning in response to chemical stimulus.

Chenopodium oil. Volatile oil which is obtained from the above-ground parts of the American wormseed plant. It is a pale yellow liquid and is used as an anthelmintic. It may be used for the treatment of ascariasis in calves hut is not given to calves suffering from diarrhoea or constipation.

Cherelle. Small and immature pods of cocoa.

Chernozem. Soil characteristic of some grassland vegetation in warm areas having moderate rainfall; dark in colour because of a high content of organic matter.

Chersophyte. (GK *chersa,* dry places: *phyton, plant*) A plant which grows on waste land or on shallow soil.

Chestnut soil. Zonal group of soil which are having a dark brown surface horizon which grades below into lighter coloured soil and finally into a horizon of lime accumulation developed under mixed tall and short grasses in a temperate to cool and subhumid to semi-arid climate. They are lound on the arid side of the chernozem soils into which they grade.

Chewing cattle-louse. The only biting species of cattle lice which concentrates on the withers and around the mot of the tail.

Chewing the Cud. An activity of Ruminants, involving the mastica-tion for the second time of food which has previously been swallowed and passed in to the first stomach or Rumen.

Chicory. Roots of *Cichorium intybus* used for blending and mixing with coffee to the extent of 40 per cent. Chicory (*Cichorium intybus*) is a hardy perennial that was brought to North America from Europe in the 1700s, and is now well-established across the continent. Though chicory has a variety of uses, it's best known for its association with coffee. At many points through history, coffee has become unavailable or too costly. During these times, people have often turned to roasted chicory as a substitute. Folks also used to make coffee from roasted acorns, yams and a variety of local grains. Anything was better than going without! The root of the chicory plant is long and thick, like the tap-root of the dandelion. When dried, roasted and ground, it makes an excellent substitute for coffee. There is no caffeine in chicory, and it produces a more 'roasted' flavour than coffee does. Many coffee pröducers offer blends with up to 30% chicory, which cuts down on the caffeine content of your cup. But many folk enjoy a cup of 'coffee' made entirely from ground, roasted chicory. Another *perk* about chicory is that it's more soluable in water than coffee, which means you use a lot less of it when brewing. Very economical for someone on a tight budget. Chicory also offers extra health benefits that you wouldn't normally get from your cup of coffee. It is reported to help cleanse the blood and improve the health of your liver. The young leaves can be used in salads, and the root can also be boiled and eaten like a vegetable (it's related to endive and radicchio). It's also grown for cattle food in Europe. The flowers are blue-purple, and will open and close at precisely the same time every day.

Chilling requirement. (1) Cold period which is required by certain plants and plant parts so as to break physiological dormancy or rest. The chilling requirement is generally expressed in terms of the required number of hours at 70C or less. (2) The number of hours at or below a specific temperature. necessary for uniform bud break following winter dormancy.

Chilling unit. A period of time at or below a specified threshold temperature that has the maximum effect toward fulfilling the chilling requirement of a given plant.

Chimera/chimaera. (1) A plant having tissues of different genetic composition adjacent to each other. (2) A fusion of unrelated tissues such as in all variegated leaves; a man-made fusion of tissue caused by grafting. (3) Plant, or part of a plant which is composed of two or more genetically distinct tissues and is growing adjacent to each other as parts of the composite plant. Plants having variegated foliage, as in coleus, dahlia, citrus, vitis, chrysanthemum are examples of chimeras.

Chip budding. (1) Chip budding is one of the primary grafting methods used for the asexual propagation of woody plants used to produce named cultivars (varieties) or clones of many ornamental trees, shrubs, fruit and nut trees. It is one of the two most popular budding systems used in nursery production, and it is the only budding system that can be done on rootstocks (stocks) that have either active or dormant vascular cambiums. (2) Method of budding which can he carried out at times when the bark is not slipping. It involves removing a chip of hark from a smooth place between nodes near the base of the stock and replacing it with another chip of the same size and shape from the budstick having a hud of desired variety. The chips in both stock and the hudstick should be cut in the same manner. The chip hud is then wrapped to seal the cut edges and to hold the bud piece tightly into the stock.

Chip. A fruit basket made from interwoven thin strips of wood.

Chirping. Production of sound by an insect, generally by rubbing one part against another.

Chisel. (1) A chisel is a tool for carving and/or cutting a hard material such as wood, stone, or metal. A chisel, typically made of hardened or tempered steel, or more rarely, common steel, consists of a sharpened end (called the blade) attached to a straight handle. The handle and blade of some types of chisels are made in one piece. In use, a worker forces the chisel into the material to cut the material. The driving force may be manually applied or applied using a mallet or hammer. In industrial use, a hydraulic ram or falling weight ('triphammer') drives the chisel into the material to be cut. A gouge, one type of chisel, is used, particularly in woodworking and sculpture, to carve small pieces from the material. Gouges are most often used in creating concave surfaces. A gouge has a 'U'-shaped cross-section. (2) A farm implement used to break through and shatter compacted or otherwise impermeable layers of soil.

Chisel plough. A type of very heavy cultivator, with large section tines, the points of which incline forward, and are drawn through the soil at a depth greater than in normal ploughing. The underlying layers are burst without subsoil being brought to the surface.

Chitin. A nitrogenous polysaccharide (Carbohydrates) found in the exoskeleton of insects and in the cell walls of many fungi. It provides mechanical strength and resistance to chemicals. In biology, chitin (pronounced keye-tin) is one of the main components in the cell walls of fungi, the exoskeletons of insects and other arthropods, and in some other animals. It is a polysaccharide, made out of units of acetylglucosamine (more completely, N-acetyl-D-glucos-2-amine). These are linked together in ^{2}-1,4 fashion, the same as the glucose units that make up cellulose. So chitin may be thought of as cellulose, with one hydroxyl group on each monomer replaced by an acetylamino group. This allows for increased hydrogen bonding between adjacent polymers, giving the material increased strength. The strength and flexibility of chitin is the reason it is the material of choice for surgical thread. "Chitin" and "chiton" (a marine animal) both derive from the same Greek word meaning "tunic", referring to the hardness of the shell. In the honeybee the color of chitin may be yellow, golden, brown or black. From chitin *chitosan* can be produced.

Chitinase. An enzyme which hydrolyses chitin.

Chitting house. A building in which trays of potatoes are stored in stacks during the winter and stimulated to sprout before plant-ing, by the provision of controlled heating to prevent frost damage and regulate the rate of growth, and movable fluores-cent lighting.

Chlordane. A persistent, highly toxic, organochlorine insecticide, used to control earthworms in turf.

Chlorella. Any single-celled, green, freshwater alga of the genus *Chlorella*, 3–10 micrometres in diameter, which can increase its weight by four times in 12 hours. Nutritive content: 50% protein, 20% fat, 20% carbohydrate, 10% phosphate, calcium, and trace elements.

Chlorinated hydrocarbons. A group of insecticides which mainly act on the central nervous system. The tend to be persistent in the environment and can build up of toxic levels in the body fats of other animals higher up food chains. They include D.D.T., Aldrin, Dieldrin and Endrin amongst other. (Organo Phospho-rus Compounds). Also known as organochlorines, these synthetic organic compounds contain chlorine. They tend to be persistent in the environment and to biomagnify in the food chain. Chlorinated hydrocarbons that are pesticides include DDT, aldrin, dieldrin, heptachlor, chlordane, lindane, endrin, mirex, hexachloride, and toxaphene. Most chlorinated hydrocarbon pesticide uses have been canceled because of their persistence, propensity to bioaccumulate, and toxicity to nontarget species.

Chlorine test (live stock). 'This test finds use for determining mas-titis. The test is based on the fact that normal milk is having a chlorine content varying from 0.08% to 0.14% and that the chlorides of the blood escape into the milk and the chlorine content of the milk has been raised with the inflammation of udder. It is a rapid test which should be run by a dairyman, but it should be used with a second test for confirmation of the findings.

Chlorine. Plant food element which has been absorbed by plants as chloride (Cl-) ion, exact role in plant nutrition not known. Chlorine compounds used as disinfectants and deodourants.

Chlorocruorin. Green respiratory pigment which occurs in blood plasma of certain worms.

Chlorophenoxy herbicides. A class of pesticides that includes 2,4-D. They mimic plant hormones. Uses of some have been canceled because of concerns about adverse health effects.

Chlorophyll. Chlorophyll is the green pigment in leaves. Chlorophyll pigment is produced in the chloroplasts of leaf, stem and flower cells. Chlorophyll pigment dominates all other pigments present including the xanthophylls and carotenes. Chlorophyll captures the sun's energy which is used to manufacture sugar from carbon dioxide and water. Chlorophylls are constantly being "used up" and thus are continually manufacture by the plant, provided the environmental conditions are appropriate. (2) A group of green molecules used to convert light to energy.

Chloroplast. A chlorophyll containing organelle in a plant cell, the site of photosynthesis.

Chlorosis. A condition in plants resulting from the failure of chlorophyll to develop caused by a deficiency of an essential nutrient. Leaves of chlorotic plants range from light green through yellow to almost white.

Chlorosis. (1) A condition of plants characterised by normally green parts becoming pale green or yellow (chlorotic). It is due to the prevention of Chlorophyll formation, caused by various factors including lack of light, magnesium or iron deficiency, and excess calcium (when called Lime chlorosis). Also called green sickness. (2) The yellowing of plant tissue due to nutrient deficiencies or disease. (3) Yellowing caused by loss of or reduced production of chlorophyll (4) Yellowing of foliage due to a loss or breakdown of chlorophyll. Chlorosis may result from disease or infestation, poor growing conditions, or lack of nutrients.

Chocolate spot. A fungi disease *(Botrytis cinerea* and *B. fabae)* of beans, particularly when winter sown. It causes the develop-ment of chocolate coloured spots on the leaves, often quite rapidly, in warm wet conditions following late spring frosts. The shoots become black and die.

Choke. A fungal disease *(Epichloe typhina)* of cocks foot, timothy and other grasses, characterised by fungal cylinders around the stems and leaves, affecting seed production.

Cholinesterase inhibitors. A class of chemicals that includes numerous insecticides, such as parathion or carbaryl. They inhibit an enzyme found in animals that regulates nerve impulses. Cholinesterase inhibition is associated with a variety of acute symptoms such as nausea, vomiting, blurred vision, stomach cramps, and rapid heart rate.

Chopper spreader. A mechanism sometimes attaches to the straw outlet of a Combine Harvester which chops the straw into short lengths and spreads it over the field.

Chopping out. The removal of unwanted plants in a crop.

Chorion. The outermost membrane around a mammalian foetus (contiguous with the uterus surface), and around a developing chick in an egg.

Chromoplast. A coloured plastid containing pigments other than chlorophyll, often yellowish or red. *Chupon* Vertical stems or shoots of cocoa.

Chromosome. (1) A thread-shaped body, comprising mainly D.N.A. and protein, carrying the Genes, and present in the nucleus of a cell. (Diploid, Haploid, Meiosis, Mitosis). (2) A unit of heredity material in all living cells.

Chronic hunger. Long-term hunger caused by endemic problems of availability and access, rather than by temporary emergencies.

Chronic toxicity. The capacity of a substance to cause long-term or delayed adverse health effects. For example, a cancer resulting from exposure to a carcinogen may not appear for years or decades.

Chrysanthemum. *Dividing*: Chrysanthemums should be divided every three to five years to avoid overcrowding and maintain adequate flowering. Divide plants in the spring as new growth appears. Dig entire clumps and separate plants with a sharp knife or spade. Remove dead and diseased plants. Replant as soon as possible in a loose, well- drained, organic soil. *Pinching*: Chrysanthemums should be pinched to develop a compact plant that won't collapse or fall when flowering. An actively growing plant during spring and summer, properly fertilized and watered, will require 4 to 6 pinchings. Start pinching plants when shoots are 8 inches long. Pinch plants back by half. New shoots should develop within weeks. When these shoots are 8 inches long, pinch them back by half. As shoots reach 8 inches in height, prune back by half. No pinching should occur after July 15th in order to develop a compact, blooming plant. Late pinching (after July 15) may remove flower buds. *Planting*: Chrysanthemums should be planted in the spring as soon as the ground thaws. Transplants are place at ground level; roots should be an inch or two below ground. Potted plants can alsc be purchased in fall for color. Establishment before winter isn't always guaranteed. Make sure the soil is properly prepared. Water twice weekly until plants are established if rain isn't provided.

Ciliate. (1) Margined with hairs *don* Obsolete term for scion. (2) Hairy on the margin, like the eyelids. (3) With short fine hairs as on a leaf or petal margin.

Cinchona. Any of a group of tropical American shrubs or trees belonging to the madder family. The drug quinine is produced from the bark of some species, and these are now cultivated in India, Sri Lanka, the Philippines, and Indonesia. (Genus *Chinchona,* family Rubiaceae.)

Cineraria. Cinerarias appear as small, brightly and unusually colored daisy plants. Blooms are long-lived, provided the soil is kept moist and temperatures on the cool side. Evening temperatures of 45-50 degrees F. and day temperatures of 60-65 degrees F. are ideal. Cinerarias are grown as an annual and should not be looked at as a houseplant. Once plants are finished blooming, discard.

Circling disease. Specilic bacterial disease which effects mammals including man and birds. It is characterized by meaningo-encephalities and sometimes abortion in domestic ruminants and human beings. It is caused by *Listeria monocytogenes.*

Circumutation. An irregular spiral movement of a plant which takes place from unequal growth rates.

Citizen suits. Refers to the provisions often found in environmental laws which authorize any citizen to bring a lawsuit to enforce the provisions of an environmental law.

Citrange. An intergeneric hybrid between *Poncirus trifoliata* and *Citrus sinensis*, used as rootstock

Citrate soluble. Refers to that part of the total phosphoric acid in a fertilizer which is insoluble in water hut soluble in a neutral solution of citrate of ammonia. It is mostly in the form of dicalcium phosphate, $CaHPO4$.

Citrus molasses. Product which is obtained from the juice of citurs wastes.

Cladophyll (Cladode). A stem that resembles a leaf in appearance.

Clarifying agents. Substances used to remove haziness or sediment produced by oxidative deterioration in fruit juices, wines, beer etc.

Clasper. Onc of a pair of appendages at the tail end of a caterpillar by which it clings to leaves, twigs and other surfaces.

Clasping. A stalkless leaf, with the base partly surrounding the stem.

Class I equivalency. The amount of less productive land in a water district receiving Bureau of Reclamation water (Classes 2, 3, and 4) that would be necessary to be equivalent in productive potential to Class I land. This equivalency rating is made to adjust the number of acres that may be irrigated (see acreage limitation) so that less productive lands are equivalent in productive potential to 960 acres of Class I land.

Class I land. Under reclamation law, Class I land is defined as irrigable land within a particular agricultural economic setting that is productive enough to yield the highest level of suitability for continuous, successful irrigation farming, and has the highest relative productive potential as measured in net income per acre.

Classified pricing. The pricing system of federal milk marketing orders, under which milk processors pay into a pool for fluid grade (Grade A) milk; its value is based on how the milk ultimately is used. Milk used for fluid (Class I) consumption receives a higher price than milk for processed (Class II, Class III, Class IIIa) dairy products.

Clat. To remove wool from around the udder and the inside of the thighs of a ewe prior to lambing.

Claw. The constricted petiole-like base of petals and sepals of some flowers. A claw is a curved pointed growth found at the end of a toe or finger, or in arthropods, of the tarsus. The claws of arthropods are sometimes called pincers. In tetrapods, claws are made of keratin, and consist of two layers. The unguis is the harder external layer which consists of keratin fibers perpendicular to the direction of growth arranged in layers at an oblique angle; and the subunguis is the softer, flaky underside layer whose grain is parallel to the direction of growth. The claw grows outward from the nail matrix at the base of the unguis, and the subunguis grows thicker while travelling across the nail bed. The unguis grows outward faster than the subunguis to produce a curve, and the thinner sides of the claw wear away faster than their thicker middle, producing a more or less sharp point. Tetrapods use their claws in many ways, commonly to grasp or kill prey, to dig, and to climb and hang. A nail is homologous to a claw, but is flatter and has a curved edge instead of a point. A nail that is big enough to walk on is called a hoof. A talon is the claw of a bird of prey. Every so often, the growth of claws stops and restarts, as does hair. In hair, this results in the hair falling out and being replaced by a new one; in claws, this results in an abscission layer, and the old segment breaks off. This process takes several months for human thumbnails; cats are often seen working old unguis layers off on wood or on boards made for the purpose. Many predators have retractile (retractable) claws that can partially hide inside the animal's limb.

Clay minerals. The particles comprising clay consist of layers of hydrated aluminium and magnesium silicates forming a crystal lattice. Various types of clay mineral exist, e.g, kaolinite and montmorillonite, having different layer structures, and differing abilities to absorb and retain water, and to absorb and exchange cations (Cation Exchange). The agricultural productivity of a soil is related to the nature and content of the clay minerals in it and to its ability to retain (and subsequently release for crop use) cations, particularly those applied in the form of fertilizers.

Clay pan. (1) In common usage, it refers to a dense, heavy subsoil, more or less impermeable to roots (2) A dense and heavy soil horizon which underlies the upper part of the soil profile, hard when dry and stiff or plastic when wet.

Clay pots. Familiar red clay flower pots, which are long used from growing young plants.

Clay soils. Soils containing large amounts of clay, often called heavy soils because they require much more effort (i.e, greater tractor power) to plough and cultivate than sandy or light soils. They are sticky and swell when wet and are liable to poaching, particu-larly in winter. They retain moisture well but become hard, shrink and crack in dry weather. Artificial drainage of clay soil is often carried out, particularly for arable cropping. (Land Drainage, Loam).

Clay. A constituent of Soil comprising extremely finegrained par-ticles, less than 0.002 mm in diameter, characterized by high moisture retention. Sticky and plastic when wet, hardening and shrinking on drying. (Clay Minerals, Clay Soils. Loam, Sand, Silt).

Clean corn. A sample of wheat, barley, etc., not mixed with the seeds of other plants.

Clean cultivation. (1) Cultivation which is carried out to prevent the growth of all vegetation except the particular crop desired. (2) Gardening practice of removing all weeds at all times. (3) Periodic soil tillage to eliminate all vegetation other than the crop being grown.

Clean land. Land free from weeds.

Clean tillage. Periodic soil tillage which is carried out to eliminate all vegetation other than the crop being grown.

Clearcut harvest. A harvest and regeneration technique removing all the trees (regardless of size) on an area. Clearcutting is commonly used with shade-intolerant species such as Douglas fir or lodgepole pine, which require full sunlight to reproduce and grow well. Clearcutting produces an even-aged stand of trees.

Cleft craft. A graft involving the inseration of the scion into a wedge-shaped cut in the centre of the stock.

Cleft grafting. A grafting method where the scion piece is placed in a split (cleft) formed on a stock plant (See Wedge grafting)

Cleft. Divided to or about the middle into divisions.

Cleistogamous. (1) A small, closed self-fertilized flower, usually near the ground. (2) Describes a small, closed self-fertilized flower, usually near the ground. (3) Flowers which fertilise themselves without opening.

Cleistogamy. (1) Close fertilization within the unopened flower due to self-pollination. (2) Self pollinating nature of closed flowers.

Cleistothecia. Cleistothecium (singular). The characteristic fruiting structure of powdery mildew fungi. This is a tiny sphere usually 0.1 to 0.2 mm in diameter. At maturity the cleistothecia liberate ascospores from microscopic sacs (asci) that develop within the sphere. Cleistothecia are at first colorless, then yellow, brown, and finally black in most species.

Clematis. Any of a group of temperate woody climbing plants with colourful showy flowers. They belong to the buttercup family. (Genus *Clematis*, family Ranunculaceae.) The wild traveller's joy or old man's beard (*Clematis vitalba*) is the only native British species, although many have been introduced and many popular garden hybrids have been bred.

Climacteric fruits. Fruits exhibiting increased rate of respiration before full ripening e.g. mango; banana.

Climacteric. The maximum respiration rate of a fruit, reached just before full ripening

Climate. The aggregate of weather conditions over a long period of time. Climate in a narrow sense is usually defined as the "average weather", or more rigorously, as the statistical description in terms of the mean and variability of relevant quantities over a period of time ranging from months to thousands or millions of years. The classical period is 30 years, as defined by the World Meteorological Organization (WMO). These quantities are most often surface variables such as temperature, precipitation, and wind. Climate in a wider sense is the state, including a statistical description, of the climate system.

Climax ecosystem. The *ecosystem* that will eventually overtake a certain area of the globe, perhaps after the landscape has moved through a succession of other ecosystems, and will persist indefinitely thereafter, ending the process of succession. Also called the dominant ecosystem.

Fig.: Cleistogamous

Climax vegetation. The most fully developed natural vegetation the climate can sustain.

Climber. (1) A plant which raises its foliage by supporting itself on surrounding objects, either by twining or coiling them (bittersweet), by the aid of tendrils (greenbrier, Virginia creeper) or aerial roots (ivy), or by scrambling over them without either coiling or having such specialized organs of attachment (rose). (2) A vine-like or rambling plant that raises its foliage by supporting itself on surrounding objects, either by twining or coiling around them, by the aid of tendrils or aerial roots, or by scrambling over them. (3) Annual or perennial herb or shrub with special structures to climb over a support.

Clinal variation/cline. With a spectrum such as blue to green in spruces.

Clod. Mass of soil which is produced by ploughing or digging and usually slaken easily with repeated wetting and dying in contrast to a ped which is a natural soil aggregate.

Clonal propagation. Asexual propagation of plants from a single plant, resulting in true-to-type and uniform progenies.

Clonal seed. A seed collected from a clone.

Clonal selection. Refers a method of selection of desirable clones from the mixed population of a vegetatively propagated crop.

Clone. (1) Cloning, in biology, any organism whose genetic information is identical to that of a "mother organism" from which it was created. In botany and horticlture, a clone is an individual plant produced from a mother plant by some form of vegetative reproduction. (2) A group of genetically uniform plants, vegetatively propagated from a single parent. CO_2 compensation point The concentration of carbon dioxide at which photosynthesis balances respiration and there is no net CO_2 flux. In C_3 plants this is about 50 ppm, whereas in C_4 plants it is almost zero. (3) A group of plants all having an identical genetical make-up, being the descendants of a single parent. In plant breeding a clonal stock of plants can be vegetatively raised by taking cuttings from a single parent plant. (4) A line of plants propagated from a single mother plant, identical to each other and to the parent.

Closed-formula mixed fertilisers. The fertiliser grade has been disclosed on each bag of such fertiliser mixture but the ingredi-ents or straight fertilisers used in formulating the mixture have not been disclosed. In India, fertiliser mixtures sold to the cultivators are usually of the closed formula type.

Clover hay. Hay derived from a clover crop. Also hay from a temporary Ley as opposed to that derived from permanent grassland (Meadow Hay). Also knwon as seeds hay.

Clover rot. A fungal disease (*Sclerotinia trifoliorum*) of clovers and related plants, characterized in its early stages by leaves being covered with small brown spots. Plants later wither and die.

Clover sickness. A condition of a field on which clover has been grown for too long and which is heavily infected with Eleworms and fungus diseases.

Clover. A large genus (*Trifolium*) of leguminous plants- character-ized by trefoil leaves, having small flowers with tight roundish heads, and producing one-to four-seeded pods. Root nodules contain bacteria capable of nitrogen fixation. A number of species are grown for fodder or are sown with a mixture of grass seeds to produce Leys, *e.g.*, Aslike Clover (*T. hybridum,*) Wild White Clover or Kentish Clover (*T. repens*), Purple or Red

Clover *(T. pratense)*. Clovers usefully knit the sward together and help keep out weeds, especially in longer leys.

Club root. (1) A fungal disease *(Plasmodiophora brassicae)* of crucif-erous plants, particularly swedes, turnip, rape, cabbage and other Brassicas, which causes swelling and distortion of the roots. Growth gets stunted and leaves become pale green. The spores can remain alive in the soil for several years and are particularly active in acid and wet soil conditions. Also known as Finger-and-Toe. (2) A fungal disease causing swollen roots.

Clump-forming grass. Grass that grows as distinct plants that get larger over time. Also called *bunchgrass*. Contrast with *sod-forming grass* (also called *spreading grass*), which expands using running roots that create new plants from existing ones.

Cluster. (1) The leaves, flowers or fruits growing from a common point on the stem or shoot of a plant. (2) The four teat cups of a Milking Machine applied to a cow's udder.

Clustered. (1) Multiple leaves, flowers or fruits seemingly arising from a common juncture; crowded, cannot easily determine if they are alternate or opposite in orientation. (2) Said of leaves, crowded so as not to be clearly opposite or alternate, also said of the whorled condition.

Clutch. A batch of eggs laid by a bird. Also the brood of chicks hatched from such eggs.

Coated seed. When seeds are first treated with strong water stable adhesives, they bind to the seed coat and adhere articulate matter of various kinds tightly to the seed to better withstand the hazards of sowing.

Cobweb theorem. An economic theorem that producers of a com-modity expect the price in the next period of production to be the same as current prices and thus plan production accordingly. It is most appropriate in the case of non-storable commodities.

Cocaine. An alkaloid obtained from the leaves of *Erythroxylum coca.* It finds use as a local anesthetic and mydriatic. Its solution in water is sometimes applied to soothe the pain in eyes.

Coccidioidal granuloma. Refers to the non-contageous fungus dis-ease of animals, which is characterized primarily by multiple pulmonary granulomas and dissemination to other organs. *Coccidioides immitis* is the casual organism.

Coccidiosis. An intestinal disease of livestock and poultry caused by various microscopic protozoan parasites of two genera *Eimeria sp.* and *Isospora sp.*, belonging to the Order Coccidia. Charac-terized by diarrhoea and emaciation.

Coccus. (1) A spherical bacterium, as distinct from rod-shaped or vibrio (spiral-shaped) bacteria. (2) A portion of dry fruit contain-ing one seed when it breaks up.

Cocksfoot. A perennial grass *(Dactylis glomerata)*, with characteris-tic one-sided flower-heads of rounded or oval individual flower spikelets on wiry branches, and dull green to deep blue leaves. Deep rooting, very high yielding used extensively for pasture. It is a common grass. It is widely distributed in Europe, Asia and North Africa. It grows to 20–120 centimetres tall, with a distinctive tufted flowerhead ten to fifteen centimetres long, and leaves twenty to fifty centimetres long and up to one and a half centimetres broad. It is usually treated as the sole species in the genus *Dactylis*, but is commonly divided into several regional subspecies; some botanists treat some of these as distinct species, or at the lower rank of variety.

Cocoon. A protective silky sheath spun by many insect larvae (e.g., the Caterpillars of butterflies and moths) in which they pupate (Puppa). Also a similar sheath or capsule in which spiders and earthworms lay their eggs.

Codlin(g) moth. (1) A small moth the caterpillars of which eat the inside of growing apples and cause them to fall prematurely. (2) Small moth that is problem in fruits and vegetables.

Cod-liver oil. The oil obtained from the livers of codfish, rich in vitamins A and D, valued for feeding to young animals to improve condition. It is particularly helpful in preventing rickets developing.

Coffin-joint. Joint located between the second and third phalangel bones of the hoof.

Coital vesicular exanthema. An acute contageous disease of cattle characterized by fever, rhinitis, tracheits and considerable loss of condition and milk yield.

Colchicine. (1) A highly poisonous alkaloid, originally extracted from plants of the genus Colchicum (Autumn crocus, also known as the "Meadow saffron"). Originally used to treat rheumatic complaints and especially gout, it was also prescribed for its cathartic and emetic effects. Its present use is mainly in the treatment of gout. (2) A poisonous (alkaloid) drug obtained from Colchicum autumnale, used for inhibiting spindle formation during mitosis at/lUmetaphase for doubling of chromosome number.

Cold brooding. Also known as hay-box brooding, It is a simple method of rearing of chicks when it is not too cold, by employing an ordinary basket or small box lined thickly with fine hay or old flannel strips of feathers and provided with small hole or outlet to allow passage to chicks.

Cold hardiness zone. Refers to the average annual minimum temperature for a geographic area. Temperature is only one factor that influences a plant's winter hardiness. The USDA Hardiness zone map was revised in 2003. Much of Colorado falls into Zone 5, with Zone 4 for mountain areas and Zone 6 in warmer locations around the state.

Cold storage. (1) An insulated storage using mechanical refrigeration to maintain a stable low temperature for long term storage of agri-horticultural products (2) The preservation of perishable produce (*e.g.*, apples, meat, etc.), usually in bulk, in refrigerated chambers. The low temperature conditions inhibit the multiplication of bacteria and fungal sproes, etc.

Cold test. A germination test in which the seeds are planted for a period in cool, moist unsterilized soil before they get transferred to a higher temperature. This is done for studying the effect of possible unfavourable planting conditions while determining seedling vigour.

Cole crops. (1) Cabbage and its various allies. (2) Representatives of the species *Brassica oleracea*.

Coleoptile. A protective sheath covering, the first leaf of a grass seedling. The pointed protective sheath covering the emerging shoot in monocotyledons such as oats and grasses. Early experiments for phototropism using plants with coleoptiles suggested that plants grow towards light because cells on the darker side elongate more than those on the lighter side. In the 1880's it was found that coleoptiles only bend towards the light when their tips are exposed. Therefore the tips must contain the photoreceptor although the bending takes place lower down. Chemical messengers called hormones are sent down the dark side of the shoot. The hormone was later called auxin and is now known to be indoleacetic acid (IAA). When light strikes the coleoptile tip from one side, auxin produced in the meristem is transported to the shaded side as it decends the coleoptile.

Coli septicaemia. A bacterial infection *(Escherichia coil)* of poultry. It is usually secondary to other respiratory diseases and is characterized by inflammation of the air sacs which often contain purulent material. Young birds between 6 and 10 weeks of age are particularly susceptible, mainly in winter in condi-tions of poor ventilation or overcrowding.

Colibacillosis. Disease which is caused by *Excherichia coil.* It is found in the intestine, where it lives in symbiosis.

Colic. Abdominal pain and associated symptoms which results from any one numerous disordres of the abdominal cavity.

Coliform bacteria. Those bacteria commonly present in animal intestines and passed out in faeces *(e.g. Escherichia coli).* The presene of such bacteria in milk indicates faecal contamination.

Coliform test. Test which is used for estimating coliform bacteria in a sample of milk. The presence of such bacteria is an index of unhygienic method of milk production.

Colitis. Inflammation of the large intestine which is often caused due to diarrhoea, constipation, etc.

Collagen. A proteinous substance, the principal fibrous constituent of skin, tendon, ligament and bone. When boiled it produces gelatin.

Collar pruning. Pruning of a mature tea bush by cutting close to the soil. thereby removing all the frame, leaving only the central stem (See Rejuvenation pruning). from a complete inflorescence or from several flowers as of mulberry and pineapple

Collateral. Said of extra or supernumerary buds that are inserted on either side of a normal axillary bud. Said of buds that grow side by side.

Collective fruit. Single fruit which is formed from several flowers.

Colony brooders. Brooders which are used for rearing chicks in lots from about 60 to a few hundred, with kerosene oil, coal, wood, oil, gas or electricity supplying the fuel.

Color bowl. A small, round container placed on a table top to house compact, annual plants.

Color charts. Hue chips used to simulate colors in the field, as *flowers* may lose or change color(s) when dried. Examples of such charts are the *horticulture* Society Color Chart, Nickerson Color Fan, and Horticultural Color Chart.

Colostrum. The milk given by cows and other mammals during the first few days after giving birth. It is particularly rich in proteins including Antibodies, and has a higher fat content than normal milk. Also called beestings or beistings.

Colouring agents. Substances added to food products to make them attractive and appetising

Coloury. A descriptive term applied to dried hops having a distinctive green tint.

Colza. Colza oil is a non-drying oil obtained from the seeds of *Brassica campestris,* var. *oleifera,* a variety of the plant which produces Swedish turnips. Colza is extensively cultivated in France, Belgium, the Netherlands and Germany; and, especially in the first-named country, the expression of the oil is an important industry. In commerce, colza is classed with rape oil, to which both in source and properties it is very closely allied. It is a comparatively inodorous oil of a yellow color, having a specific gravity varying from

0.912 to 0.920. The cake left after expression of the oil is a valuable feeding substance for cattle. Colza oil is extensively used as a lubricant for machinery, and for burning in lamps.

Comb honey. Honey stored by bees in the cells of freshly built broodless combs.

Combine (crop science). A self-propelled or tractor-drawn machine which cuts, threshes, and cleans the standing crop which moving across the field. It is adapted to harvesting all the small grains, soybeans, grain sorghums, peanuts, beans, etc. In some areas, the crop is cut and placed in windrows by a swather (windrower), and a combine with a pickup attachment gathers the grain and threshes it at a later date.

Combine drill. A Drill which sows seeds and fertilizer simultane-ously, usually down the same seed-tube, although there are separate hoppers for seed and fertilizer. Combine drills are difficult to clean out effectively.

Combine. A self-propelled machine for harvesting grain and other seed crops. In one operation, it cuts, threshes, separates, and cleans the grain and scatters the straw.

Combiner. Machine which undertakes the functions of a reaper, thresher and winnower. Its main functions are as follows : (i) Cutting the standing crop (ii) Feeding the cut crops to threshing unit (iii) Threshing the crop (iv) Cleaning the grains from straw and (v) Collecting the grains in a combiner.

Combining ability. The relative ability of an inbred line or a clone, when crossed to another inbred line or clone, to transmit desirable trait(s) to its cross.

Command area. The land area that can be irrigated by canal or from a dam or an irrigation project

Commercial feed. Materials distributed for use as feed or for mixing in feed for animals other than man.

Commercial floriculture. Area of horticulture related to the commercial production and marketing of cut flowers, flowering pot plants, foliage plants and bedding plants.

Commodity certificates. Payments issued by the Commodity Credit Corporation (CCC) in lieu of cash payments to participants in farm subsidy or agricultural export programs. Holders of certificates are permitted to exchange them for commodities owned by the CCC. Certificates were used not only to compensate program beneficiaries but also to reduce the large, costly, and price-depressing commodity surpluses held by the CCC during the mid-1980s.

Commodity distribution. Direct donation of food products by the federal government to needy persons, schools, and institutions. Commodities are either entitlement or bonus. Bonus commodities can be received when they are available from surplus stocks purchased by the Commodity Credit Corporation under its price support program or the Agricultural Marketing Service under its surplus removal program (Section 32 of the Agricultural Adjustment Act of 1935).

Commodity exchange authority. A former regulatory agency of USDA established to administer the Commodity Exchange Act prior to 1975; the predecessor of the Commodity Futures Trading Commission.

Commodity exchange. An organization operating under a set of bylaws aimed at promoting trade in one or more commodities by providing services and rules for the conduct of trade.

Commodity programs. This term is usually meant to include the commodity price and income support programs administered by the Farm Service Agency and financed by the Commodity Credit Corporation. The commodities now receiving support are: (1) those included in the production flexibility contract payments program, specifically wheat, feed grains, cotton, and rice; (2) those eligible for nonrecourse marketing assistance loans, soybeans and minor oilseeds; (3) those under marketing quota limits, peanuts and tobacco; (4) sugar and milk. A broader term that includes these programs and others is farm programs.

Commodity promotion programs. Programs that advertise and promote an agricultural commodity or product without reference to the specific farmer, brand name, or manufacturer. Producers can and do organize voluntary commodity promotion programs, but most are operated under the authority of either federal or state laws, frequently with the objective of requiring that all members of the industry participate. At the federal level, the programs are authorized by law, implemented by industry groups (after USDA review, rulemaking and approval), and financed by assessments (also called check-offs) of industry members such as producers, importers, and/or handlers. In the past, Congress enacted separate laws permitting producers of specifically-designated commodities to create such programs. The FAIR Act of 1996 also gives USDA general authority to create programs for any commodity at the request of a group of producers. In early 1999, 12 federal promotion programs were fully operational: beef, cotton, dairy products, eggs, fluid milk, honey, mushrooms, popcorn, pork, potatoes, soybeans, and watermelon. In addition to the federally authorized programs, there are between 300 and 350 state-legislated promotion programs covering about 80 farm commodities. Nine out of ten U.S. farmers contribute to one or more of these efforts, which, collectively, raise and spend hundreds of millions of dollars annually.

Common Horticulture Policy (CHP). The set of legislation and practices jointly adopted by the nations of the European Union (EU) in order to provide a common, unified policy framework for agriculture. Its stated purposes are to increase farm productivity, stabilize markets, ensure a fair standard of living for farmers, guarantee regular supplies, and ensure reasonable prices for consumers. The CAP rests upon four basic principles: common import restrictions, common financing, common pricing, and common treatment of surpluses.

Community forestry. Forest managed for the interest of local community or village, run almost exclusively by them

Community. (1) A term applied to any assemblage of plants making up a distinct vegetation type, e.g., deciduous woodland. (2) A short-term for the European Economic Community.

Compacted surface. Compacted surface or crust on soil which is somewhat loose underneath but does not break regularly during ploughing.

Companion crop. (1) Any subsidiary crop grown in association with a main crop. (2) Two crops grown together in the same field, one of the crops, and often both, benefiting from the presence of the other. The term is particularly applied to crops grown together for seed production which have similar cultivation and manage-ment needs and whose seed can be separated after harvesting (e.g., clover and ryegrass, the clover facilitating soil Nitrogen Fixation, beneficial to ryegrass growth, whilst the ryegrass assists in the harvesting of the clover).

Companion plant. A plant making an showy, contrasting display or providing pollination

Comparative advantage. Refers to the economic theory that in international trade it is more advantageous for a country to devote its resources not to all lines of production in which it may have superiority (least cost production), but to those in which its relative superiority is greatest. Two countries may find trade mutually profitable even if one of the countries could produce all goods at lower cost than the other.

Compatibility. (1) In sex cells, the ability to unite and form a fertilized egg that can grow to maturity. (2) Congenial condition for successful graft union between stock and scion.

Compatible varieties. (1) Varieties that will set seed when cross pollinated (2) Varieties that will make a successful graft union when inter-grafted.

Compensation point. The light intensity at which the photosynthetic intake of carbon dioxide is equal to the respiratory output of carbon dioxide.

Compensatory amounts. Monetary Compensatory Amount.

Compensatory growth. The phenomenon of animals turned out to pasture after a period of energy restriction exhibiting greater growth rate than animals which have not been so restricted. It is due to the greater feed intake following the period of energy restriction.

Competition (Interference). The process in which two individual plants or two populations of plants interact such that at least one exerts a negative effect on the other

Competition. Demand of organisms within a community for the same substance. Plants may compete for light and nutrients; animals for food and shelter. Competition usually causes the displace-ment of one organism by another.

Competitive advantage. A situation in which one country, region, or producer can produce a particular commodity more cheaply than another country, region or producer.

Complement. A constituent of blood serum, the presence of which is required for an Antibody to kill an Antigen.

Complete diet. Winter feed for livestock comprising a mixture of Concentrates and bulk forage fed as a complete mixture, the animals usually receiving no other feed.

Complete fertilizer. (1) A fertilizer containing nitrogen, phosphorus, and potassium, the three nutrients in which plants are most commonly deficient. (2) A plant food with all three of the primary elements: nitrogen, phosphorus and potassium.

Complete flower. (1) A flower having pistil(s), stamens, petals and sepals. (2) One that has corolla, calyx, stamens and one or more pistils. (3) Complete flowers contain both the male and female parts (stamen and pistil) as well as sepals and petals.

Complex fertilisers. Commercial fertilizers having at least two or more of the primary essential nutrients. When such fertilisers have only two of the primary nutrients they are known as

Complimentary imports. Agricultural import items not produced in appreciable commercial volume in the United States, such as bananas, coffee, rubber, cocoa, tea, spices and cordage fiber.

Component crops. Individual crops of a cropping system.

Composite seed sample. Seed sample composed of mixture of different subsamples taken from various parts of a seed lot.

Composite. Compound; the common name of a member of the Compositae.

Compost activator, Compost accelerator. A chemical substance which promotes decomposition and fermentation of decaying plant remains in a compost heap.

Compost. (1) A manure derived from decomposed plant remains, usually made by fermenting waste plant material (e.g.. straw, mowings, etc.) in heaps, usually in alternate layers with added lime, nitrogen and water. Compost is usefully used in greenhouses, to enrich soil, either dug in or as a surface Mulch. Straw compost is valuable in building up poor soils for intensive vegetable cropping and assists in maintaining soil structure on heavier land. (2) A product obtained by the controlled decomposition of organic matter having a C:N ratio of about 10:1. (3) An organic soil amendment resulting from the decomposition of organic matter.

Composting citrus peels. Citrus peels may take several years to decompose in a compost pile due to their high concentration of citrus oils. The smaller the pieces, the faster the decomposition.

Composting. The controlled biological decomposition of organic material, such as sewage sludge, animal manures, or crop residues, in the presence of air to form a humus-like material. Controlled methods of composting include mechanical mixing and aerating, ventilating the materials by dropping them through a vertical series of aerated chambers, or placing the compost in piles out in the open air and mixing it or turning it periodically.

Compote. Fruits cooked in sugar syrup, retaining their shape. A stemmed container used for arranging flowers. A compote is a sweet cooked preparation of whole or cut fruit (such as apples, pears, cherries, strawberries, plums) and sugar, usually more liquid in consistency than jams, jellies or preserves. Compotes may also contain spices. Known in Polish and Russian as kompot, it is a traditional drink in Eastern European countries. In the mid-1980s, 60 percent of beverages consumed by an average Pole consisted of compote and other home-made concoctions. Today that number has dropped to 30 percent. In some homes tea with lemon may be an alternative for compote (in many Eastern European countries tea is drunk not only after meals, but also with meals). Note that you are unlikely to find compote in Polish restaurants but it is often available in smaller Russian restaurants. It is considered a typically homemade beverage in Poland. Also note that in Poland nowadays compote and tea are being more and more frequently supplanted by fruit juices.

Compound. (1) A leaf composed of two or more distinct leaf blades (leaflets), each attached to a petiole-like structure (rachis) or directly to the top of the petiole; compare with simple. (2) A quality of leaves. Leaves that are compound are in an array of small, symmetrically-arranged leaflets on each stem. In contrast, a plant with simple leaves has one leaf per stem.

Compound fertilizer A containing a mixture of two or three of the major plant nutrients (i.e. Nitrogen, Phosphorus and Potassium). Many proprietary compound fertilizers are pro-duced with differing nutrient ratios to suit specific crops and soils. Mostly produced in easily stored granulated form, but powders and less concentrated liquids are available. Also called mixed fertilzer.

Compound layering (Serpentine layering): Layering method where a flexible branch is alternately covered and exposed along its length, resulting in several new plants.

Compound leaf. (1) A leaf composed of two or more leaflets e.g. ferns. (2) A leaf of two or more leaflets, in some cases (Citrus) the lateral leaflets may have been lost and only the terminal leaflet remain. Ternately compound- when the leaflets are in threes; palmately compound- when three or more leaflets arise from a common point; Pinnately compound- when the leaflets are arranged along a common rachis or if only three are present at least the terminal leaflet is petioled; Odd-pinnate- if a ternimal leaflet is present and the total number of leaflets is an odd number; Even-pinnate- if no terminal leaflet is present and the total number of leaflets is an even number. (3) A leaf of two or more leaflets, in some cases the lateral leaflets may have been lost and only the terminal leaflet remain. (4) One divided into separate leaflets, e.g., Clover. (Simple Leaf)

Compressibility of soil. The susceptibility of a soil to compact under a heavy load (i.e. a tractor).

Concave. A semi-circular grating in a Threshing Machine, encirclind the drum, against which the grain is dislodged by the rotating Beaters on the drum.

Concentrates. A term for a variety of animal feedingstuffs with a high food value relative to volume, with low fibre content, some rich in Protein, others rich in Carbohydrates or Fat. Concentrates include cereal grains and their by-products, leguminous seeds, oil seeds, oil Cakes and Meals, and various animals by-products. Mainly supplied by compound feedingstuff manufactuers as expertly balanced formulations, often computer assisted, for all types of farm stock. Ingredients may vary in relation to current prices and availability and Trace Elements, Vitamins and Min-erals are usually included to produce complete foods. Some farms mix their own concentrates using home-grown cereals.

Concrete (Floral concrete). The substance containing odoriferous principles of commercial flowers like jasmine, rose and tuberose obtained by solvent extraction

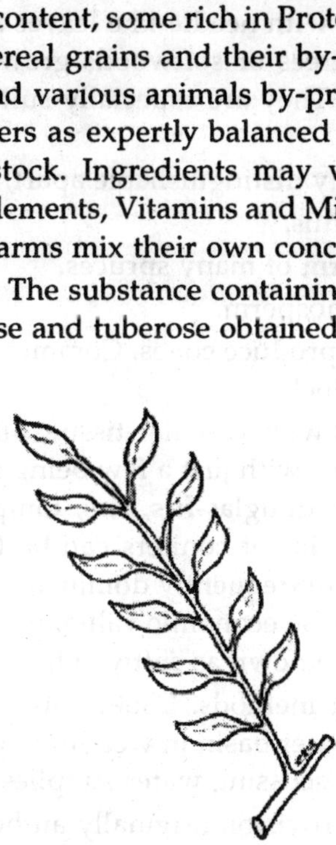

Fig.: Compound leaf

Condiments. These are substances of plant origin with pronounced odour and piquancy, used for seasoning food and stimulating appetite (See Spices and condiments).

Conditional registration. Under special circumstances, the Federal Insecticide, Fungicide, and Rodenticide Act (FIFRA) permits registration of pesticide products that is "conditional" upon the submission of additional data. These special circumstances include a finding by the Environmental Protection Agency that a new product or use of an existing pesticide will not significantly increase the risk of unreasonable adverse effects. A product containing a new (previously unregistered) active ingredient may be conditionally registered only if the EPA finds that such conditional registration is in the public interest, that a reasonable time for conducting the additional studies has not elapsed, and the use of the pesticide for the period of conditional registration will not present an unreasonable risk.

Conditioner (of Fertilisers). Material which is added to a fertiliser to prevent caking and to keep it free-flowing.

Cone. (1) The fruit of the female hop-plant, growing as a cluster of hops on a Bine. (2) The fruit of a coniferous tree, consiting of compact over- lapping scales, arranged spirally around a central axis. (3) A coniferous fruit, having a number of woody, leathery, or fleshy scales, each bearing one or more seeds, and attached to a central axis; the seed-holding structure of trees like pines and spruce.

Conelet. A young, immature first season cone, in the pines.

Confiers. A group of trees and shrubs with needle-like leaves and typically bearing cones, e.g., pine, fir, spruce. Coniferous trees or softwoods grow quickly and provide early profit where grown commercialiy. They are especially suited to exposed hili and acid conditions (Broadleaf)

Confluent. Blending together, not easily distinguishable apart, as applied to bundle-traces.

Congeneric. Belonging to the same genus.

Conical. Cone shaped, as he young form of many spruces.

Conifer. A cone bearing plant or gymnosperm.

Conifer. Woody trees and shrubs that produce cones. Common conifers include pines, firs, spruce, juniper, redwood and hemlocks.

Coniferous. Cone-bearing seed plants with vascular tissue; all extant conifers are woody plants, the great majority being trees with just a few being shrubs. Typical examples of conifers include cedars, cypresses, douglas-firs, firs, junipers, kauris, larches, pines, redwoods, spruces, and yews. Species of conifers can be found growing naturally in almost all parts of the world, and are frequently dominant plants in their habitats, as in e.g. the taiga. Conifers are of immense economic value, primarily for timber and paper production; the wood of conifers is known as softwood.

Conjunctive use. Water management methods. Usually used to describe the practice of storing surface water in a groundwater basin in wet years and withdrawing it from the basin in dry years. Often used in discussing water supplies and water conservation.

Conservation (cross) compliance. A provision originally authorized by the Food Security Act of 1985 that requires farmers who operate on highly erodible land to manage this land under an approved conservation system in order to maintain eligibility in specified federal farm programs. The FAIR Act of 1996 amended the conservation compliance provisions in several ways to provide greater planting flexibility to farmers.

Conservation easement. Acquisition of rights and interest to a property to protect identified conservation or resource values, using a reserved interest deed. Since the mid 1970s, conservation easements have been purchased to protect nearly 420,000 acres of farmland in fifteen states, primarily in the Northeast.

Conservation farm option program. A provision of the FAIR Act of 1996 authorizes a pilot program for producers who receive production flexibility payments to enter into a contract to consolidate payments at rates that are equivalent to payments that would otherwise be received from the Conservation Reserve Program, Wetlands Reserve Program, and/or the Environmental Quality Incentives Program in exchange for implementing practices to protect soil, water, and wildlife.

Conservation plan. A combination of land uses and farming practices to protect and improve soil productivity and water quality, and to prevent deterioration of natural resources on all or part of a farm. Plans may be prepared by staff working in conservation districts and must meet technical standards. For some purposes, such as conservation compliance, the plan must be approved by the local conservation district. Under the 1996 FAIR Act, conservation plans for conservation compliance must be both technically and economically feasible.

Conservation practice. Any technique or measure used to protect soil and water resources for which standards and specifications for installation, operation, or maintenance have been developed. Practices approved by the Natural Resources Conservation Service are compiled at each conservation district in its field office technical guide.

Conservation Technical Assistance (CTA). CTA has been the central activity of the Natural Resources Conservation Service since it was established in 1936. NRCS field staff help landowners and farm operators plan and implement soil and water conservation and water quality practices. The most common use of this program in recent years has been preparing and updating conservation compliance plans. In FY1993, CTA assisted 1.2 million farmers and serviced 62 million acres.

Conservation tillage. Any tillage and planting system that leaves at least 30% of the soil surface covered by residue after planting. Conservation tillage maintains a ground cover with less soil disturbance than traditional cultivation, thereby reducing soil loss and energy use while maintaining crop yields and quality. Conservation tillage techniques include minimum tillage, mulch tillage, ridge tillage, and no-till.

Conservation. (1) The optimum rational use of natural resources and the environment, having regard to the various demands made upon them and the need to safeguard and maintain them for the future. (2) The protection of the soil against erosion or loss of fertility. (3) The preservation of grass as Hay and Silage and of fodder crops for winter feed. Also applied to the preservation of certain growing crops *in situ* for later use, e.g., frost- hardy Thousand Headed Kale kept through the winter for folding in spring. (4) The management of human and natural resources to provide maximum benefits over a sustained period of time (see sustainable agriculture). In farming, conservation entails matching cropping patterns and the productive potential and physical limitations of agricultural lands to ensure long-term sustainability of profitable production. Conservation practices focus on conserving soil, water, energy, and biological resources. Contour farming, no-till farming, and integrated pest management are typical examples of conservation practices. (5) Sustainable use of natural resources, such as soils, water, plants, animals and minerals

Conservatory. A greenhouse of the tropics which protects plants from fierce sunlight and strong wind

Conserving use acreage. Farmland diverted from crop production to an approved cultural practice that prevents erosion or other degradation. Though crops are not produced, conserving use is considered an agricultural use of the land.

Considered planted. Refers to a provision of the Agricultural Act of 1949 that was used to implement the base acreage and yield system for the 1991-95 crops, a provision that was suspended by the FAIR Act of 1996. Under previous law, crop acreage bases were, in general, calculated as a 5-year average of planted and considered planted acreage. Acreage considered planted includes acreage idled under production adjustment programs or for weather-related reasons or natural disasters; acreage devoted to conservation purposes or planted to certain other allowed commodities; and acreage USDA determines is necessary for fair and equitable treatment.

Consociation. Climax plant community which is characterized by a single species.

Consolidation. In horticulture and other economic sectors, consolidation usually is a reference to the trend from numerous smaller-sized operations toward fewer and larger ones. Consolidation can lead to higher concentration. See industrialization.

Constipation. The term used for the retention of faeces in the intestines because of difficult evacuation. It may develop due to feeding of coarse, dry, or indigestible feed or due to over-feeding or from overeating.

Consultative Group on International Agricultural Research (CGIAR). An informal association of 56 public and private organizations that support 16 international agricultural research centers. All but three of the research centers are located in developing countries. The cosponsors of CGIAR are the Food and Agriculture Organization of the United Nations, the United Nations Development Programme, the United Nations Environment Programme, and the World Bank (where the CGIAR Secretariat is headquartered). CGIAR's mission is to promote sustainable agriculture in developing countries with the goal of creating food security, alleviating poverty, and preserving natural resources.

Consumer Subsidy Equivalent (CSE). A measure of the value of monetary transfers to consumers resulting from agricultural policies in a given year. If negative, it measures the implicit tax imposed on consumers by agricultural policies. The main component of the CSE is market transfers due to market price support to producers. The CSE can be measured in money terms, in money terms per unit of production, or in percentage terms. See producer subsidy equivalent (PSE).

Consumer-oriented agricultural products. One of three broad categories of agricultural products used by the Foreign Agricultural Service to report export and import data under its BICO system. (The others are bulk commodities and intermediate agricultural products. Consumer-oriented agricultural products are high value products that usually (but not always) are those ready, or easily made ready, for immediate use by consumers. Notable examples are snack foods, breakfast cereals, bakery mixes, eggs and products, dairy products, fresh or processed red meats and poultry, fresh or processed fruits, vegetables, nuts, pet foods, wine, and beer.

Consumptive water requirement. The amount of water potentially which is needed to meet evapo-transpiration needs of the plant so that the plant does not suffer in its growth through short supply of water.

Consumptive water use. Water removed from available supplies without return to a water resources system, e.g., water used in manufacturing, agriculture, and food preparation. Crop consumptive water use is the amount of water transpired during plant growth plus what evaporated from the soil surface and foliage in the crop area.

Contact herbicides. Herbicides which kill only those parts of a plant with which they come into contact and used mostly to control annual weeds when seedlings. They have very little residual effect and are normally applied just before crop sowing.

Contact placement. Drilling of seed and fertilizer together while sowing.

Contagious. Contagious disease which gets conveyed from a sick animal to another by actual contact.

Container garden. Includes all shapes and sizes of containers, usually sitting on the ground, sometimes massive enough to hold all types of plant species including woodies, perennials, and bulbs.

Contaminants. Those factors affecting genetic and physical quality of the seed. For example of types, foreign pollen, other crop plants, weed plants, plants affected by designated diseases, weed seeds etc., are regarded as contaminants.

Continental climate. A climate having large daily and annual ranges of temperature

Continous ploughing. In this method, first the headland gets marked and the first ridge is set up at three quarter of land width from the side. The outer ridges are set at full width over the field. The operator starts ploughing between the first ridge and the sideland, continue to turn left and cast in the three quarter land until a quarter land width of ploughing gets completed on each side.

Continuous brooders. Brooders which are used for rearing chicks in large number with coal, gas or oil suppling fuel, used in big commercial enterprises.

Continuous cropping. Growing of crops one after the other without seasonal fallowing.

Continuous drying. Refers to the process of drying with a continuous flow of grain and air, in contrast to back operation.

Continuous flow driers. The grain is conveyed through the drier in a continuous flow in a shallow bed, either at rightangles to the flow of high temperature air (cross-flow type) or in the opposite direction (counter flow type). *Storage Driers.* These are involving slow drying with large quantities of warm air until equilibrium is reached. They include bins or silos with perforated floors (in-bin-type) or buildings with flat concrete floors overlaid by ventilating airducts (on-floor type). The rate of drying varies from 6% per day in silos to 6% per hour in small driers. The safe, long term storage of grain requires a moisture content of 14%.

Continuous layerage. Refers to an asexual reproductive process of plant propagation in which the whole branch except the tip has been buried 8-10 cms of earth and allowed to develop roots while still attached to parent.

Continuous. Said of pith which is solid; not spongy, chambered or interrupted by cavities.

Contour border irrigation. Refers to the method of irrigating gently sloping fields, the whole area being divided into strips by ridges along the contours and cross ridges. The ridges confine the water to a particular strip till it is completely full, before, it is allowed to flow to the next lower strip. Also called contour check irrigation.

Contour cropping. Cultivation of crops along the contours of a slope.

Contour farming. Field operations (such as plowing, planting, cultivating, and harvesting) at right angles to the natural slope to reduce soil erosion, protect soil fertility, and limit water runoff. Contour strip farming is a kind of contour farming in which row crops are planted in strips, between alternating strips of close-growing, erosion-resistant forage crops. Refers to the method of cultivation wherein operations including sowing have been carried out along the contour. It reduces runoff, conserves more moisture for crop production, reduces soil losses and increases crop yield.

Contour furrows. Furrows which are ploughed on the contour on pasture of range land to prevent soil loss and allow water to penetrate the soil. Sometimes used in planting trees or shrubs on the contour.

Contour planting. Planting crops along the contour.

Contour ploughing. Ploughing on the level following the contour, as distinct from up and down the slope. This practice is an effective method of soil erosion control.

Contract acreage. Enrolled 1996 commodity base acreage under the FAIR Act of 1996 for wheat, feed grains, upland cotton, and rice (generally fixed for 1996 through 2002). A farmer may voluntarily choose to reduce contract acreage in subsequent years. Land leaving the CRP may be entered into a production flexibility contract if the land was previously commodity base acreage.

Contract commodity. The commodities previously eligible for deficiency payments and now eligible for production flexibility contracts under the FAIR Act of 1996: wheat, corn, sorghum, barley, oats, rice, and upland cotton.

Contract for future sale. A sales contract under which a farmer agrees to deliver products of specified quality and quantity to a buyer for a specified price within a prescribed time frame. Contract sales are a growing practice, recently accounting for 86% of poultry, more than 50% of fruits, and 43% of milk. The benefits to processors are greater uniformity and predictability resulting in lower costs of grading, processing, and packing. The benefits to farmers are more stable income from a guaranteed market and price, and possibly access to a wider range of production inputs and advanced technology.

Contract production. A form of vertical integration where a firm commits to purchase a commodity from a producer at a price formula set in advance of the purchase.

Contract sanctity. The concept that U.S. agricultural products already contracted to be exported should not be subject to government cancellation because of short supply, national security, and/or foreign policy reasons. The FACT Act of 1990 provides for contract sanctity by prohibiting the President from restricting the export of any agricultural commodity already under contract to be delivered within 270 days from the date the embargo is imposed, except during national emergency or war.

Contract. Written or oral agreement spelling out the parties' understanding of how a commodity is to be produced and/or marketed, including specifications for quantity, quality, and price. Marketing contracts are commonly used for crops, while production contracts are more prevalent in the livestock industry. Contracts contrast to cash markets. Cash markets continue to dominate the agriculture sector, accounting for almost 70% of farm commodity sales in 1997. However, contracting could likely continue to grow as a risk management tool for farmers and a coordination tool for processors. Futures contracts provide a way to manage price risk that typically do not involve actual delivery of commodities.

Control, biological. The employment of the enemies and diseases of a pest for the purpose of maintaining adequate control. The artificial application of biotic control.

Controlled atmospheric storage. Cold storage in which the concen-trations of atmospheric gases have been adjusted to extend the storage life of fresh produce. Usually oxygen has been lowered and carbon dioxide raised.

Conventional agriculture. Generally used to contrast common or traditional agricultural practices featuring heavy reliance on chemical and energy inputs typical of large-scale, mechanized farms to alternative agriculture or sustainable agriculture practices. Mold-board plowing to cover stubble, routine pesticide spraying, and use of synthetic fertilizers are examples of conventional practices that contrast to alternative practices such as no-till, integrated pest management, and use of animal and green manures.

Conventional tillage. Tillage operations considered standard for a specific location and crop and that tend to bury the crop residues; usually considered as a base for determining the cost effectiveness of erosion control practices. See no-till farming.

Converted wetland. Under the swampbuster program, these are wetlands that were drained or altered to improve agricultural production after December 23, 1985, the date swampbuster was enacted. On lands with this designation, no drainage maintenance and no additional drainage are allowed.

Conveyance loss. Water loss in pipes, channels, conduits, ditches by leakage or evaporation.

Conveyors. Equipment which is used for moving the seeds in seed testing laboratory.

Cool season vegetables. Vegetables whose growth and productivity are favoured by low temperatures; most of them have the ability to withstand frost; also known as winter vegetables

Cool-season grass. A grass that greens up and grows during the spring, sets seed in early summer, then goes dormant until fall, when it begins growing again. A second dormancy may occur during cold winter weather.

Cooperative extension system. A federal-state-local cooperative education system that provides continuing adult education based on the academic programs of the land grant colleges of agriculture and their affiliated state agricultural experiment stations. The system employs approximately 32,000 people located on land grant campuses and offices in virtually every county in the nation. About half of Extension's education programs focus on agriculture and natural resources, one-quarter on youth development (including the vocational 4-H program), and the balance on home economics and community resource development work.

Cooperative. An enterprise or organization owned by and operated for the benefit of those using its services. In agriculture, such an organization is owned and used by farmers mainly to handle the off-farm part of their businesses — buying farm supplies, marketing their products, furnishing electric and telephone service, and providing business services —at cost. Essential features are democratic control, limited return on capital, and operation at cost, with distribution of financial benefits to individuals in proportion to their use of the services made available by the cooperative (called patronage refunds). In 1997, there were 3,884 farmer cooperatives in the United States. As a variation from the traditional design, so-called "new generation cooperatives" are characterized by limited membership, require substantial investment, and include delivery contracts. Producers are increasingly using this model to create their own value-added business enterprises. The Rural Business-Cooperative Service (RBS) assists in forming new cooperative businesses and improving the operations of existing cooperatives through technical assistance, research, information products. Cooperatives are afforded certain antitrust exemptions by the Capper-Volstead Act.

Copper fungicides. Copper compounds which are used for killing fungi causing plant disease. They act by contact, killing only those spores or other parts of disease agents with which they are brought in to close contact, popular examples include copper fungicides like Bordeaux mixture; copper oxychloride, cuprous oxide, perenox perelan (a dusting powder based on copper).

Coppice farming. Practice of intensive exploitation of land capability by regular harvesting of trees which produce numerous shoots from the stump after cutting. It is a traditional method of woodland management, by which young tree stems are cut down to a low level, or sometimes right down to the ground. In subsequent growth years, many new shoots will grow up, and after a number of years the cycle begins again and the coppiced tree or stool is ready to be harvested again. Typically a coppice woodland is harvested in sections, on a rotation. In this way each year a crop is available. This also has the side-effect of providing a rich variety of habitats, as the woodland always has a range of different aged stools growing in it. This is beneficial for biodiversity. The cycle length depends upon the species cut, the local custom, and the use to which the product is put. Birch can be coppiced for faggots on a 3 or 4 year cycle, whereas oak can be coppiced over a 50-year cycle for poles or firewood. The shoots (or "suckers") may be used either in their young state for interweaving in wattle fencing as is the practice with willows, or the new shoots may be allowed to grow and mature into large poles, as was often the custom with hardwood such as oaks or ashes. This practice creates long, straight poles which are better for working than naturally grown trees which have bends and forks. Coppicing may also be practiced to encourage specific growth patterns, as with cinnamon trees which are grown for their bark.

Coppice growth. Arising from sprouts at the stump; bushy.

Coppice. (1) A small wood of small trees and underwood cut periodically, essentially comprising sprouts from cut stumps. (2) Growth arising from sprouts at the stump, bushy.

Coppicing. The cutting back of trees or shrubs close to the main trunk or ground level resulting in the production of a number of vigorous new shoots (See Pollarding)

Copra. The mature endosperm (kernel) of coconut when dried to about six per cent moisture content

Cord (forestry). A stack of wood that has a gross volume of 128 cubic feet. A standard cord measures 4 feet x 4 feet x 8 feet.

Cordate. A leaf base that extends in a curving, dual lobe-like fashion below the top of the petiole; heart-shaped.

Cordial. A sparkling, clear, sweetened fruit juice from which pulp and other insoluble substances have been completely removed. A cordial is any invigorating and stimulating preparation; as, a peppermint cordial. The term derives from obsolete medical usage, as various beverages were concocted which were believed to be beneficial to one's health, especially for the heart (cordialis, Latin). Alcoholic cordials are also known as liqueurs. In the United Kingdom and Australia, cordial is an extremely sweet (usually entirely artificial) non-alcoholic fruit flavoured drink concentrate that is mixed with water to taste. This cordial is said to have some medical properties for stomach problems. Squash is sometimes called a cordial. A cordial is also a type of candy, in which a fruit filling is place within a chocolate shell. A well known confectionary of this type is the cherry cordial.

Cordiform. Heart-shaped or formed.

Cordon. A fruit tree trained and closely pruned to grow as a single stem, supported by wire frames or walls.

Cordoned. Having been espaliered.

Corking. The forming of heaps or cocks of a loose cut crop for drying in the field.

Corky-ridged. With elongated warts or ridges on the bark.

Corm. (1) A short, swollen underground stem, carrying buds, acting as an organ for storage and vegetative preproduction. Similar to a bulb but the food is stored in the swollen stem as distinct from the scales of a bulb. Examples, include crocus and gladiolus. (2) A solid bulb-like underground stem not differentiated into scales, often depressed-globose in form, bearing scale-like buds on the surface, usually tunicated. Ex. Gladiolus, crocus (3) A solid, swollen part of a stem. This is typically underground. Plants with corms include *Crocus* and *Gladiolus*. (4) Fleshy underground stem that resembles a true bulb, which produces roots, leaves and flowers during the growing season. (5) The underground swollen base of stem axis enclosed by scale leaves.

Cormel. (1) A small corm arising from the base of a parent corm. (2) A tiny corm (fleshy stem base) around a large mother corm. (3) Miniature conn arising from a parent corm

Corms. Corms are swollen underground stem tissue. The basal plate is located on the bottom of the corm; roots develop from this plate. Cormels are the small corms that are produced to the side of the corm each year. Cormels usually take two to three years to develop. Examples include:

- Crocus
- Erythronium
- Freesia
- Gladiolus
- Tritonia

Corn celt. That area of the United States where corn is a principal cash crop, including Iowa, Indiana, most of Illinois, and parts of Kansas, Missouri, Nebraska, South Dakota, Minnesota, Ohio and Wisconsin.

Corn cockle. A tall weed (*Agrostemma githago*) of cornfields. Hairy, long-stalked, with much divided leaves and single pale purple flower, with woolly sepals. Corncockle (*Agrostemma githago* - also Corn cockle and Corn-cockle) is a slender pink flower of European corn fields. In the 19th century, it was reported as a very common weed of wheat fields and its seeds were inadvertently included in harvested wheat seed and then re-sown the following season. It is very likely that until the 20th century, most wheat contained some corncockle seed. It is now present in many parts of the temperate world as an alien species, probably introduced with imported European wheat. It is known to occur throughout much of the USA and parts of Canada, parts of Australia and New Zealand. In parts of Europe such as the UK, intensive mechanised farming has put the plant at risk and it is now uncommon or local. This is partly due to increased use of herbicides but probably much more to do with changing patterns of agriculture with most wheat now sown in the autumn as winter wheat and then harvested before any corncockle would have flowered or set seed. It is a stiffly erect plant up to 1 metre tall and covered with fine hairs. Its few branches are each tipped with a single deep pink to purple flower. The flowers are scentless, are 25 mm to 50 mm across and are produced in the summer months - May to September in the northern hemisphere, November to March in the southern hemisphere. Each petal bears 2 or 3 discontinous black lines. The five narrow pointed sepals exceed the petals and are joined at the base to form a rigid tube with 10 ribs. Leaves are pale green, opposite, narrowly lanceolate, held nearly erect against stem and are 45 mm to 145 mm long. Seeds are produced in a many-seeded capsule. All parts of the plant are reported to be poisonous. It can be found in fields, roadsides, railway lines, waste places, and other disturbed areas.

Corn gluten. A byproduct of wet milling of corn. Corn gluten is used as a medium-protein (20-24%), medium-fiber (10%) feedstuff. The Eurõpean Union is the major market for U.S. corn gluten feeds.

Corn. A general term for cereal crops (barley, maize, oats, rye and wheat) grown for their edible seeds or grain. The grain itself is also called corn. In America the term is restricted to maize. In Scotland and Ireland the term is generally restricted to oats, whilst in England it is mainly applied to wheat.

Corn/hog ratio. See hog/corn ratio, and feed ratio.

Corner plants. grow well over the container's edge and benefit from a corner position where they have maximum elbowroom.

Corolla. (1) The combined term for floral petals. (2) The inner floral envelope, consisting of petals. (3) The petals and normally colored portion of a flower; the inner floral envelope, consisting of petals. (4) A crown; an appendage or extrusion that stands between the corolla and stamens; an out gowth of perianth tissue in the "cup" of Narcissus, or of the androecium in milkweeds.

Corona. An appendage between the petals and stamens e.g. cup of the daffodil.

Corona/crown. A fused flower tissue as in the cup of a Daffodil; the point at which a plant's roots and top join.

Corporate farm. A form of farm ownership which is a separate legal entity from the owners of the farm. Changes in the tax law in the 1970s encouraged the incorporation of farms as corporate tax rates declined while individual tax rates rose, mainly because of inflation. The 1992 Census of Agriculture reports that less than 4%, or nearly 73,000, of the 1.925 million farms in the nation were corporate farms. By contrast, more than 1.653 million (86%) were individual or family-owned operations and 186,000 (10%) were partnerships.

Corrosive sublimate. Bichloride of mercury which is a violent poison, ft has been highly effective as an antiseptic,, germicide, and disinfectant. As an antiseptic would wash and in dressings, it is used in 1:1000 to 1:5000 solutions; as disinfectant in 1:1000 to 1:5000 solutions. Corrosive sublimate solution has to be stored in wooden, glass or earthenware as it corrodes metals.

Cortex. The rind or outer covering of a stem; technically it lies outside of the vascular bundles, which constitute the stele.

Corymb. A more or less flat-topped indeterminate inflorescence whose outer flowers open first. Ex. Viburnum

Cosmetic appearance. "The exterior appearance of an agricultural commodity, including changes to that appearance resulting from superficial damage or other alterations that do not significantly affect yield, taste, or nutritional value." The Agricultural Marketing Service sets grades and standards for many agricultural commodities. Some consumer and environmental groups have argued that some of these standards are harmful because they encourage excessive pesticide use merely to make fruits and vegetables "attractive." Agricultural interests disagree, countering that consumers prefer blemish-free produce and that cosmetic standards are no less important than other grading factors.

Cosmetic pruning. Minor pruning of dead flowers and weak or crossing shoots to keep a plant tidy and balanced in shape.

Cost/benefit analysis. A quantitative and sometimes qualitative evaluation of the costs which would be incurred by some action (such as building a dam, or implementing an environmental regulation) versus the overall benefits to society of the proposed action. See risk-benefit analysis.

Cost-containment (for WIC). Refers to statutory provisions in the Child Nutrition Act of 1966 that require state agencies to contain WIC program costs, particularly with respect to the cost of infant formula sold through the program. See competitive bidding and sole source bids.

Cotton boll. Fruit of cotton plant is termed as boll.

Cotton competitiveness provisions. Provisions added by the Food, Agriculture, Conservation, and Trade Act of 1990 to the cotton program designed to keep U.S. cotton price competitive in domestic and export markets. Sometimes referred to as the "three-step competitiveness"

Fig.: Cotton boll

provisions. Step 1 is the discretionary authority for USDA to reduce the adjusted world price (used in the cotton marketing assistance loan program) when world prices are declining to near the adjusted world price, but U.S. prices are higher than world prices. Though rarely used, the Step 1 adjustment is intended to make marketing loans more effective in keeping U.S. cotton globally competitive. Step 2 payments, sometimes referred to as the "user marketing certificate program," are made to U.S. cotton users and exporters when U.S. prices are higher than world prices. Step 2 payments are intended to bridge price gap and keep U.S. cotton competitive. Step 3 mandates the opening of a "special import quota" when the differential between the higher U.S. price for cotton and the lower price for foreign cotton extends for a specified length of time. Its purpose is to allow imports to enter, acting to lower U.S. prices to bring them more in line with world prices. Step 3 quotas were in effect in April-May 1995, from late October 1995 through early May 1997, and were triggered in late February 1999. A step 3 quota cannot be established if a limited global quota for upland cotton is in effect, which operates differently and is triggered when other price conditions are met.

Cottonseed cake, cottonseed meal. An animal feedingstuff made from the seeds of the cotton plant after the extraction of oil, mainly used for adult ruminants. It has a high fibre content, is poor in calcium, with low protein quality. It is yellow-coloured due to the presence of gossypol which is toxic to young pigs and poultry, and is available either in decorticated or undecorticated forms.

Cottony. Much the same as woolly, with white hairs.

Cotyledon. A seed leaf, forming part of the embryo of seeds. The first leaf to develop when a seed germinates, initially lacking chloro-phyll. Monocotyledonous plants have one cotyledon in each seed whilst dicotyledonous plants have two types of plants are also distinguished by their leaves. The former possess long, narrow leaves with parallel veins (e.g., grasses) whilst the latter have broad leaves with branching xeins (e.g., broadleaf weeds such as dock, fat hen, etc.) (2) The primary leaves of the embryo, present in the seed.

Cotyledons. Embryonic seed leaves that stores nutrients to sustain the growth of a newly germinated plant

Coulter face. The side of a slice cut by a coulter ploughing, and then turned over by the mouldborad to become the front of the slice.

Coulter. The part of a plough which makes the vertical cut in the soil, usually mounted in front of the share, attached to the beam. There are three main types; knife, disc and skim coulters. Knife coulters were a feature of horse ploughs. Flat, circular, freely rotaing disc coulters are used on most tractor-drawn ploughs. They rotate as the plough moves, separating a slice of soil from unploughed land, which the share and mouldboard then turn to produce a clean furrow slice. Skim coulters are attached to the beam, either in front of knife coulters or behind disc coulters. Each acts like a small digger and turns a small furrow slice from the coulter of the main furrow into the furrow bottom, thus completely burying crop residues or manures also called colter.

Coumarin. A naturally occurring plant growth inhibitor.

Counter irritant. An agent causing a skin irritation which is intended to relieve another irritation of an organ below the skin. Mustard plaster is somtimes used in the treatnent of pneumonia.

Counterfire. Intentionally burning a bell of trees towards an advanc-ing forest fire so that both the fires may get extinguished when they meet.

Countertrade. A trade transaction of goods and services without the exchange of money. Forms of countertrade include barter, buy-back or compensation, counter-purchase, offset requirements, swap, or triangular trade.

Country schedules. The official schedules of subsidy commitments and tariff bindings as agreed to under GATT for members countries.

County committees. Panels of three to five farmers, elected by other farmers, to oversee the local operation of commodity programs, credit, and other programs of the Farm Service Agency. County committees, established by the Soil Conservation and Domestic Allotment Act of 1935, are so named because they have overseen USDA field offices for farmers that once existed in most rural farm counties throughout the United States. Today, the committees often oversee activities in multi-county areas, due to USDA reorganization and consolidation of its field office structure into a network of about 2,500 field service centers. The committees are responsible for hiring and supervising the County Executive Director (CED), who manages the day-to-day activities of the field service center and its employees. The director and most county office staff legally are employees of the farmer-elected committees rather than the federal government, although their salaries come from federal funds.

County extension agent. A worker who is jointly employed by the county, State Cooperative Extension Service, and the U.S. Department of Agriculture's Extension Service, to bring agricultural and homemaking information to local people and to help them resolve farm, home and community problems. Also called extension agent, farm and home advisor, agricultural agent, extension home economist, and 4-H or youth agent.

Cover. Vegetation or other natural shelter serving to conceal wildlife from predators.

Cover crop. (1) A close-growing crop, planted primarily as a rotation between regularly planted crops, or between trees and vines in orchards and vineyards, to protect soil from erosion and improve it between periods of regular crops. (2) A crop grown alone or in between rows of crops to reduce soil erosion, conserve nutrients, and supplement organic matter.

Cover, ground. Any vegetation which produces a protecting mat on or just above the soil surface. In forestry, low-growing shrubs and herbaceous plants under the trees.

Crabtree effect. The continued improvement in the feed-use effi-ciency of an animal under feed restriction, so that the feed supply has to he repeatedly reduced further in order to maintain a flat growth rate.

Cradle. A set of fingers or light rods in a light frame set into a scythe (a cradle-scythe), used to cut com, laying it evenly with each stroke.

Craft. A class of work in agriculture such as milk production, machinery operation and maintenance, or plant nursery practice.

Crawler tractor. Tractor which has endless chain or track in place of pneumatic wheels. This is also known as *track type tractor* or *chain type tractor.*

Cream separator. A machine which uses centrifugal action to separate milk into lighter cream for butter making and Skim milk or separated milk containing the heavier particles. Either hand-operated or power-driven. Farmhouse butter making is now uneconomic except for home consumption, and cream separa-tors are now rarely seen.

Creeper. A plant which spreads along the ground surface or other substrata.

Creeping. (1) Growing and spreading over the ground. (2) Prostrate and spreading over the ground.

Crenate. Having rounded teeth at the leaf margins. (3) Scalloped; with rounded teeth. (4) Scalloped; with rounded teeth.

Crenate-serrate. Having a mixture of blunt and sharp teeth.

Creolin. Refers to an antiseptic, germicide, deodorant and disinfec-tant which is used for the control of certain external parasites of animals. It is a dark-brown liquid which consists of high boiling coal-tar, phenols and oils. It forms milky emulsion when diluted with water.

Creosote. An oil-based preservative used to treat wood used in gardens.

Cress. Cress is one of those leafy greens that makes you sound snooty and rich when you order a soup, salad, or sandwich. However, it is easy and quick to grown (12 to 20 days from seed), and provides vitamins and minerals. How it became associated with afternoon teas and bridge luncheons is not fully known. Cress, as it is sometimes called, is a hardy salad/sandwich green that prefers growing in cool seasons - early spring and late fall. Sow 10 to 15 seeds per foot of row every 2 weeks starting as soon as the ground can be worked. Plantings sown after May 31 usually bolt or produce a flower head quickly. However, additional plantings can resume after Labor Day and continue until Thanksgiving if the ground is not frozen. Later plantings may not grow as fast and could be killed by severe temperatures.

Crimp. (1) To crush grass with a crimper. (2) In general terms the waviness of wool. One of the waves in wool fibres. Measured as the number of crimps per unit length and an important quality in assessing manufacturing value.

Crimper. A machine with ribbed rollers between which grass is crushed during haymaking, allowing sap to be removed by the action of sun and air. Similar to a Roller Crusher.

Crisped. Wavy on the margin; short and curly when applied to pubescence.

Criss-cross breeding. A system of pig breeding in which the same two breeds are used as alternative sires on home-produced, cross-bred gilts. By comparison, rotational crossing involves the use of boars from three or four breeds used in strict rotation.

Critical control point. An operation (practice, procedure, process, or location) at or by which preventive or control measures can be exercised that will eliminate, prevent, or minimize one or more hazards. Critical control points are fundamental to Hazard Analysis and Critical Control Point (HACCP) systems, which are now being adopted by the food industry to prevent health hazards in the food supply.

Critical habitat. Under the Endangered Species Act, critical habitat is an area essential to the conservation of a listed species, though the area need not actually be occupied by the species at the time it is designated. Critical habitat must be designated for all threatened and endangered species under the Act (with certain specified exceptions). The areas may be federal or nonfederal land, but only the federal government is required to protect it. A federal agency with whom a landowner is dealing must ensure that its actions (which may include giving a loan, increasing irrigation flows, etc.) do not adversely modify these areas.

Critical level. Concentration of a nutrient element below which deficiency symptoms may develop or a response to addition of the nutrient may be observed.

Critical state. Unique conditions of pressure, temperature and composition where in all properties of coexisting vapour and liquid become identical.

Critical temperature for water. Temperature at which the volume of water and the volume of saturated vapour formed from it are equal. This is about 365°C.

Critical temperature. Approximate temperature of the air when the heat lost by the animal body through physical means such as evaporation just balances the heat produced as a result of internal work and tissue oxidation, and for which further tissue-oxidation is necessary to maintain the normal temperature of the body.

Critical velocity. Of a liquid is that velocity of flow above which the flow ceases to streamline.

Crochets. Any small hook-like organs, particularly those on the prolegs of caterpillars.

Crocks. Broken pieces of earthen pots.

Crop acreage base. A crop-specific measure equal to the average number of acres planted (or considered planted) to a particular program crop for the previous five years. The sum of the crop acreage bases for all program crops on a farm may not exceed the farm acreage. The acreage base was used in determining the number of acres a farmer, under an acreage reduction program, had to remove from normal crop production and devote to conserving uses in order to be eligible for USDA price and income supports. The FAIR Act of 1996 suspends the base acreage provisions of the permanent law.

Crop bound. A bird with its crop obstructed by material which cannot pass on to the stomach and gizzard, e.g. feather, straw, wool, etc.

Crop cafeteria. The demonstration of identified efficient crops / varieties in an agrometerological region/zone offering an opportunity to the farmer to choose a suitable crop or crop combination commensurating with the available resources and requirements.

Crop canopy. The overhead cover or horizontal projection of a crop formed by its leaves, branches etc. which can influence the penetration and interception of radiant energy.

Crop competition. The tendency of plants of the same or different species growing together to strive for and capture common resources like light, water, nutrients and space, resulting in reduced growth and/or altered form of the crop(s).

Crop cover. Subsidiary crop of low plants which has been introduced in a plantation to afford cover between or below the main crop; also any crop which is used to protect land from erosion. A cover plant is one suitable for use as a cover crop.

Crop duration. Days taken by a crop from germination to maturity.

Crop insurance. Insurance that protects farmers from crop losses due to natural hazards. Hail and fire insurance are offered through private companies without federal subsidy. A subsidized multiperil federal insurance program, administered by the Risk Management Agency, also is available to most farmers. The program is authorized by the Federal Crop Insurance Act (which is actually title V of the Agricultural Adjustment Act of 1938), as amended. Federal crop insurance is available for about 60 different crops, although not all insurable crops are covered in every county. With the amendments to the Federal Crop Insurance Act made by the Federal Crop Insurance Reform Act of 1994, USDA is authorized to offer basically "free" catastrophic (CAT) coverage to producers who grow an insurable crop. Farmers must sign a waiver foregoing any federal disaster assistance if they decline CAT coverage. For an additional premium, farmers can buy additional coverage beyond the CAT level. Crops for which insurance is not available are protected under the Noninsured Assistance Program (NAP). Federal crop insurance is sold and serviced through private insurance companies. A portion of the premium is subsidized by the federal government, as well as the administrative and operating expenses of the private companies. The Federal Crop Insurance Corporation reinsures the companies by absorbing the losses of the program when indemnities exceed total premiums. Several revenue insurance products are available on major crops as a form of additional coverage.

Crop reports. Reports compiled by the National Agricultural Statistics Service (NASS) on various commodities that are released throughout the year. Information in the reports includes estimates on planted acreage, yield, and expected production, as well as comparison of production from previous years.

Crop residue management. Use of the non-commercial portion of the crop for soil protection and improvement.

Crop residue. That portion of a plant, such as a corn stalk, left in the field after harvest. Crop residues are measured for farmers who use conservation tillage to implement their conservation plans to meet conservation compliance requirements. These farmers are required to maintain a minimum level of crop residue to be in compliance. Under revisions to the conservation compliance program in the FAIR Act of 1996, farmers are allowed to use third parties, certified by USDA, to measure levels of crop residue.

Crop Revenue Coverage (CRC). A form of revenue insurance that protects a producer's revenue for an insurable crop whenever low prices, low yields, or a combination of both causes revenue to fall below a guaranteed level selected by the producer. It differs from other revenue insurance programs by allowing producers to use the higher of the planting price or the market price in determining a target level of revenue.

Crop rotation. (1) The rowing of different crops, in recurring succession, on the same land in contrast to monoculture cropping. Rotation usually is done to replenish soil fertility and to reduce pest populations in order to increase the potential for high levels of production in future years. (2) Definite succession of crops following one another in a specific order.

Method and Purpose: Crop rotation is the practice of growing two (or more) dissimilar type of crops in the same space in sequence. It is one component of polyculture. Examples of types of crops that can be rotated include cereals and legumes, and deep-rooted and short-rooted plants. Crop rotation is widely practiced in vegetable production, where it is possible to grow a cool-season crop (such as lettuce) in the spring, follow with a warm-season crop (such as tomatoes) in the summer, and then grow a winter crop (such as mache) harvested in early spring. Crop rotation is less common with field crops because the timing is difficult to manage, particularly since most crop-growing regions experience a good deal of rainfall during the late spring and early summer season, which interrupts fieldwork. Crop rotation avoids a decrease in soil fertility, as growing the same crop repeatedly in the same place eventually depletes the soil of various nutrients. A crop that leaches the soil of one kind of nutrient is followed during the next growing season by a crop that returns that nutrient to the soil, for example, rice followed by wheat, or rice followed by cotton. If crop rotation is done properly, farmers can keep their fields under continuous production, without a need to let them lie fallow or to apply artificial fertilizers, both of which can be expensive. Legumes, plants of the family Fabaceae, for instance, have nodules on their roots which contain nitrogen-fixing bacteria. It therefore makes good sense agriculturally to alternate them with cereals (family Poaceae) and other plants that require nitrates. A common modern crop rotation is alternating soybeans and maize (corn). In subsistence farming, it also makes good nutritional sense to grow beans and grain at the same time in different fields. Crop rotation is also used to control pests and diseases that can become established in the soil over time. Families of plants tend to have similar pests. By regularly changing the planting location, the pest cycles can be broken or limited. This principle is of particular use in organic farming, where pest control is achieved without synthetic pesticides, herbicides and fungicides. Some plants follow one another favorably and others do not which brings risks of yield losses, development of diseases and weeds. The choice of the rotation depends on the nature of the soil, which determines the species of plants that one can cultivate, but also on animal breeding, which plays a role in the choice of the rotation (fodder crops, straw). Finally a good rotation takes into account the proportion of natural meadows.

The choice of the rotation is also determined by the economical aspects (local needs, market accessibility).

Crop scouting. Precise assessments of pest pressure (typically insects) and crop performance to evaluate economic risk from pest infestations and the potential effectiveness of pest control interventions. Scouting is usually sold as a commercial service to farmers.

Crop season. The most favourable weather condition to get better yield

Crop share rent. In contrast to cash rent, the tenant farmer pays the landlord a share of the crop. This arrangement puts the landlord, like the tenant operator, at risk from variation in yields and prices. For the farm operator, crop share rent is a mechanism for sharing risks with the landlord. In relation to commodity programs for supporting prices and farm incomes, cash rent landlords do not have a beneficial interest in the commodity and are not eligible for some benefits compared to crop share landlords that do have a beneficial interest in the crop.

Crop timing. Regulating the harvesting time of crops by way of adjusting the planting time, pruning or suitable treatments.

Crop year. The year in which a crop is produced. This contrasts with the marketing year, which is the 12-month marketing period that begins with harvest.

Crop yield index. Refers to a measure of comparison of the yields of all crops on a given farm with the average yield of these crops in the locality. The relationship has been expressed in percent-age. It is a suitable method under Indian conditions where diversified type of arming has been followed.

Crop. (1) Plants, carefully selected and developed over many years, sown on cultivated land to produce food for man and animals or raw materials, e.g., barley, potatoes. The term is also applied to plants which are not sown but come up naturally in cultivated land from wild seed, e.g., a crop of thistles. (2) To mow, cut, reap or gather a cultivated crop. (3) The total quantity produced, cut or harvested from cultivated land. (4) To bite off in eating, e.g., grazing sheep will crop a ley. (5) The expanded part of a bird's oesophagus or gullet, in which food is stored before being digested. Also called the craw. (6) A hunting whip with a loop as distinct from a lash. (7) Plants sown and harvested by man for economic purposes.

Cropland. Land used primarily for the production of row crops, close-growing crops, and fruit and nut crops. It includes cultivated and noncultivated acreage, but not land enrolled in the Conservation Reserve Program. Approximately 382 million acres of cropland, including 50 million acres of irrigated land, was in use in the United States during the most recent national resources inventory, conducted in 1992. Cropland is 30% of all non-federal rural lands. In 1996, the value of production from cropland was about $108 billion.

Cropping intensity. Refers to the number of crops which are raised during the year.

Cropping pattern. Yearly sequence and spatial arrangements of crops and fallow, on a given area.

Cropping system. Cropping pattern and its interaction with farm resources, farm enterprises and available technology.

Cross compliance. A no longer used requirement that a farmer who participates in a price support program for one crop must also participate in price support programs for other crops grown on the same farm. See conservation compliance.

Cross cultivation. Means the working of land by ploughing the second time at right angles to the first.

Cross pollination. Transfer of pollen from the anther of one plant to the stigma of another plant (unless of the same clone)

Cross protection. One strain of a virus which is infecting plant tissues and potect them from infection by other strains of the same virus.

Cross-breeding. The mating of animals (or plants) of different breeds, but both of which are purebred, in order to combine the-best characteristics of the two breeds. The progeny known as crossbreds, possess hydbrid vigour, often expressed in females as increased mothering ability. They are sometimes crossed with a male of a third breed to introduce further qualities.

Cross-fruitfulness. Interfruitfulness; the ability of a variety of fruit plant to mature fruit following pollination by another specific variety.

Crossing. Another name of hybridization. It is defined as the artificial cross-pollination between genetically unlike plants.

Crotch. (1) Any angle formed between two woody branches, or that point where the main trunk divides into branches. (2) Common joint of large branches from which sometimes more than one lateral branch arises.

Crown bud. A flower bud, whose development has ceased, resembling a crown.

Crown budding. Replacing the undesirable and susceptible crown of a high yielding tree with a resistant clone by budding.

Crown craft. A type of graft in which the scion is fixed into a vertical slit in the bark and sap wood of the stock. Also called a rind graft.

Crown grafting. A graft union made at the crown of a plant on an established rootstock.

Crown slips. Shoots growing on the crown of pineapple (between fruit and crown), used for planting after rooting.

Crown vetch. Crown vetch, is not a true vetch, though it is a legume. It was or is commonly used by transportation departments to control erosion on banks and interchanges. Crown vetch is drought and cold hardy, disease resistant and seems to grow on the worst possible soils. However, this pink flowering perennial plant tends to become weedy, and allows a perfect habitat for burrowing animals, which isn't the best thing on banks. Plants don't respond to close mowings. Seeds should be sown at the rate of 1/2 to 3/4 pounds per 1000 square feet.

Crown. (1) The junction between stem and root near the ground level. (2) The tree-top. (3) Leafy top of pineapple fruit used for vegetative propagation. (4) Part of a rhizome with a large bud suitable for propagation. (5) The top pan of a plant, such as the top branches of a fruit it tree or the leaves of certain root crops, e.g., sugar beet. (6) The junction of the root and stem of a plant. (7) The perennial root-stock of certain plants e.g., rhubarb. (8) The part of a budded plant when the scion and understock meet. (9) On a clumping plant, the point at which the main stem (or stems) and the roots join together.

Crucifer. A plant belonging to the cabbage family *(Cruciferae)* characterized by four petals and sepals, both arranged in the form of a cross.

Cruciferous. Of or pertaining to the mustard family, Cruciferae.

Crush spread. In the soybean futures market, the simultaneous purchase of soybean futures and the sale of soybean meal and soybean oil futures to establish a processing margin. See gross processing margin.

Crush. Fruit beverage containing at least twenty five per cent fruit juice or pulp and fifty five per cent total soluble solids.

Crushed grain. Cereal grains crushed in a crushing Mill before feeding to livestock to assist in their digestion. Also called rolled grain.

Crushing mill. A machine which crushes cereal grains before feed-ing to livestock, by rolling it between two iron cylinders rotating

Cryo-preservation. Storage of cells, tissues, organs and seeds at super-low temperature.

Cryptogam. A plant reproducing by spores such as most ferns and algae.

Crystallised fruit/vegetable. Candied fruit/vegetable when covered or coated with crystals of sugar, either by rolling in finely powdered sugar or by allowing sugar crystals to deposit on them from a dense syrup.

CTC method. Crushing, tearing and curling method of processing tea.

Cuber. A machine which forces meal, usually binding it with molas-ses at the same time, through circular holes to produce extruded pencils of meal which are chopped into cubes or pellets of various lengths ranging from 3.25 mm for poultry to 13 mm for cattle.

Cucumber. A creeping plant cultivated under glass and in the open where there is no frost risk, both as a market garden and vegetable garden crop. Used as a salad ingredient and for pickling.

Cuke. Refers to the seeds of various plants, mainly tropical (e.g., groundnut, cotton, soya bean, coconut, linseed and palm nut kernel) which is compressed (to extract oil for cooking, soap manufacture, etc.) into flat slabs, or straights, rich in Protein, having oil and fibre in verying amounts, used as cattle food.

Culamintha alpina. A hardy aromatic plant, lorming a thick carpet of coliage and carrying sprays of violet flowers all through the summer.

Culinary. Suitable for cooking or processing in some form.

Culm. (1) The flowering stem of a grass. (2) An upright flowering stem from a rhizome; stem of grasses and sedges. (3) The joined stem of a grass, which is usually hollow, except at the nodes.

Cultigen. (1) A plant species or race which has arisen in the course of cultivation. (2) Any plant arising from cultivation and not the wild.

Cultivar (cv.). (1) A contraction of the phrase 'cultivated variety1 which designates a group of plants which are significant in agriculture, forestry or horticulture. (2) A cultivar is a cultivated selection that can be propagated reliably in a prescribed manner. This may be by seed, by grafting or it may be vegetatively propagated, i.e, be a clone. The word cultivar is a portmanteau coined from "cultivated" and "variety". Cultivars may be either particularly desirable selections from populations of a single species, or hybrids between species. Sometimes a cultivar can be placed within a species, but this is not required; in popular genera (such as Dahlia, Hosta or Rosa) the breeding lines are so complex that it would be impossible to ascribe most cultivars to any particular species. (3) A variety of a cultivated plant. (4) The name officially given for all cultivated varieties of plants. (5) A cultivated variety; does not occur naturally. (6) Short for "cultivated variety," or human-produced hybrid. You know the plant is a cultivar if part of the name is in single quotes, capitalized and not Latinized. For example, Sedum spectabile 'Brilliant.' (7) Shortened form of *culti*vated *variety*. A group of plants, selected for particular attributes, that are clearly distinct, uniform, and stable and that retain these characteristics when propagated by appropriate means.

Cultivar group. An association of related cultivars with a species, subspecies, or variety due to common origin or groups of traits.

Cultivation sheet. The term used for the record which furnishes the complete history of the individual crop grown on the field with its cost of cultivation from preparing the land to harvesting the crop and all other extra charges upto sale, revealing at the end total expenses of the crop, yield, and net profit or net loss.

Cultivation. Cultivation (a term which also has broader meanings related to the raising of plants in general) is the agricultural preparation of the soil to receive seeds. Primary tillage loosens the soil and mixes in fertilizer and/or plant material, resulting in soil with a rough texture. Secondary tillage produces finer soil and sometimes shapes the rows. It can be done by a using various combinations of equipment: plow, disk plow, harrow, dibble, hoe, rotary tillers, subsoiler, ridge or bed forming tillers, roller. Tillage can also mean the land that is tilled. Modern agriculture has greatly reduced the use of tillage for weed control. Through the use of herbicides and no-till planters crops can be grown for several years without any tillage. This practice has greatly reduced soil erosion and improved the quality of the environment by reducing the total tonnage of soil that is washed into streams and blown into the air to produce dust. Tillage and its effects are now a modern *agricultural science* carefully researched by universities across the planet which have disproven many of the ideas put forth by the passages quoted below. The following text is taken from the *Household Cyclopedia* of 1881: Tillage is an operation whereby the soil is either cleared from noxious weeds, or prepared for receiving the seeds of plants cultivated by the husbandman. When this operation is neglected, or even partially executed. the soil becomes foul, barren, and unproductive; hence, upon arable farms, tillage forms the prominent branch of work; and, according to the perfection or imperfection with which it is executed, the crops of the husbandman, whether of corn or grass, are in a great measure regulated. Tillage, in the early ages, was performed by hand labor; but, in modern times, the plough has been the universal instrument used for executing this necessary and important branch of rural work. In no other way can large fields be turned over because the expense of digging with the spade, the only other method of turning over the ground, would much exceed any profit that can be reaped. Stones lying above or below the surface are the most formidable obstruction to perfect tillage. On stony ground, the work is not only imperfectly executed, but in many cases the implement is broken to pieces, and a considerable portion of time lost before it is repaired and put in order. The removal of stones, therefore, especially of such as are below the surface, ought to be a primary object with every agriculturist; because a neglect of this kind may afterwards occasion him considerable loss and inconvenience. To drain the ground, in other words, to lay it dry, also facilitates tillage exceedingly; for ploughing cannot be performed with advantage where either the surface or subsoil is wet. The only sure and certain way by which the soil is cleaned or rendered free of weeds, is by ploughing in the summer months, when the ground is dry, and when, by the influence of the sun and air, the weeds may be destroyed with facility. Seldom at any other period is the soil much benefitted by ploughing, unless so far as a seed-bed is thus procured for the succeeding crop; and though the situation or state of the ground, when these intermediate ploughings are bestowed, is of importance in judging of their utility, yet the radical process of summer fallow cannot, by any means, be altogether dispensed with. Though, if the winter and spring ploughings are executed under favorable circumstances, and plenty of manure is at hand, it may be delayed for a greater number of years than is otherwise practicable, if good husbandry is to be maintained. Without summer fallow, or, which is the same thing, without working the ground in the summer months, perfect husbandry is unattainable on all heavy or cold soils, and upon every variety incumbent on a close or retentive bottom. To keep his land clean will always be a principal object with every good farmer; for if this is neglected, in place of carrying rich crops of grain or grass, the ground will be exhausted by crops of weeds. Where land is foul, every operation of husbandry must be proportionately noneffective; and even the manures applied will, in a great measure, be lost. The necessity of summer fallow depends greatly upon the nature and quality of the soil; as, upon some soils, a repetition of this practice is less frequently required than upon others. Wherever the soil is incumbent upon clay or till, it is more disposed to get foul, than when incumbent upon a dry gravelly bottom; besides, wet soils, from being ploughed in winter, contract a stiffness which lessens the pasture of artificial plants, and prevents them from receiving sufficient nourishment. When land of a dry gravelly bottom gets foul, it may easily be cleaned without a plain summer fallow; single crops, such as turnips, etc., may be substituted in its place, which, when drilled at proper intervals admit of being ploughed as often as necessary; whereas wet soils, which are naturally unfit for carrying such crops, must be cleaned and brought into good order by frequent ploughings and harrowings during the summer months.

Cultivator, trailed type. Having a main frame which is carrying a number of cross members to which tynes are fitted. At the forward end of the cultivator, there exists a hitch arrangement for hitching purpose. A pair of wheels are fixed in the cultivator. The lift has been operated by both wheels simultaneously so that draft is even and uniform. The height of the hitch has to be adjusted so that main frame becomes horizontal over a range of depth setting.

Cultivator. (1) A machine used to till the upper portion of the soil, primarily used to destroy weeds or form a moisture retaining mulch. (2) An implement having number of tynes which are attached to a frame. It is used for tilling the soil between standing rows of crops. It stirs the soil and is able to break the clods. The tynes fitted on the frame of the cultivator have been found to comb the soil deeply in the field.

Cultural methods. Practices used to enhance crop and livestock health and prevent weed, pest or disease problems without the use of chemical substances; examples include the selection of appropriate varieties and planting sites; selection of appropriate breeds of livestock; providing livestock facilities designed to meet requirements of species or type of livestock; proper timing and density of plantings; irrigation; and extending a growing season by manipulating the microclimate with green houses, cold frames, or wind breaks.

Culture. Bacteria growing on or in a substances which provides suitable feeding conditions.

Culvert. A channel for carrying water under a road, railway.

Cuneate. (1) Having a very narrow leaf base. (2) Wedge-shaped with an essentially straight side, the structure attached at the narrow end.

Cupped. A flower with depressed or concave center.

Curd. Milk thickened or coagulated by acid (Rennet) in cheese making, and distinguished from the more watery whey.

Curing. Natural or artificial aging of the plant produce brought about by some (dry or wet) heat treatment.

Curring. A drying process by which most of the moisture in the harvested leaf has been removed. Different methods of curing include air curing, rack curing, smoke curing, pit curing, ground curing, sun curing. It is process by which the harvested produce of tobacco is made ready for the market.

Cusec. Refers to the quantity of water which is flowing at the rate of one cubic foot per second. One cubic foot of water weighs 6.24 Ibs i.e., 6.24 gallons. One cusec flowing for one hour will be equal to 62.4 X 60 X 60 lbs=101 tons approximately.

Cuspidate. With an apex somewhat abruptly and concavely constricted into an elongated sharp-pointed tip.

Custom feeders. Producers who provide the service of feeding animals (e.g., cattle, hogs) they do not own, in return for a fee paid by someone else (such as a packer) who does own the animals. Custom feeding potentially provides packers with more control over supplies and prices of animals. Custom feeding is a form of vertical integration.

Custom inoculation. Refers to the inoculation of fanner's seeds, usually by machine not longer than 3 or 4 weeks in advance of planting.

Cut. (1) A term used to described the steepness of the crest of a Furrow slice. (2) A term used to described the mowing of grassland for silage.

Cuticle. (1) Cuticle is also used as a term for the exoskeleton, outside of epidermis of many invertebrates. In botany the cuticle is the waxy covering produced by the epidermal cells of leaves to protect the plant from excessive water loss. The cuticle is thicker in plants living in dry climates than in those from wet climates, and tends to be thicker on the top of the leaf. The cuticle is mostly composed of cutin and waxes. "The waxy sheet of cuticle also function in defense, forming a physical barrier that resists penetration by virus particles, baterial cells, and the spores or growing filaments of fungi". (Freeman, 2002). Cutin, as a structural component of the cuticle, is covered with cuticular and epicuticular waxes, a mixture of hydrophobic materials containing C26 to C36 aliphatic compounds. (2) The waxy covering on plant parts like leaves and fruits, which prevents excess moisture loss

Cutin. Refers to a fatty substance which is formed from a mixture of complex derivatives of fatty acids and impregnated the outer wall of the epidermis of plants and forms a separate layer, the cuticle, outside this wall. It is relatively impermeable to water and gases, restricting water loss.

Cuttage. (1) Plant propagation by cutting. (2) The process of rooting cuttings. (3) Method of plant propagation by means of cuttings.

Cutter bar. A mechanism on a Binder, Combine Harvester or Mower which cuts the crop and consists of a fixed finger bar (Fingers) and reciprocating knives. The knives of a mower operate at a faster speed than those on binders and combines since grass is more difficult to cut than the dry straw stems of cereals.

Cutter blower. Forage Blower.

Cutting. (1) A cut section of material removed from a living plant which, when placed in a suitable rooting medium, will produce roots and give rise to a new plant. Cuttings are used extensively in the propagation of certain types of plant, e.g., Chrysanthemums, Geraniums, etc. (Refer to Graft, Budding, Layering). (2) Any stem or vegetative part of a plant used for propagation. (3) The portion of a plant cut and removed for rooting, in asexual propagation.

Cutworm. A caterpillar of various noctuid moths, active at night only, attacking roots and stems and named after its habit of eating completely through young plant stems at ground level. Also called surface caterpillar.

CVM. Center for Veterinary Medicine.

Cyanamide (CaCN2). Calcium cyanamide, an Artificial Fertilizer, converted by soil water into Ammonia. Also called nitrolime.

Cyclamen. Cyclamen are traditionally sold during the Christmas season and into the winter as a florist plant. Most will bloom from mid-November until spring, provided conditions are satisfactory. Most of the blooms are either white, red, pink or a lavender and appear to hang upside down over the heart-shaped mottled green/gray foliage. Cyclamen prefer cool temperatures and bright indirect light. Ideal daytime temperatures are 60-65 degrees F. with night temperatures around 50 degrees. An east window provides adequate light. High humidity during the winter indoors is also crucial. To maintain humidity, fill a large plate or broad, shallow pan or tray with water. Set the cyclamen on an inverted dish, just out of the water. Pebbles could be placed in the plate, pan or tray with the pot setting on the pebbles. Plants prefer to be kept moist. Most are planted in a peat moss soil that dries quickly. Plants can wilt quickly. Make sure pots have drainage holes. Repot if drainage holes are not present. Avoid watering the crown or center of the plant. Bud

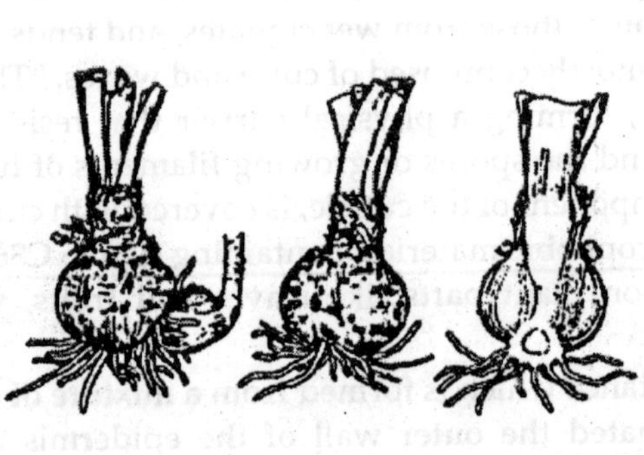

Fig.: Cuttage

blasting or aborting as well as yellowing leaves result from hot and dry conditions, lack of water or insufficient light. After flowers start fading, gradually withhold water. When the foliage is withered, remove the "bulb" (actually a tuber) from the soil, clean off all soil from it and store it in unmoistened peat moss or vermiculite in a plastic bag at 50 degrees F. Replant in good potting soil in May or June, keeping the upper half of the tuber above the surface. Grow the plant in a cool, bright, protected spot outside, with partial shade during the hottest part of the day, and with the pot sunk in a bed of moist peat moss. Water adequately and feed about twice a month with a complete liquid fertilizer. Bring indoors before cold weather, and provide full sun and the temperature suggested above.

Growing from Seed: Growing cyclamen from seed is discouraged, even though this is the only method used by professional growers. Germination is slow and erratic, and 9 to 15 months are needed to produce full sized blooming plants, even under the best greenhouse conditions.

Cyclic. Having floral parts in whorls.

Cycloheximide. An inhibitor of protein biosynthesis.

Cyme. (1) A branding, relatively flat-topped flower cluster whose central or terminal flower opens first, forcing development of further flowers from lateral buds. (2) A more or less flat-topped determinate inflorescence whose outer flowers open last. (3) An inflorescence in which primary axis bears single terminal flower that develops first, the arrangement being repeated by axis of secondary and higher orders

Cyst. (1) A membrane secreted around the resting stages of many animals during their development, e.g., that enclosing Eelworm eggs in the soil. (2) A swelling or hollow tumour containing soft or watery matter as distinct from pus.

Cytogenetics. Science dealing with the study of the biological systems using the combined approach of cytology and genetics.

Cytokinins. Group of plant growth substances which promote cell division and exert other growth regulatory functions e.g. zeatin, kinetin, benzyl adenine, 2iP

Cytology. (1) Branch of biology dealing with the structure, function, properties, physiology, development and reproduction of cells. (2) The study of cells and often chromosomes.

Cytosol. (1) Soluble part of cytoplasm which is left after the particles got separated, e.g., by centrifugation. (2) The cytosol (as opposed to *cytoplasm*, which also includes the organelles) is the internal fluid of the cell, and a large part of cell metabolism occurs here. Proteins within the cytosol play an important role in signal transduction pathways, glycolysis; also, they act as intracellular receptors and form part of the ribosomes, enabling further protein synthesis. In prokaryotes, all chemical reactions take place in the cytosol. In eukaryotes, the cytosol contains the cell organelles. In plants, the amount of cytosol can be reduced due to the large tonoplast (central vacuole) that takes up most of the cell interior volume. The cytosol is not a "soup" with free-floating particles, but is highly organized on the molecular level. The cytosol also contains the cytoskeleton. This is made of fibrous proteins (microfilaments, microtubules, and intermediate filaments), and (in many organisms) maintains the shape of the cell, anchors organelles, and controls internal movement of structures, e.g., transport vesicles. As the concentration of soluble molecules increases within the cytosol, an osmotic gradient builds up toward the outside of the cell. Water flows into the cell, making the cell larger. To prevent the cell from bursting apart, molecular pumps in the plasma membrane, the cytoskeleton, the tonoplast or the cell wall (if present), are used to counteract the osmotic pressure.

D

D Trials. Series of large plot trials which are carried out for comparison of varieties using farm methods of seeding and harvesting.

Dairy. (1) A farm building, often a single room, in which milk is cooled and temporarily stored prior to collection for transportation to a commercial dairy, and in which it may be treated and made into cream, butter and cheese—a practice now be-coming rare on farms. Milking equipment is usually washed and kept in the dairy. (2) A commercial processing plant to which fresh farm milk is transported by tanker (Bulk Milk) for treat-ment and distribution as bottled milk, and for the production of cream, butter, cheese, yogurt and other products for retailing. Some dairies are purely milk bottling plants whilst others are mainly creamers.

Damping off. (1) A seedling disorder caused by microorganisms. (2) Decay of young seedlings at ground level following fungal attack. Often the result of soil borne diseases or over watering.

Dark seed. Seed which will germinate, only when kept in the dark, when other conditions would normally favour germination.

Dark storage. Storage of foliage plants, cut flowers etc. in darkness during transit (from harvest to end use)

Data call-in. A part of the Environmental Protection Agency Office of Pesticide Programs (OPP) process of developing key required test data, especially on the long-term, chronic effects of existing pesticides, in advance of scheduled Registration Standard reviews. Data call-in from manufacturers is an adjunct of the registration standards program intended to expedite re-registration.

Day neutral plants. Plants in which flowering is not affected by day length.

Day-old-chick. One that is depatched to the buyer within 24 hours of hatching.

De-acclimation. De-hardening or the loss of adaptation to a climate.

Dead furrow. Refers to the open trench which is left in between two adjacent strips of land after finishing the plough.

Dead-head. To prune the spent flowers or the unripe seed pods from a plant.

Dead heading. (1) Removal of dead flowers or the unripe seed pods from plants. (2) The processing of removing old or spent flowers to prevent them from setting seed. This will promote continued bloom. (3) The removal of old blossoms to encourage continued blooming or to improve the appearance of the plant.

Dead seed. Seeds which are not capable of germinating.

Dead-in-shell. In general terms a fertile egg in which the developing embryo has died. In specific terms, embryo death occurring between the 14th and last day of incubation.

Deadstock. The equipment required to operate a farming enterprise, such as tractors, implements, hay racks, and hurdles etc.

Deadweight. The weight of a dressed carcase.

Deassimilation. Refers to the utilization of food by a plant.

Debearder. Preconditioning machine which is used in seed testing laboratory to debeard barley, breakup lucerne seed pods, breakup grass seed doubles, separate and polish vegetable seeds, decor-ticate sugar beet seeds and remove extra glumes and hulls.

De-blossoming. Removing blossoms from a plant.

De-budding. Removal of floral or vegetative buds.

De-bunching. The removal of the bunch soon after emergence to speed up growth of the follower

Decalcification. (1) Means the removal of calcium carbonate or calcium ions from soil by leaching, a natural process in soil formation. (2) Also refers to the removal of calcium from the bones of animal in negative calcium balance.

Decarboxylase. An enzyme which is able to liberate carbon dioxide from the carboxyl group of a molecule.

Decidua. A deciduous species in a normally evergreen genus.

Deciduous plants. Trees and shrubs that drop their leaves before entering onto the rest period. *Decortication:* Removal of bark or outer covering of plant or plant produce.

Deciduous tree. A tree that loses its leaves or needles during the fall and winter.

Deciduous. (1) Shedding leaves during winter or dormant season. (2) The falling of plant parts at the end of a growing period. (3) A descriptive term of plants, particularly trees, which shed their leaves annually, usually in the autumn. (Evergreen) (4) A plant that loses its leaves in the autumn; falling or not persistent as in maple or oak trees each fall; leaves or stipules which do not persist in a green condition throughout the winter. (5) Having leaves that fall at the end of a growing season. (6) Non evergreen woody plants. Leaves drop each fall. Influenced by temperature and light. (7) Plants that shed all their leaves annually, usually in the fall. (8) Said of leaves or stipules which do not persist in a green condition throughout the winter; usually falling in the autumn. (9) Perennial plants whose leaves die all at once (and usually fall) at the end of each growing season, to be replaced by new leaves at the next growing season. Most deciduous plants are broad-leaved, though a few, such as *Larix laricina* (Tamarack), have needles. Plants whose leaves live year-round are *evergreen*.

Decortication. The removal of husks from seeds. Oil seeds are frequently decorticated before oil extraction. Cattle Cake de-rived from oil seeds, *(e.g.,* Cottonseed Cake) may be either decorticated or undecorticated (the husks having been left on).

Decoupling. The concept of separating federal farm payments from the requirement that farmers produce specified program crops and/or divert land from production. A chief goal of decoupling is to remove a seemingly inherent contradiction in traditional policy: asking farmers to reduce production, while implicitly encouraging more output by tying their benefits to each unit produced. The decoupling concept was first introduced during debate over policy options in the 1985 omnibus farm bill, and was effectively implemented by policy changes made by the FAIR Act of 1996.

Decumbent. Reclining on the ground with the tip ascending; trailing stems on the ground with lateral shoots upright

Decurrent habit. Plant form typical of angiosperm trees and shrubs which are more round-headed and spreading, lacking a main leader to the top of the plant

Decurrent. Continued down the stem in a ridge or wing, as applied to leaf-bases.

Decussate. Situation where opposite leaves at any two consecutive nodes on a stem are oriented at 90 degree angles to one another

De-differentiation. Process in which the inter-organ, inter-tissue and inter-cellular relations are broken, leading to the formation of a mass of parenchymatous cells, called callus.

Deduplication. Chorisis, *i.e.*, increase in parts of floral whorl which occurs due to division of its primary members.

Deep percolation loss. Water percolating downwards through the soil beyond the reach of plant roots.

Deep percolation. Refers to the moisture which penetrates below the depths from which it may be used by plants. It represents that part of the water absorbed which exceeds the field capacity of the soil within the depth of root development.

Deferred grazing. Means withholding animals from a pasture be-yond the normal beginning of the grazing season.

Deferred pricing. A cash forward contract that provides for determining price by formula at a later date. This also may be called "booking the basis," when the formula sets price relative to a futures price.

Deficiency deseases. Diseases of animals and plants due to an inadequacy of one or more essential food substances, such as a Vitamin, mineral Element, Protein of Amino Acid, etc., *e.g.*, Grey Leaf, Brown Heart.

Deficiency payments. Direct government payments made to farmers who participated in an annual commodity program for wheat, feed grains, rice, or cotton, prior to 1996. The crop-specific deficiency payment rate was based on the difference between the legislatively set target price and the lower national average market price during a specified time. The total payment was equal to the payment rate, multiplied by a farm's eligible payment acreage and the program payment yield established for the particular farm. In the latter years of the program, farmers could receive up to one-half of their projected deficiency payments at program signup. If actual deficiency payments, which were determined after the crop year, were less than advance deficiency payments, the farmer was required to reimburse the government for the difference, except for zero, 50/85-92 payments. The FAIR Act of 1996 eliminated deficiency payments and replaced them with production flexibility contract payments.

Defleating. The cutting down of leaves that could scar the fruit. leaves with leaf spot and leaves that have collapsed.

Deflocculation. The dissociation of large particles into smaller ones. Particularly applied to soils and the disintegration of crumbs under wet conditions.

Defoliant (Defoliator). Chemical used for inducing leaf fall, usually to facilitate harvest. (2) A chemical that removes leaves from trees and growing plants; regulated as a pesticide.

Defoliate. To strip off leaves.

Defoliate/defoliation. (1) Loss of leaves such as premature summer leaf drop; casting off or falling off of leaves. (2) Casting off or falling off of leaves. (3) The removal of leaves by hand or using chemicals or by withholding water.

Deforestation. The process of clearing forests.

Dehanding. Removal of the false hand (hand with undeveloped fingers) and / or the last apical hand of banana at the time of bagging.

Dehisce. To open spontaneously when ripe

Dehiscence. The bursting or splitting of fruit (to emit the seeds) or anther (to emit the pollen grains)

Dehiscent fruits. A type of dry fruit in which the carpels split along definite sutures at maturity.

Dehiscent. (1) A descriptive term for fruits which spilt open to release the seeds, *e.g.*, peas. (2) Opening to discharge the seeds or pollen.

Dehorning. The removal of an animal's horns to prevent it wounding others. The best method is to disbud the young animal, but horns can also be removed under anaesthetic by electric saw. Also called dishorning.

Dehulling. Refers to the removal of the outer covering from intact grains of seeds.

Dehydration. Removal of water by sun drying or by artificial heat under controlled conditions of temperature humidity, and air flow.

Dehydro- freezing. A process of freezing food where freezing is preceded by partial dehydration.

Delaney clause. The Delaney Clause in the Federal Food, Drug, and Cosmetic Act (FFDCA) states that no additive shall be deemed to be safe for human food if it is found to induce cancerin man or animals. It is an example of the zero tolerance concept in food safety policy. The Delaney prohibition appears in three separate parts of the FFDCA: Section 409 on food additives; Section 512, relating to animal drugs in meat and poultry; and Section 721 on color additives. The Section 409 prohibition applied to many pesticide residues until enactment of the Food Quality Protection Act of 1996 (P.L. 104-170, August 3, 1996). This legislation removed pesticide residue tolerances from Delaney Clause constraints.

Delayed pricing. A type of deferred pricing that provides for transfer of title before the price is determined and final settlement made. Contracts including this feature are sometimes called "price-later" contracts.

Deliquescent. Breaking up into fine branches.

Delivery point. A location where a commodity can be delivered to fulfill a futures contract.

Demography. Refers to the study of numbers of organisms in a population and their variation with time.

Denaturing. The treatment of wheat grain so as to make it unusable for human consumption either by staining with a suitable dye or by the addition of fish oil to give a smell to it. Wheat, when denatured, can be used in animal feed, but not for flour-milling.

Dendrochronology. Means the dating of trees by observation of annual growth rings in the trunks.

Dendroclimatology. Determination of past climates by studying of relative widths of annual rings.

Denitrification. The breakdown of nitrates and nitrites by denitrify-ing bacteria in the soil in anaerobic conditions with the release of gaseous nitrogen.

Density crop. Compactness of tree stocking which is expressed as a decimal co-efficient, by taking normal numbers, basal area or volume as unity and assessed by comparison with yield table figures.

Dent corn. Horticulturally produced variety or com which is characterized by a depression in the end of the dry kernel.

Dentate. (1) Having marginal sharp teeth which are roughly perpendicular to the margin having male and female flowers on separate plants like Holly. (2) Toothed leaf margin.

Dentition. The cutting or growing of teeth. Also the arrangement and number of teeth.

De-potting. Removing the plant from the original container mainly for repotting.

Depreciation. A figure included in the annual farm accounts as an expense indicating the amount by which various assets such as machinery and implements, etc., have lost value with ageing.

Depressed. Flattened, as if compressed somewhat.

Depth wheel. A wheel on an implement used to control the depth of working.

Derris. An insecticide powder from the root of the tropical plant, *Derris elliptica.* Used against fleas, lice and Warbles. *Derris elliptica* is a large climber that is mainly cultivated in the tropics for its roots, a source of the insecticide Rotenone. It is known to be invasive in Fiji and western Polynesia (PIER 2000). In Hawai'i, it forms a dense canopy and smothers vegetation, fences, forest, pastures, and farm land. There are few locations of this aggressive invader on Maui. The full potential range is not known and it is uncertain whether this species will invade the rain forest or be restricted to the lowlands. Island wide eradication of this species is still possible and will probably only become harder with time

Desert. Places where the annual rainfall is often less than 250 mm and with extreme temperature fluctuations.

Desertification. Land degradation in arid, semi-arid, and sub-humid areas resulting from various factors, including climatic variations and human activities.

Desiccant. (1) A chemical agent that absorbs moisture; desiccants can be used to control certain insect pests or mildew, and also to dry foliage before harvest (as with potatoes). Desiccants are regulated as pesticides under the Federal Insecticide, Fungicide, and Rodenticide Act. (2) To dry thoroughly to remove moisture from an object to below normal level.

Desiccation. The process of drying up of tissues due to excess water loss or due to conditions that prevent the entry of water to the roots / shoot.

Designated diseases. Plant diseases which have been specified for seed certification of a given kind and in those regard certifica-tion standards must be met.

Designated uses of water. Water uses identified in state water quality standards that must be achieved and maintained as required under the Clean Water Act. Uses can include cold water fisheries, public water supply, irrigation, etc.

Designated weeds. Species of weeds have been specified for seed certification of a given kind and in those regard certification standards must be met.

Desired future condition. Used to describe the future condition of federal forests and rangeland resources that meet management objectives. Desired future condition is based on ecological, social, and economic considerations during the land and resource management planning process. Desired future condition is usually expressed as ecological status or management status of vegetation and desired soil qualities.

Desired plant community. The plant community that has been determined through a land use or management plan to best meet the plan's objectives for a site. A desired plant community is consistent with the site's capability to produce the required resource attributes through natural succession, management intervention, or a combination of both.

Dessicant. A chemical applied to a crop causing the foliage to dry up so that harvesting can commence.

Detassel. Means the removal of young tassels in maize plants so as to render the plants fully female, specially in hybrid maize seed production.

Determinant inflorescence. Inflorescence in which the apical flower blooms first.

Determinate. Said of an inflorescence when the terminal flower opens first and the prolongation of the axis is thereby arrested.

Dethatch. Process of removing dead stems built up beneath lawn grasses.

Detritus food chain. Food chain in which microorganisms are able to absorb and break down the energy-rich compounds synthe-sized by the primary producers.

Devernalization. Negation of a vernalizing stimulus by temperatures above a critical level.

Dewberry. Dewberries, like loganberries and boysenberries are a trailing type of blackberry. Plants grow horizontally, compared to the erect habit of most blackberries. To grow, most plants are tied to supports or wires to limit their trailing habit. Fruits are sweeter, ripen sooner and are in looser clusters. Dewberries are not winter hardy and need protection from cold temperatures. Plants should be untied from supports and lain horizontally after the ground starts freezing.

Cover canes with straw or loose hay at least a foot or more. Prune dewberries in a similar fashion to blackberries. The most common form of loganberries are thornless. Flowers tend to be more pink than white. Dewberries, like most trailing blackberries, are difficult to grow in Illinois. Choose a protected location and expect a crop every five to ten years. Be proud if you succeed more often.

Dew-claw. The rudimentary fifth digit protruding from the heel of cattle, pigs and dogs.

Dew-point (Dew point temperature). The temperature at which condensation will start and 100 per cent relative humidity results in a given mixture of air and water vapour.

Dextrins. Refer to the products which are obtained by the partial chemical breakdown of starch by the action of heat, acids, enzymes, or a combination of these agents.

Dhaincha (Ind.). *Sesbania cannabina* (Retz.) pers. (*S. aculeata* Pers var. *cannabina* Baker). Family Papionaccae. Ranks next to sannhemp as green manure crop, a legume which has been adapted to a wide range of soil and climatic conditions, growing well on soils too wet, too dry, or too salty for most green manure crops. It is grown as green manure mostly in low lands, and planted on paddy bunds or in high lands for seed production.

D-horizon. Refers to a sub-surface layer or rock stratum which is neither parent material nor the source of parent material for the soil.

Diallel crossing. (1) Crossing of a number of genotypes in all possible combinations. (2) Means type of crossing in which each of a number of males is made to cross to each of a number of females so as to compare the breeding value of males.

Diaphragmed. Said of pith which is solid with transverse bars of denser tissue at intervals between the nodes or at the nodes.

Diara cultivation. Cultivation of vegetables in the river beds, making use of the organic sediments and subterraneous moisture.

Diarrhoea. Waterly faeces which are caused by internal parasites, bacterial infection, excessive feeding of green succulent or other feed. Simple diarrhoea could be treated with catharitics like castor oil.

Diastase. An enzyme produced in germinating seeds and by the pancreas which converts starch to sugar.

Dibber, dibble. (1) A pointed hand tool used to make holes in the ground for seeds or plants. (2) To sow seeds, or plant seedlings, in holes made with a planting peg. (3) A stick or tool that makes planting holes in soil or media. The surname DIBBLE originated in medieval times in England for better identification of each individual. Prior to that time a person was known only by the single name which was given at birth. Later it became necessary to distinguish one man named John or Thomas from others with the same name so a descriptive word would be given to each individual. There were four major sources from which these additional names arose: a person's occupation, a location, a descriptive nickname, and baptismal from the name of the father. Of these, the name Dibble began as a baptismal name and was used to further identify one who descended from Dibb, from Dibald or Theobald (people, bold). In the beginning, the name Dibble belonged only to the person it described. Generations later it became hereditary and was passed down from a father to his children to become their family name. Like most names, Dibble has undergone some spelling changes since it originated. The name Dibble was likely to be written down and spelled as to what the writer thought he heard with the various accents of the times like Dible, Dibel, Dibell, Deeble, Dybol, Dibbs, Dibbles, etc. The Dibble name may even be spelled like Thibodeau or Thibault (French) in other countries. Genealogy research found the earliest "Dibble" ancestors originating from Somerset, England in the early 1500s. Other ancestors spelled Deeble or Deble may have come from St. Germains, Cornwall, England in the late 1400s. There were many families with the Dibble surname located in different towns within the county of Somerset, which is located in southwest England. Unless a Dibble individual was a prominent person or a criminal the only mark he was likely to leave for posterity was the supposedly compulsory record of his baptism, marriage or burial written in a church register by the incumbent of the local parish church or his clerk. There were a very large number of parishes in Somerset but only about half of them had church registers prior to dates we may be researching and many have huge gaps of missing records. Only some of these records actually survived the church fires and various wars. It was an age of successive outbreaks of plague, famine, war, iconoclasm and sackings of clergy, and many church records were lost forever. It is very rare indeed to be able to follow any Dibble individual through baptism, marriage and burial in the church registers. It is nearly impossible to be convinced that a Dibble baptism followed by a Dibble marriage about 25 years later refers to the same Dibble individual. After researching most of the church records, wills and other written documents available we have not found the missing link to who might be the "original Dibble". The internet and email has helped all genealogy researchers worldwide combine their efforts only to find the same dead end - incomplete and missing records. Genealogy research has proven that there were many Dibble families in southwest England. One can only guess that the surname Dibble had been used there for quite some time and that we were much more likely to have descended from Saxons than from French immigrants. So even though we have not identified the original Dibble, where he was born, married or died, we do know that if we search far enough back in time we will find that we are all related.

Dibbling. (1) A method of sowing by placing crop seeds in the holes manually by using a dibbler, maintaining specific spacing and number of plants between the rows and within the rows. (2) Refers to the method of sowing the crops by using manual labour or a dibbler where in specific spacing and number of plants are maintained between the rows and within the rows.

Dichotomous. Forked in pairs. A dichotomous key is an important tool used by biologists to identify plants. The user of a dichotomous key must answer a series of questions about the plant he/she is trying to identify. Some of the questions are easy to answer but some take knowledge of plant biology or special vocabulary. This activity teaches you how to use a dichotomous plant key and learn some of the terms necessary to identify plants. Click on one of the plants above to begin.

Dicliny. A condition in which male and female organs are separate and in different flowers.

Dicot. A plant of the angiosperm group having two cotyledons.

Dicotyledon (Dicot). Class of plants having embryos with two cotyledons.

Dieback. (1) As ugly as it sounds, it's when limbs of trees or parts of plants die, usually a sign of something such as insects, poor drainage or extreme cold damage. (2) Death of shoot tips such as by winter cold or chemical damage. (3) Progressive dying from the extremity of part of the plant.

Dietary supplements. Defined by law as a vitamin, mineral, herb, botanical, an amino acid, a dietary substance added to supplement the diet to increase the total dietary intake, or concentrate, metabolite, constituent, extract or combination of any ingredients described above. Supplements cannot be represented for use as a conventional food or sole item of a meal or the diet; and they must be labeled as dietary supplements. Regulatory authority rests with the Food and Drug Administration. There are concerns about the degree of regulation by FDA for safety and claims made for the products by manufacturers.

Differential assessment. Basing the valuation of farmland for property tax purposes on its use for farming rather than on its "highest and best" use, such as potential development. Types of differential assessment include preferential assessment, deferred taxation, and restrictive agreement.

Digestibility (Indicator Method). A quick method which is used for determining the digestibility of animal feed without measuring either the feed intake or faces output. This is carried out by using an indicator with the feed to be tested and then finding out the ratio of the concentration of the indiator to that of a given nutrient in the feed and also in the faces resulting from feed.

Digestible crude protein. A measure of the protein value of a Feedingstuff, determined by a Digestibility Trial. It is more useful than Crude Protein or True Protein content in formulating the diets of Ruminants since they can vary considerably in digestibility and composition.

Digestible energy value. The amount of utilisable energy in an animal food after deduction of the energy value of undigested faecal residues. (Digestibility Trial)

Digestion coefficient. The percentage digestibility of a constituent of a foodstuff (e.g., Crude Fibre, Protein) calculated from the results of a Digestibility Trial.

Digger. A type of mouldboard, abruptly concave and twisted, that produces a rough well-broken furrow slice, as distinct from the continuous smooth slice of the Lea Plough. Used for deep ploughing and popular for general work.

Digitate leaf. Leaf with many finger like leaflets radiating from a central point, or joined only at the base

Digitate. Spreading like the fingers, the leaflets leaving the petiole at one point as applied to compound leaves. The same as palmate.

Dihybrid. Plants that differs by two pairs of genes.

Dimorphic. Exhibiting two distinct forms.

Dimorphism. The occurrence of two types of leaves, flowers or other structures on a single plant or on different plants of the same species.

Dingy. Neither white nor brighly colored, as applied to pubescence.

Dioecious. (1) A species having unisexual flowers, each sex confined to a separate plant. (2) A term applied to plants having the male and female flowers borne on different individual plants of the species e.g., willow. (3) Condition in which staminate and pistillate flowers are produced on separate plants. (4) Dioecious translates from Latin to mean "two houses." Dioecious plants have flowers of only one sex per plant, such as ginkgo and holly. (5) Species in which male and female reproductive organs are found on separate plants. Most hollies (genus *Ilex*) are dioecious; female plants will only bear fruit when a male grows nearby, and male plants do not bear fruit.

Dioxin. Any of a group of toxic chlorinated compounds known chemically as dibenzo-p-dioxins (or the most toxic of these compounds, 2,3,7,8 tetrachloro dibenzo-p-dioxin). They are produced inadvertently as a by-product of chemical production or combustion and are widespread pollutants in the environment.

Dip slope. The long gentle slope of a line of hills following the inclination of the rock strata, as distinct from the shorter, steeper Scarp Slope.

Dip. (1) A proprietary chemical which is diluted in a recommended volume of water in a Dipping Bath for Dipping animals, mainly sheep. Dips containing Dieldrin were more advantageous than arsenic and sulphur based dips as they persisted in the fleece. They significantly reduced strike in sheep. Dieldrin was banned from use in sheep dips in 1966, being highly poisonous, and a possible meat contaminant. Nowadays dips mainly contain organo-phosphorus chemicals. (2) A Dipping Bath.

Diploid. (1) Having two set of chromosomes. (2) A nucleus, organism, or generation that has two sets of chromosomes. (3) A term applied to living cells having two sets of chromo-somes in the nucleus, present in pairs. Characteristic of most animal and plant cells. In reproduction, the pairs separate in the process of meiosis producing gametes which are haploid, having only one set of chromosomes.

Direct drilling. The sowing, by drill, of seeds direct into a field, without previous cultivation and usually following the applica-tion of weed control sprays. The lack of tillage reduces the effectiveness of soil-acting herbicides and encourages certain weeds in winter-sown crops, and intermittent ploughing is now advised.

Direct reseeding. Sowing grass seeds (usually a Seeds Mixture) without a Cover Crop in a field in which the immediate past crop was grass.

Direct seeding, direct sowing. Sowing grass seed (usually a Seeds Mixture) on bare ground without a Cover Crop. Germination of seed in their final spot as opposed to in a nursery.

Dirt rating. The term used for the habit of young animals which sometimes eat litter, soil, wool and other indigestible substances that may lead to the formation of balls of foreign material in the stomach, which may lead to death due to stoppage of the exit from the stomach.

Disarticulating. Falling away by abscission, leaving a clean cut scar, as with most leaves, many flowers, some twig tips, etc.

Disbud. (1) To remove buds from a plant. (2) To remove or prevent the growth of the horn buds of calves by painting them with a caustic compound or by use of a hot iron, usually in the first 7-10 days after birth. Kids and lambs are also sometimes disbudded.

Disbudding. (1) The processing of removing extra buds to promote larger flowers. (2) The removal of vegetative or floral buds.

Disc colour. A flat circular kind of colour which rotates as the plough moves and makes a vertical cut in the soil, separating a slice of soil from unploughed land, which the share and mouldboard then turn to produce a clean furrow slice.

Disc floret. The central regular five lobbed floret of a flower-head in compositae.

Disc Harrow, Disc Cultivator. A type of harrow with a number (normally four) of sets or gangs of sharp-edged concave rotating discs, the angle of which can be varied in relation to the direction of travel, altering the severity of cultivation according to requirements. The gangs can be positioned so that disc faces in different gangs face in opposite directions. Often used on light land, and on freshly ploughed grassland or stubble since unrol-led trash is not broughl to the surface.

Disc plough. A type of plough with heavy dish-faced wheels, indi-vidually mounted, and usually adjusted at an angle (between 35 and 45°) to the line of travel, which cut and turn the soil to the side, partly inverting it. The discs are usually between 50-80 cm (20-32 in.)

in diameter and are sometimes fitted with scrapping devices to remove accumulaling mud. The discs may be plain or cut-away. The latter give more bile and cut into hard soils. Disc ploughs are used for deep, rapid cultivations on rough land but are not very common Mould board ploughs produce better quality work.

Discharge. In water resources, the term refers to the flow of surface water in a stream or canal or the outflow of ground water from a flowing artisan well, ditch, or spring. In environmental protection, the term is used synonymously with effluent or emission as a term of point source pollution release.

Discoid. The same as chambered, when applied to pith.

Disease endurance. Refers to the ability of plants to tolerate the invasion of a pathogen without showing much damage and many symptoms. Synonym, disease tolerance.

Disease escape. Refers to the abilily of susceptible planls to avoid disease attack due to some inherent qualities such as rapid growth and´early maturity or environmental conditions like changes in sowing time, growing period and method of cultiva-tion. Synonym, klendusity.

Disease resistance. Refers to the ability of plants to withstand, oppose, or overcome the attack of pathogens.

Disease vectors. Plants or animals that harbor and carry disease organisms which may attack crops or livestock.

Diseminule. A part of a plant which gives rise to a new plant (propagule).

Dish garden. (1) A miniature planting in a shallow tray. (2) The art of making miniature garden by growing delicate plants in containers such as bowl shaped drinking cups or a brandy goblet.

Disinfection. (1) The cleansing of appropriate buildings, animals, implements and utensils, etc., which may be hardbouring harmful bacteria or viruses, after an infection has existed on a farm. The process cm involve physical cleaning, the application of approved chemical solutions, steam cleaning and fumigation. (2) The cleaning of farm equipment and utensils (e.g., milking plant) using suitable detergents and disinfecting agents (e.g., sodium hypochloride) to reduce bacterial contamination. (3) The chemical treatment of seed before sowing to control fungal disease in crops, particularly cereals.

Disinfestant. Substance which is able to kill or inactivate pathogenic microorganisms in the environment or on the surface of plant parts before infection.

Disinfestation. External parasites Mechanical or physical means of hindering their development *e.g.*, cleaning the yards of all refuse, removing litter and droppings frequently etc.

Disk flower. The tubular flower in the center of the usual Compositae inflorescence.

Disk. (1) An enlarged tip, as applied to tendrils. (2) An tractor-drawn implement composed of circular plates arranged at an angle with the soil. Used to prepare the soil for seeding.

Dispersed soil. A type of soil in which the clay is readily to form a colloidal soil, usually with a low permeability, which upon drying shrinks, cracks, and becomes hard, and upon wetting slakes readily and is plastic.

Dispersing agent. Substance added in water dispersible insecticide powders and concentrate pastes to obtain uniform concentration of spray all over treated area. Gammon examples include protective colloids such as sodium carboxy and methyl celluloses.

Display index. Percentage of peak display for flowers or other showy part.

Dissected. (1) Cut, lacerated or divided in narrow, slender segments, said of fringed petals or leaves. (2) Divided in narrow, slender segments.

Dissectum. Deeply cut, as in foliage.

Dissolved oxygen (DO). The oxygen freely available in water, vital to fish and other aquatic life and necessary for the prevention of odors in water. DO levels are a critical indicator of a waterbody's ability to support desirable aquatic life. Secondary and advanced wastewater treatments are generally designed to ensure adequate DO in waste-receiving waters by removing, digesting, or oxidizing oxygen-demanding wastes (see biological oxygen demand).

Distillers' grains. A by-product of whisky manufacturing similar to Brewers' Grains consisting of the remains of malted barley. Used as an animal Feedingstuff either wet or dry, but mainly preferred dry as wet grain retains raw alcohol which may intoxicate the animals.

Distillery waste. The form used for the residue which is obtained from the fermentation and distillation of beer and distilled spirit. It contains considerable amount of potash and some ammonia recovered and used as fertilizer material.

Ditch rider (crop science). Manages water for an irrigation district. This person is responsible for delivering water through ditches to farmers for irrigation.

Ditch water. Stagnant, foul-smelling water such as that often found in ditches.

Ditch. A trench dug in the ground to drain water from farmland, usually discharging to a stream or other watercourse. The water reaching a ditch may be by surface run-off, by natural under-ground flow, or via an artificial underfield drainage system, the drains of which discharge into the ditch.

Ditcher. A machine designed to excavate, clean or repair a ditch. The Offset Cut Power Ditcher gives you the same durability and performance of our center cut ditcher, but gives you the added advantage of being able to offset the unit to either the left or right. A power ditcher is the simplest, most economical approach to field water control. The unit is designed to spread soil up to 30 ft. and dig ditches up to 12 inches deep in one pass; you get excellent performance in most soil conditions and all types of crops. The standard unit comes as shown below; available options include a ripper shank assembly and a gauge wheel assembly.

Diurnal. (1) Blossoms opening only during the day. (2) Of or during the day; not nocturnal

Divaricate. Spreading very wide apart.

Divergent. The same as spreading.

Diversified cropping. The cropping plan in which no single crop contributes to 50 per cent or more towards the annual crop production or income.

Divided. Deeply lobed when applied to leaves; separated to the base into divisions.

Dividing plants. Hostas should be divided in early spring as plants emerge from the ground. Dig the entire clump and carefully separate plants. Some plants are slow to divide, such as the large, blue colored plants. Others may need dividing every three to four years, such as those used as edging.

Dividing/division. A method of propagating plants by separating each one into two or more sections and then replanting.

Diviner. One who divines or detects the presence of water beneath the ground (Dowser)

Division. (1) A taxonomic grouping of plants belonging to similar classes. (2) The process of cutting into sections of rhizomes, stem tubers, tuberous roots etc. for propagation.

Dock. (1) A weed *(Rumex sp.)* with large leaves and a long root. (2) To remove the whole or part of an animal's tail. Low-land breeds of sheep are customarily docked as lambs to avoid the accumulation of dirt and faeces in the tail.

Dockage. A factor in the grading of grains and oilseeds; i.e., dockage in wheat is described as "weed seeds, weed stems, chaff, straw, or grain other than wheat, which can be readily removed from the wheat by the use of appropriate sieves and cleaning devices; also, underdeveloped, shriveled and small pieces of wheat kernels removed in properly separating, properly rescreening, or recleaning." The term is also used to describe the amount of reduction in price taken because of a deficiency in quality.

Doctrine of prior appropriation. Water rights doctrine adopted by most western states, giving the first person to use water from a stream the first right to such water. If the first user does not consume all of the water, then the second and later users can appropriate water for their needs. The water right is not necessarily tied to land ownership.

Domestica. A plant having been domesticated or cultivated near residences.

Dominance. Ability of an allele to express itself in the heterozygous state.

Dominant plant. The major plant in a vegetation community, usually the most numerous or the tallest, e.g., beech wood.

Door yard fruit. Fruit species grown in small numbers around homes.

Dorking. A table breed of fowl, square-bodied, variously coloured (including red, white, silver-grey and barred), with a single comb and five-clawed white legs. Hens produce white eggs.

Dormancy. (1) A state of low metabolic activity in living organisms, particularly in seeds and spores, usually associated with unfa-vourable environmental conditions (e.g., low temperature, in-adequate moisture, short day length) for growth and thought to be controlled by hormones. Stored seeds can remain dormant for years and will germinate after sowing if conditions are right. Deciduous trees lie dormant in the winter after shedding their leaves. (2) A state of suspended growth; the yearly cycle in a plant's life when growth slows and the plant rests. (3) Resting state of buds, seeds and spores that fail to grow when provided with seemingly optimal conditions.

Dormant. A biological process in which a plant ceases most growth activities and simply maintains existing tissue. Resting condition for a living plant or animal with a relatively inactive metabolism.

Dormant bud. A bud remaining inactive for an indefinite period, until stimulated into growth.

Dormant oil. Oil sprayed on deciduous trees while they are dormant. Dormant oils are used to kill overwintering insects or insect eggs on plant bark.

Dormant spore. The spore which fails to germinate under the same nutritive and environmental conditions which later allow the formation of germ tube.

Dormant spray. A spray applied to a crop during a dormant period.

Dormin. Initial name given to abscisic acid (ABA). A dormancy inducing substance which has been extracted from buds and leaves. Also known as abscisin II.

Dorsal. (1) Of or on the back or outer surface of a leaf, etc.

(2) Relating to the back or outer portion of a plant part; of or on the back or outer surface of a leaf.

Dorset wedge silage. Silage produced by filling a silo in wedge shaped layers with cut grass or other greenstuff, covering with plastic sheeting and sealing the silo to ferment.

Double chop harvester. A type of offset-trailed or semi-mounted Forage Harvester, consisting of a cutting or pick-up mechanism of the flail-type, with an auger delivering the cut crop sideways to a flywheel chopper mechanism for further chopping before it is loaded by chute into the trailer. This type of machine produces a non-uniform chop length. Also called a Flywheel Chopper.

Double cropping. (1) The practice of consecutively producing two crops of either like or unlike commodities on the same land within the same year. An example of double cropping might be to harvest a wheat crop by early summer and then plant corn or soybeans on that acreage for harvest in the fall. This practice is only possible in regions with long growing seasons. (2) Taking two crops a year in sequence from the same piece of land.

Double digging. Soil bed preparation done by two or more spading sessions, preparing the soil by systematically digging an area to the depth of two shovels.

Double fertilization. Union of two male gametes with the female gamete and polar nuclei

Double hedge planting. Pairs of row of plants form a double hedge where the spacing within the rows and between the set of two rows are close but the spacing between pairs of rows is wider.

Double leader. Two shoots competing as leaders on a tree, each trying to assert apical dominance.

Double mounting. The fitting of dual wheels to the rear axle of tractor in an effort to reduce soil compaction by increasing the contact area, and to increase traction.

Double nose. Said of bulbs with two growing apices.

Double recessive. An organism with two recessive alleles for a particular trait.

Double working. Type of grafting where the graft combination contains an interstock between the scion and the stock.

Double. (1) A flower with numerous and full petals. (2) Term to describe flowers having more than the usual number of petals.

Double-dropping. Means growing a second crop in one growing season after the first crop is harvested from the same piece of land (overlap between crops can occur).

Double-suckling. A method of feeding beef cattle in which a second calf is introduced and allowed to suckle alongside a cow's own newborn calf. The second calf is sometimes rubbed with part of the Placenta to encourage the cow to accept it as her own. The system is only used with cows producing sufficient milk to feed two calves.

Double-work. A method of ensuring a successful fruit-tree Graft betweena Scion and Stock which otherwise would be incompat-ible, by grafting between the two a scion of a variety compatible with both, called an intermediate stem piece.

Doubly serrate. Serrations bearing minute teeth on the margins.

Dowser. (1) A water divine—a person able to locate underground water using a dowsing-rod (a forked twig) which quivers when over a water source. (2) Dowsing is a controversial practice which proponents claim empowers them to find water, metals, gem stones, and hidden objects by carrying or waving a stick or other apparatus over a piece of land and watching for any movement. Repeated tests under controlled conditions have not supported this claim. Dowsers (sometimes known as diviners) may also use a forked branch of a tree, bent pieces of metal or plastic wire, or a small pendulum. Some people use no pointing device at all.

Dowsing (water witching). Using a forked stick or wire to locate underground water.

Draft of Mouldbord Plough. According to Collins, the draft of a plough on the ground has been 18%, draft due to turning furrow slice has been 34%, draft due to cutting slice is 48%. Thus it is seen that practically 50% of the total draft of the plough has been used in cutting the furrow slice.

Drag harrow. A kind of heavy harrow, with tines having curved ends, and often with wheels, capable of quite deep work. The action is similar to that of a cultivator which has fewer tines and can work deeper. Drag harrows are preferred on difficult land for seed bed preparation, bringing fewer unweathered clods to the surface. Also called a duck-food harrow.

Drain gauge. A tank placed in a ditch beneath a land under-drain outlet to catch and measure the volume of water passing through the drain.

Drain jetting. The practice of flushing drainage systems with water in areas where sedimentation and ochre build-up occurs in the pipes.

Drain. (1) An artificially dug channel or ditch to carry surplus water away from farmland. Common in the Fens. Usually maintained by an Internal Drainage Board. (2) An underground channel in a field drainage system usually comprising a piped duct of clay, porous concrete or plastic tube.

Drainage basin. The area of land that drains water, sediment, and dissolved materials to a common outlet at some point along a stream channel.

Drainage wells. Wells drilled to carry excess water off agricultural fields. Because they act as a funnel from the surface to the groundwater below, drainage wells can contribute to groundwater pollution.

Drainage. (1) Improving the productivity of agricultural land by removing excess water from the soil by such means as ditches, drainage wells, or subsurface drainage tiles. See swampbuster and wetlands. (2) The process of removal of excess water from the land.

Drawbar power. Power of the tractor, which is measured at the end of the drawbar. It is the power available, to pull loads or draw machines.

Drawbar pull. Pull exerted by a tractor at its drawbar end.

Drawbar. A metal bar at the back of a tractor to which implements are attached.

Dray. A low strong cart with no fixed sides, used for carrying heavy loads.

DRC. Dry rubber content, ascertained by coagulating a weighed sample of latex.

Dredge corn. A mixture of cereals sown together *(e.g.,* oats and barley) producing a yield of grain better than if either crop were grown separately. (Mashlum).

Dress. In general terms to clean prepare or repair. (1) To apply fertilizer to land *(e.g.,* Top Dressing.) (2) To chemically treat seeds before sowing to control fungal disease in growing crops. particularly cereals. (3) To clean corn seeds by removing Chaff, weeds and other unwanted matter. (4) To remove the tough fibres from flax. (5) To sort potatoes into various sizes. (6) To prune a hop plant by cutting the previous year's growth from the crown. (7) To recut grooves in a millstone.

Dried blood. An Organic Fertilizer derived from waste blood from slaughterhouses, dried by evaporation. It has a nitrogen content of about 13%.

Dried grass. Grass which, after cutting, has been thoroughly dried by artificial means, and is used as an ingredient in animal feeding-stuffs. It is either milled and mixed with molasses to produce cubes or pellets, or pressed directly into cobs or waters. Used for many years for poultry and pigs, and now becoming poupular as a dairy cow ration combined with a concentrate and roughage. Sometimes called dehydrated foddeer.

Drill (crop science). A farm implement for planting seeds which forms a small furrow, deposits the seed in dribbles, covers the seed, and packs soil over it. It can also deposit fertilizer, lime, or other amendments into the soil, alone or with the seed.

Drill barrow. A hand drive Seed Drill.

Drill harrow. (1) A light flexible Chain Harrow trailed behind a Seed Drill to cover the seed with extra soil. (2) A drill mounted on a Spring Tined Harrow. Also called a tine harrow.

Drill plough. An old type of trailed plough with an apparatus attached to the beam which drilled seed directly into the newly produced furrow.

Drill. (1) An implement used to sow seed in rows (Seed Drill). Seed and fertilizer may also be sown simultaneously using a Combine Drill. (2) A ridge with seed or growing plants on it *(e.g.,* turnips, potatoes), or the plants in such a row.

Drilled to a stand. Seeds sown by a Seed Drill at predetermined intervals so as to produce a crop with a specific plant population which requires no thinning. (Thin).

Drip culture (Sand culture). A modification of hydroponics where plants are kept in sand medium and watered with a nutrient solution

Drip irrigation. (1) The application of small quantities of water directly to the root zone through various types of delivery systems. (2) A method used to place irrigation water near plants' roots through pipes or tubes. This reduces water evaporation and runoff, but may not be cost effective for all crops. (3) Watering plants by small droplets over a long period of time each day.

Drip line. (1) On the ground, the line that marks the outer edge to which a tree's (or other plant's) branches spread. The drip line usually signals a change in microclimate, where the area under the tree, which sees less precipitation, sunlight, and wind and may also be subject to competition from the tree's roots (see *root zone*), meets an area that isn't sheltered by the tree. (2) The circle which would exist if you drew a line below the tips of the outer most branches of a tree or plant. (3) The edge of the foliage of trees and shrubs. Water the ground all the way out there when you water new plantings. This is where the feeder roots are.

Drought (resistance). (1) Characteristics of plants which are suit-able for cultivation in dry condition regardless of the inherent mechanism that provides resistance. One of the more important properties is the capacity to endure, without injury, an intense loss of water. (2) Refers to the relative ability to maintain growth yield under moisturestress conditions.

Drought stress. Turfgrass is considered under drought stress and in need or a thorough watering when footprints remain in the lawn thirty minutes later. Water thoroughly. Turf may go dormant and turn brown during hot conditions even though it's not drought stressed.

Drought year. When rain fall is short by more than twice the deviation, year is regarded as a drought year for a particular place.

Drought. (1) Shortage of moisture for crop growth or animal production. (2) An extended period without rain.(3). Lack of moisture that hinders plant growth. (3) A drought is an extended period where water availability falls below the statistical requirements for a region. Drought is not a purely physical phenomenon, but instead is an interplay between natural water availability and human demands for water supply.

Drum. The cylinder of a Threshing Machine or Combine Harvester bearing Beaters, which revolves and dislodges the grain from the ears of cereals. Also called a threshing drum.

Drumhead. A kind of Cabbage having a flattish round head.

Drup. A kind of fruit comprising a hard kernel containing a single seed enclosed within a fleshy pulp, *e.g.*,plum, cherry. Also called a stone-fruit.

Drupe. (1) A fleshy one-seeded indehiscent fruit with the seed enclosed in a strong endocarp. (2) A stone fruit (cherry, plum, peach). Many berry-like fruits are technically small drupes, like huckleberry. (3) A stone-bearing fruit, such as a plum.

Drupelet. A small drupe as that in an aggregate fruit as the blackberry.

Dry cured bacon. Bacon preserved by being rubbed with dry salt as opposed to being soaked in brine.

Dry farming. (1) Refers to the type of agricultural operation which involves deep cultivation of the soil to form a sufficient reservoir for the moisture as it falls, surface cultivation to prevent or reduce evaporation; selection of drought resistant crops. (2) The production of crops without irrigation on the land which receives annual rainfall of less than 500 mm.

Dry feeding. A method of feeding meal to animals in a dry state without the addition of water. Used with pigs and poultry.

Dry matter. The various mineral and organic components (*e.g.* Ash, Crude Protein, etc.) in feedingstuff derived by Proximate Analysis.

Dry measure. A system of measuring by bulk. (Bushel, Peck).

Dry period. (1) A rest period between lactations when a cow is not lactating, normally 60 days in duration. (2) A period between Lactations when a cow is allowed to rest from the strain of producing milk, usually of 6-8 weeks duration.

Dry planting (sugarcane). Refers to the practice of planting the cane in dry condition of soil and irrigating later on. This method of followed in heavy black soil.

Dry rot. (1) The decay of timber caused by fungi. The wood is destroyed and reduced to a dry, brittle, powdery mass. (2) A storage disease of potatoes caused mainly by the fungus. Spores infect the tubers via woulds and bruises caused by rough handling during harvesting. Tubers are caused to wither and wrinkle.

Drying front. Refers to the advancing zone of the bin of drying grain which includes all the grain which has just reached some arbitrarily initial moisture content, the zone between the initial and final moisture content.

Drying off. (1) The process of preparing bulbs and tubers for their necessary rest or dormant period by withholding water. (2) Gradual reduction of the volume of milk taken from a cow so as to cause it to cease lactating.

Dryland farming. (1) Farming on non-irrigated land. Success is based on rainfall, moisture-conserving tillage, and drought-resistant crops. (2) A system of producing crops in semi-arid regions (usually with less than 20 inches of annual rainfall) without irrigation. Dryland farmers often try to rebuild soil moisture by leaving the land fallow (unplanted) or mulched in alternate years, called summer fallowing.

Dry-pack storage. The storage of cut flowers in vapour-proof containers, usually at 0°C.

Dual-purpose breeds. Breeds of (a) cattle considered useful for both beet and milk production, (b) poultry considered good both for egg laying and as table birds, and, (c) pigs, the females of which are kept generally for crossing with either bacon or pork type boars.

Due restraint provision. The UR Agreement on Agriculture provision that sets a 9-year period during which green box policies are exempt from GATT challenges.

Duff (Litter). (1) The partially decomposed organic matter (e.g. leaves, flowers, fruits etc.) found beneath plants, as on a forest floor. (2) The organic layer in forests. It consists of fallen vegetative matter in the process of decomposition, including everything from pure humus below to the litter on the surface. Duff is a general, non specific term.

Duhhing. The removal of the comb of a cock. Also the trimming of part of the comb in Day-Old-Chicks. The practice reduces frost bite, and is advantageous of serious Cannibalism or pecking occurs.

Dump box. A large portable container similar to a Forage Wagon, placed adjacent to a Silo, into which trailer loads of Forage from the field are tipped, and from which the forage is delivered in an even steady flow, usually by a moving floor mechanism or elevator belt, for loading into the silo. It may also be used to deliver forage directly into a Manger.

Dumping-off. A disease of seedings caused by *Pythium* and various other fungi in excessively moist conditions. It causes wilting and roots to die.

Dung liquor. The dark coloured liquid which drains from a manure heap, containing soluble compounds of ammonia and organic matter.

Duster, power operated. It mainly consists of a power driven fan, a hopper and a delivery spout. The fan creates strong air flow which makes the dust to blow off from the hopper to a considerable distance either vertically or horizontally.

Duster, rotary type. Duster with a hand operated rotor in front of the operator. For dusting tall crops, more force of delivery is required and hence rotary dusters are preieraed.

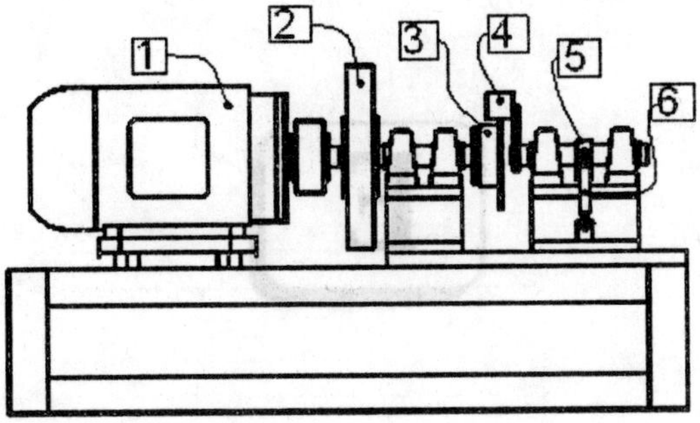

Fig.: Dynamometer

Duster. Machine to apply chemicals in dust form. Dusters make use of airstreams to carry pesticides in finely divided dry form on the plants.

Dusting. The application of insecticide or fungicide to crops in a dry, powdery state. It is less efficient than spraying, usually requiring about twice the amount of chemical per unit area.

D-value. The percentage of digestible organic matter in the dry matter of animal feeds such as hay, silage and dried grass.

Dwarf trees. Trees made smaller than the size commonly accepted as normal for the species and variety under consideration.

Dwarf varieties. Varieties bred to grow smaller than their parent plants. Dwarf plants may lose the ability of the parents to set fruit (Bailey's dwarf highbush cranberry, *Viburnum trilobum 'Bailey's Compact'*). They may not resemble miniatures of their parents (dwarf Alberta spruce, *Picea glauca var. albertiana 'Conica'*).

Dwarf. Said of a genetically smaller plant, often 1/4 or less normal size.

Dwarfing rootstock. Rootstock which has a dwarfing effect on scion

Dyke. (1) A ditch or minor lowland watercourse. (2) A raised bank of earth, stones, etc., constructed to keep a river or sea water from flooding adjoining, usually low-lying, land.

Dynamometer. An instrument which measures power. Used in agriculture to measure the resistance or Draft of any implement drawn by a tractor.

E

Ear. (1) The spike or flowering head of grass, usually applied to cereals. Also to form such ears. (2) An obsolete term meaning to plough or till.

Early bite. A sudden growth of grass which can be grazed before the main crops of grass has really begun to grow.

Earth vegetables. All forms of vegetables in which food is stored in underground parts.

Earth. (1) A fox's burrow. (2) To cover with soil. Earth-board. A Mouldboard.

Earthing up. The building up of a ridge with soil removed from adjacent furrows, particularly used to cover tomatoes during planting.

Earth-nut. Another name for round nut.

Easement. A landowner sells or surrenders the right to develop a portion of the property, usually in return for a payment or some other benefit. Some local and state governments, and land trusts, have programs to acquire development easements from landowners to prevent conversion of farmland to other uses. Since the mid 1970s, conservation easements have been purchased to protect nearly 420,000 acres of farmland in 15 states, primarily in the Northeast.

Easter lilies. Easter lilies will tolerate extremes that many other plants won't survive. You can enjoy the blossoms longer if you keep the plants out of direct sunlight or warm drafts. It's worthwhile to place the plant on an unheated but frost-free porch at night. To keep the plant looking its best, remove the lily flowers as soon as they wither and clip any leaf ends that may brown. After all the flowers have been removed, you can keep the plant in a sunny window for its pleasing foliage or remove it to a basement window until danger from frost is over. Should the plant begin to go into a rest period, the leaves will start to yellow and fall. The plant should then be kept on the dry side to discourage rots. The lily can be planted in a sunny garden spot as soon as danger from frost is past. Remove the plant from the pot by inverting it and while gently holding the top, tap the edge of the pot on a step or heavy board. Open the root ball by pulling upward and out from the center of the ball. A few torn roots are better than an undisturbed dense root mass that may not be able to establish new roots in the soil. Clumped and matted roots are more likely to die and even injure the bulb. Place the bulb a few inches deeper than it was in the pot, open the root as much as possible and work soil through them. Thoroughly water the plant. One-half teaspoon of a 10-10-10 fertilizer per gallon helps promote new top and root growth. Soon after the old tops die, new shoots will start. These will flower later, usually in July or August, if given ordinary garden care. Be sure that trees, weeds, or other plants don't shade the lily. Although many people report good results, none of the lily varieties are reliably hardy. If the ground is well drained, you can cover plants with a mulch during the cold winter months — just as is done for roses. Mulches of straw, leaves, evergreen boughs, wood chips or ground corn cobs are satisfactory. Most of the lilies are killed by exposure to winter winds and sun. Mulches limit the heaving action of the soil and, thus, prevent bulb exposure.

Easy feed. A method of livestock feeding in which the stock are allowed easy access to feed. Usually the feed is transferred mechanically to large hoppers or feeding passages.

Eatage. A obsolete term for grass used for grazing as distinct from for hay, particularly aftermath. Also the right to use aftermath for grazing.

Ecology. The study of the relationship between living organisms and their environment.

Economic botany. Branch of botany dealing with the study of various uses of plants. It includes methods for their better utilization and improvement by mankind.

Economic irrigation efficiency. Means the ratio of the actual income attained with the operating irrigation system, compared with the income expected under ideal conditions. It is relating the final output to input.

Economic protection. Protection of crops from pest damage only to the extent that the resultant economic gains exceed the costs of protection.

Economies of size. The concept that the average cost of production per unit declines as the size of the operation grows. One reason farms have been growing in size is to make more economical use of machines capable of covering more ground with less labor, to capture economies of size. Larger sized farms can typically get volume discounts on such inputs as chemicals and seed.

Ecosystem. (1) Means complete ecological system of an area, including the plant, animal and other environmental factors. (2) A functioning community of nature that includes fauna and flora together with the chemical and physical environment with which they interact. Ecosystems vary greatly in size and characteristics; an ecosystem can be a mud puddle, a field or orchard, or a forest. An ecosystem provides a unit of biological study and can be a unit of management. (3) An ecological system involving living and non-living units where a stable cyclic interchange of energy and materials occur in a defined space and time.

Ecotophyte. External plant parasite, that is one that lives outside its host.

Eco-type. Plant type or strain within a species, resulting from exposure to a particular environment

Ectocarp. The outer layer of cells of the pericarp or fruit wall.

Ectocommensal. Refers to an organism that lives on the external surface of another organism, the host, without either benefiting or injuring it.

Ectoparasite. A parasite living on the outside of its host, e.g., a Louse. Also called exoparasite.

Edaphic. Of or pertaining to the soil. Soil is unconsolidated rock particles mixed with organic matter from plant decay. Soil is vital to all life on Earth because it supports the growth of plants, which supply food and oxygen and absorbs carbon dioxide and nitrogen. Soil serves as a habitat for animal life from microorganisms to small animals.

Edaphology. The study of the influence of soils on living organism, particularly plants, including man's use of land for growing crops.

Edge plants. Drape over the edge, softening the look of the container and filling out the space between its corners.

Edging plant. A compact, short plant used to line or trim borders.

Edgings. A border of bricks, low shrubs or dwarf perennials provided to demarcate garden paths or to make an area more conspicuous.

Edible landscape. A landscape planted with vegetables and fruit plants.

Effective ground-water velocity. Actual or field velocity of ground water which is percolating through waterbearing material. It is a measured by the volume of ground water passing through a unit cross-sectional area divided by the effective porrosity. It is also called actual velocity : field velocity.

Effective rainfall. Precipitation which falls during the growing period of the crop and is available to meet the evapotranspiration requirements of crops. It does not include precipitation lost through deep percolation below the root zone or through surface runoff.

Effective rooting depth. Soil depth from which the full grown crop extracts most of the water needed for evapo- transpiration.

Effluent limitation. An Environmental Protection Agency "standard of performance" reflecting the maximum degree of discharge reduction achievable by the best available technology for various categories of sources of water pollution. These categories include feedlots, grain mills, and several kinds of food processing.

Effluent. (1) Waste, usually liquid, released or discharged to the environment. Generally the term refers to point source discharges of sewage or contaminated waste waters into surface waters. (2) Liquid, solid or gaseous water material, e.g., Slurry, indus-trial waste sewerage.

Egg weight grades. There are seven weight grades under E.E.C. regulations for Class A and B eggs. Grade 1 are 70 g or over, Grade 7 are under 45 g.

Elaiosome. An outgrowth from the sufrace of a seed, having fats or oils. They are sometimes attractive to ants, hence aiding in seed-dispersal.

Elderly and disabled. For food stamp purposes: "elderly" persons are age 60 or older; and "disabled" persons are beneficiaries of disability-based governmental assistance, such as social security disability payments and certain veterans disability payments.

Electronic colour sorter. A machine which finds use in seed testing laboratory for separating seeds on the basis of difference in colour. Each seed is viewed by photoelectric cells in com-parision with a selected background.

Electronic leaf. Device in a mist system to switch on the mist as and when the cuttings/ plants become dry

Electrostatic seed separator. A machine which is used for separat-ing the seeds on the basis of their differential electrical proper-ties. The seed are separated by allowing them to fall free in space, and then using an electric field for deflecting some seeds from their normal flight path.

Elevator digger. An implement for harvesting potatoes. There are three kinds; trailed, semi-mounted, and fully mounted. They lift potatoes from the field rows and separate the soil from the crop on a chain web and deposit them in a row on the field for collection. The Potato Harvester is essentially an improved elevator digger bearing a platform for the pickers to stand on, and a system of elevators delivering the potatoes to a sorting table, and conveying them to the side for collection and trans-port. Some complex modern machines incorporate electronic automatic sorting and rquire only one operator.

Elevator. (1) A tall warehouse facility that uses vertical conveyors to raise or elevate grain, generally owned privately or by an agricultural cooperative, where grain is stored before being marketed. The term elevator often refers to any grain storage facility, even if the grain is not elevated. The country elevator is where a farmer delivers grain; a terminal elevator is a major transshipment facility; while an export elevator is at a port facility. (2) A machine for raising a cut crop, Grain, or Bales to the top of a Silo, storage bin, or bale stack, etc., usually for storage. Various designs, exist. Grain elevators usually consist of a continuously moving belt bearing buckets, or an endless chain bearing flights or steps (operating like a moving staircase). Elevators for hay or straw bales often comprise a Continuous moving chain bearing spikes or cross slats.

Elite plants. Plants which in trials and tests have been proved to be genetically superior to other plants of the same species.

Elite variety. An improved variety developed by plant breeders and released to farmers due to it superiority in at least one respect.

Elutriation method. Type of mechanical analysis of soil where the separation of different soil fractions is carried out by washing out particles in a rising current of water (particles whose settling velocity is equal to or less than the velocity of the rising water will be taken away by the current).

Eluviation. Process of removal of organic material in solution or in suspension from the soil by percolating waters.

Emasculation. Removal of stamens of self fertile plant to aid self sterility.

Embden. A heavy white breed of goose, with blue eyes.

Emblements. Crops raised by cultivation (e.g., wheat, potatoes) as distinct from the fruits of trees or grass.

Embryo. (1) A young animal or plant in its earliest stages of development after fertilization, e.g., a young animal in its mother's womb, or the innermost part of a plant seed. (2) The rudimentary plant within the seed. *Embryo culture In vitro* culture of isolated mature or immature embryo.

Emergency Conservation Program. A program administered by the Farm Service Agency to help farmers to rehabilitate farmland damaged by natural disasters by sharing in the cost of rehabilitation.

Emergency feeding agency. This refers to an organization serving the food needs of the poor and unemployed that is designated by a state as eligible for commodities and administrative support to distribute commodities or operate a meal service program under the Emergency Food Assistance Program.

Emission. Waste released or emitted to the environment. The term is commonly used in referring to discharges of gases and particles to the atmosphere, i.e., air pollutants, and also is used in referring to particles or energy released radioactively. Sometimes the term is used broadly, encompassing any pollutant discharge.

Enation. Outgrowth on a plant surface.

Enclosure. An area of land bounded by fences, hedges or walls.

Endangered species. (1) A species likely to become extinct. (2) Species of animals or plants likely to go extinct in the foreseeable future unless current trends are altered. They are listed by regulation under the Endangered Species Act and assigned the Act's highest level of protection. Only scientific factors may be taken into account in deciding whether to list a species as endangered, though economic factors may be taken into account at other stages of the Act. See also threatened species. For the legal definition, see Section 3 of the Act.

Endemic. A descriptive term for plants or animals naturally occurring in a particular area. Also called indigenous.

Endocarp. (1) The inner layer of cells of the pericarp or fruit wall. (2) The stony part of a drupe or pome. (3) Endocarp is the hard inner layer of the pericarp of some fruits that contains the seed. It is especially, but not exclusively noticeable, in members of the subfamily Prunoideae including sloes, damsons, apricots, peaches, plums and cherries. In nuts, it is the stony layer that surrounds the kernel of pecans, walnuts, almonds etc. and which is removed when eating.

Endocrine disruptor. A chemical agent that interferes with natural hormones in the body. Hormones are secreted by endocrine glands (such as the pituitary, thyroid, pancreas, ovary, and testis), are transported through the body in the bloodstream, and regulate body growth and metabolism, other endocrine organs, and reproductive functions. There is emerging concern that endocrine disruptors may be causing human health or ecological effects, such as abnormal thyroid function, decreased fertility, and alteration of immune and behavioral function. This concern arises from demonstrated instances (an example is the ability of diethylstilbestrol (DES) to disrupt female reproductive function throughout the lifespan in laboratory animals and humans) and the fact that hormones are biologically active at very low concentrations (at parts per billion or less), so low levels of disruptors may similarly be biologically active. In amendments to the Safe Drink Water Act and the Federal Insecticide, Fungicide, and Rodenticide Act in 1996, Congress directed the Environmental Protection Agency to study endocrine disruptors. The outcome of this research will be of consequence to agriculture because some pesticides and animal growth stimulants have been hypothesized to act as endocrine disruptors.

Endoparasite. A parasite living on the inside of its host, e.g., Tapeworm.

Endosperm. (1) Food storage tissue surrounding and nourishing the embryo in plant seeds. By the time the seed is fully developed the embryo may have completely absorbed the endosperm in some seeds, e.g., peas and beans, whilst in others part may remain and be absorbed after germination, e.g., wheat. Cereals are grown for the endosperm content of the grain, which is a valuable food source. (2) Triploid food storage tissue of angiosperm seed resulting from triple fusion.

Endospermic seeds. Seeds with dominant endosperm.

Endrin. A very toxic insecticide. A poly-chlorinated hydrocarbon.

Energy cropping plantation. Growing of crops like casuarina and eucalyptus for energy purpose

Enteritis. Inflammation of the intestines. Enteritis is the inflammation of the small intestine (inflammation of the large intestine is termed colitis, while enterocolitis refers to an inflammation of both the large and small intestine.)

Enterotoxins. Products of bacterial growth that cause inflammation of the intestinal tract, a form of food poisoning. Certain kinds of staphylococci can produce food poisoning owing to enterotox-ins if they gain access to improperly refrigerated milk or soft foods such as custards, gravy and creamed meats. Enterotoxins are heat-resistant to the extent that they are not destroyed by pasteurization.

Entomology. (1) Branch of agricultural science which is dealing with all aspects of insect activity including their structure, behaviour, their place in nature and form, whether they are harmful or useful *i.e.*, their economic importance. It also helps us to devise suitable control measures. (2) The study of insects and their environments.

Entomophilous plants. Plants which are pollinated by insects.

Entomophilous. Insect-pollinated. Entomophilous is a form of *pollination* whereby *pollen* is distributed by *insects*, particularly *bees, moths,* and *beetles.* Entomophilous species frequently evolve mechanisms to make themselves more appealing to insects, e.g. brightly colored or scented *flowers, nectar,* and appealing shapes and patterns. Entonophilous species include the *sunflower, orchid,* and *cycad.*

Environment (Ecology). External surroundings in which an organism lives. It is influenced and determined by the interactions of climate, water, light etc. and other living organisms. It affects the growth, development, and behaviour of the organism.

Environment. The totality of the surrounding external conditions—biological, chemical, and physical—within which an organism, community, or object exists. The term is not exclusive in that organisms can be and usually are part of another organism's environment. Thus one can speak of the environment as that within which humankind lives, i.e., separate and external; or, one can speak of humankind as a component of the environment.

Environmental equity justice. Equal protection from environmental hazards for individuals, groups, or communities regardless of race, ethnicity, or economic status. This applies to the development, implementation, and enforcement of environmental laws, regulations, and policies, and implies that no population of people should be forced to shoulder a disproportionate share of adverse impacts of pollution.

Enzymes. Proteins present in living organisms which act catalytically to promote chemical changes whilst remaining unchanged them-selves, e.g. Alpha Amylase.

Ephemeral(s). Persisting for one day only, of short duration. Plants that emerge and bloom during one season, then die back for the remainder of the year. Many spring ephemerals bloom in woodlands, including trillium and ladyslipper.

Epicarp. Outer layer of fruit wall.

Epicormic shoots. A cluster of shoots, derived from dormant adventitious buds, on a main stem or branch after a wound or cut has been made.

Epicotyl. That portion of an embryo or seedling above the cotyledons.

Epidermis. Outermost cell layer of young plants and some parts of older plants (like leaves and fruits)

Epigeal. Plants in which the cotyledons appear above the soil surface.

Epigeous germination. Germination pattern of seed where the hypocotyl elongates and raises the cotyledons above ground level.

Epigynous. Condition in which the perianth and stamens of a flower are attached above the ovary

Epinasty. Condition in which the upper surface of a structure, such as a leaf, grows more rapidly than the lower surface, causing a downward curvature.

Epiphyte. (1) A plant growing on another without being a parasite (orchid) or on a rock or tree trunk (moss, lichen). Contrasted usually with plants rooted in the soil; sometimes parasites. (2) A plant that grows in the air or on another plant but gets its nourishment from the air and rainfall. A plant growing on another without being a parasite or on a rock or tree trunk. (3) Nonparasitic plant growing on another plant, absorbing atmospheric moisture with the help of velamen roots.

Epiphytotic. Means the sudden and destructive development of a plant disease on extensive area. It is known as epidemic in human disease and both the words have been used interchangea-bly by plant pathologists.

Episodic growth. Growth occurring in spurts or flushes, as against continuous growth

Epizootic. Enzootic Disease. An epizootic is the nonhuman equivalent of an *epidemic*, meaning that large numbers of animals are infected with a disease. As in epidemics, high *population density* is a major contributing factor to epizootics, with *aquaculture* being one notable example of an industry plagued by disease because of the large number of fish in a small area. New agricultural research aims to harness the microbes that hold the potential to cause epizootics and use them in pest control technologies. Due to the pathogens specificity to a hosts, the risk to humans and other non-target species in the environment is minimal and the capacity for control has proven to be quite effective.

Equilibrium moisture content. Moisture content of the product which is in equilibrium with the moisture content of the surroundings or hygroscopic equilibrium.

Equine malarial fever. Infectious viral disease of equines which is characterized by a variable course and is usually associated with intermittent fever, progressive weakness, oedema and anaemia of a progressive or transitory type. It is caused by a filterable virus.

Equivalent acidity. Pounds or kgs of pure calcium carbonate which is required to neutralize the acids produced in the soil from the quantity of fertilizer used, on the basis of which acid forming nature of fertilizer is decided; 1.6 Ib in urea, 2.6 Ib in calcium sulphate nitrate, 5.1 Ib in ammonium sulphate.

Equivalent basicity. In basic residue of any fertilizers measured in terms of calcium carbonate required to neutralize it. Term is used for expressing the basicity of a fertilizer.

Eradication. (1) The term used for the method of disease control in which the pathogen is eliminated after it has established. (2) Absoluble destruction, by burning or deep ploughing etc.

Eremophyte. Desert plant.

Ergosterol. Plant sterol which is acting as a precursor for vitamin D in animals, and is converted to this vitamin by ultraviolet radiation.

Ergot. A fungal disease *(Claviceps purpurea)* of grasses, particu-larly cereals and especially rye. It attacks the seed heads and destroys the grain. Animals can suffer serious poisoning (ergo-tism) by eating diseased grain or products derived from such grain (e.g., bread, maize meal). In man it causes the condition known as St Anthony's Fire.

Erosion (erodibility) index (EI). The erosion (sometimes called erodibility) index is created by dividing potential erosion (from all sources except gully erosion) by the T value, which is the rate of soil erosion above which long term productivity may be adversely affected. The erodibility index is used in the conservation compliance and Conservation Reserve Programs. For example, one of the eligibility requirements for the CRP is that land have an EI greater than 8.

Erosion (normal). The wearing away of the land surface, particularly soil, by running water, ice and wind, etc. Ploughing up and down slopes can lead to gully formation, and light and friable soils can blow away in strong winds when exposed, usually in the spring. Wind erosion is a serious problem in the Fens.

Erosion spalsh. Erosion which is preceded by the destruction of the crumb structure by rain-drops or drip.

Erosion. (1) The wearing away of the land surface. Unconsolidated materials, such as soil, erode more rapidly than consolidated materials, such as rock. The most common causes of erosion are wind and moving water. The susceptibility of soil to erosion is quantified by the erosion index. Water causes sheet, rill, and gully erosion. (2) The wearing away, washing away, or removal of soil by wind, water or human activity.

Erosion-resistant crop. A crop because of its dense foliage, root system etc. providing effective protection against soil loss by erosion.

Erratic bearing (Irregular bearing). Inconsistent and unpredictable flowering and fruiting behaviour.

Escape. A garden plant that has been seeded or spread into the wild.

Escherichia coli. A bacterium found in the alimentary canal of most mammals. Certain strains cause disease, e.g., Coli Septicaemia in poultry and diarrhoea in various animals. Usually abbreviated to *E.coli.*

Esculentus. Means 'edible' (used in botanical names).

Espalier. (1) A lattice-work, usually of wood, to train trees on. Also fruit trees so-trained, usually so that the branches grow horizontally. (2) Any plant trained lattice fashion in one plane; woody plant trained in a lattice-like or fence-like pattern on some support. (3) Plants trained to grow flat against a wall. (4) Process of training a tree or shrub so its branches grow in a flat pattern. (5) To train a plant along a railing or trellis so that the plant grows against the rail or trellis.

Essential amino acids. Refer to those amino acids which are not synthesized in the body in sufficient quantities to support opti-mum metabolism.

Essential oil. Volatile aromatic extracts prepared by steam distillation of ground spices.

Estate. (1) The total property in single ownership, often applied to the property of large landowners, including farms cultivated land, woodland, houses, etc. (2) A property where rubber, tea. coffee etc. are cultivated.

Estrone. Female sex honnone occurring in the ovary, in the placenta and in urine of pregnant animals, women and plants etc.

Estrus (heat). The recurrent, restricted period of sexual receptivity in livestock. Non-pregnant cows and heifers usually come in heat 18 to 21 days following their previous estrus.

Ethylene C_2H_4. (1) Growth hormone which regulates fruil ripening various aspects of vegetative growth, and is also important in the abscission process. (2) A gaseous plant growth substance, regulating vegetative growth, fruit ripening and abscission of plant parts. (3) Naturally occurring gas to ripen fruit.

Etiolation. (1) A condition of plants when grown in darkness, characterized by a general pale yellowness due to lack of chlorophyll, and by long internodes and small leaves. (2) Development of plants or plant parts in darkness, resulting in elongated stem, small leaves, and reduced chlorophyll content. (3) Stretching of a plant and loss of color due to a lack of needed light.

Eutrophic. A descriptive term for nutrient or soil solution with nutrient concentrations optimal or nearly optimal for animal or plant growth.

Eutrophication. The process by which a body of water acquires a high concentration of plant nutrients, especially nitrates or phosphates. This nutrification promotes algae growth that, when it dies, can lead to the depletion of dissolved oxygen, killing fish and other aquatic organisms. While eutrophication is a natural, slow-aging process for a body of water, human activities can greatly accelerate the process.

Evaporation. Process by which water returns to the air.

Evaporimeter. An instruement which is used for the measurement of the amount of moisture which air will take up by evaporation. The quantity of water lost is usually recorded from an exposed pan (Open pan evaporimeter) or tank. Due allowance is made for water added by precipitation.

Evapotranspiration. (1) The loss of water from the soil both by evaporation and by transpiration from the plants growing in the soil. (2) The loss of water from an area (e.g. a field) by evaporation at the soil surface and by plant transpiration. Potential Evapolranspiration is the maximum transpiration pos-sible under given weather conditions with a low-growing crop which shades the soil completely and has an adequate water supply. (3) Process of loss of water from a plant's tissue and soil.

Even span greenhouse. A greenhouse whose roof-slopes are of equal length and angle.

Evergreen plants. (1) Retain some leaves longer than one growing season so that leaves are present throughout the year. Seasonal drop of some of the oldest interior leaves is a natural part of the life cycle. (2) A tree that retains some or most of its leaves, or needles, throughout the year. (3) A plant that retains all or most of it's foliage throughout the year. (4) A plant which keeps its leaves throughout the year. Mainly confierous trees and shrubs. (5) Holding or having green foliage through the winter. (6) Plants that have persistent green leaves for two or more growing seasons. (7) Plants whose leaf cover remains alive year-round, though individual leaves may die and fall. Includes species, such as *Rhododendron*, whose leaves go dormant and change color at the end of the growing season, then green up again for the new season. Other evergreens, such as *Pinus banksiana* (Jack Pine), discard batches of leaves periodically. Evergreens may have needles (pine and spruce, for instance) or "broad" leaves (holly and rhododendron). Perennial plants whose leaves all die at once (and usually fall) at the end of each growing season (i.e. maple trees) are *deciduous*.

Everlasting(s). A flowering plant that retains colors when cut and dried. The flowers of these plants keep their good looks long after they've been picked.

Eviscerated. Having had the viscera or bowels removed.

Excavated. Hollowed out, as applied to pith, making the stem fistulous.

Excess land. Irrigable land, other than exempt land, owned by any landowner in excess of the maximum acreage limitation (ownership entitlement) under the applicable provision of reclamation law.

Excised embryo test. (1) Germination test of seeds whose embryos require long periods of after-ripening. (2) This is used to test the germination of seeds of woodyshrubs and tree whose embryos need long periods of after-ripening before germination will occur. In this test, the embryo is excised from the rest of the seed and germinated alone. A viable embryo will either germinate or show some indication of germination, while a nonviable embryo becomes discoloured and deteriorates. The excision is carried out care-fully to avoid any possible injury to the embryo. Procedures for germinating excised embryos are similar to those for germinat-ing intact seeds.

Excurrent habit. The cone-shaped form typical of many conifers and a few angiosperms with a clearly identified central leader and growing point at the top of the plant

Exoenzyme. An enzyme which is produced by a plant and exuded into the surrounding area where it breaks down potential food material which can then be assimilated by the plant.

Exogenous. Arising from the external tissues.

Exotic plant. A newly introduced plant, not native to a place.

Exotic species. (1) A species that is not native to a region. (2) A non-native plant or animal species introduced by humans, either deliberately or accidentally.

Exotic. (1) A descriptive term for plants or animals introduced into an area or country from outside or abroad, not naturally occurring there. (2) Said of a plant not native; rare or tropical plants.

Exotoxin. Substance having a pronounced toxicity and is produced by some pathogenic bacteria, which as soon as it is formed in the cytoplasm of the living pathogenic germ, filters out through the membrane and spreads in to the surrounding medium.

Expeller process. A process for removing the oil from oilseeds in which the seeds are forced along a tapering tube, squeezing out the oil. The residues, containing usually less than 5% oil, may be used as animal feeds.

Explant. (1) Any removed portion of a plant used for tissue culture. (2) Excised plant part used for tissue culture.

Explantation. The removal of living tissue from an organism and its culturing in an artificial medium.

Explosive fruit. Fruit which bursts suddenly and violently, scattering seeds over a considerable area

Extensive farming. A method of fanning in which a large amount of land is used to raise stock and produce crops, yields usually being about average as distinct from Intensive Farming.

Extensive production. Crop production using large area of land with minimum labour and inputs.

External seed treatment. Treatment of seed for external seedborne fungal disease, by which fungicide like formalin sulphur, cop-per carbonate, organomercurials like Agrosan and ceresan, is throughly mixed with seed, preferably in a rotating drum so as to cover the seed surface with a layer of the fungicide. In the case of formalin, it gets diluted with an equal quantity of water and sprayed over a thin layer of seed and the seed heaped and kept covered with canvas overnight.

External seed-borne disease. Plant disease, in which the disease organism in its living state or its spores have been present superficially on the seed without any organic union with it, or as mycelium in superficial infection. It is controlled by application of fungicides to the surface of the seed.

Extra-long staple (ELS) cotton. Also called American Pima, this cotton has a staple length of 1-3/8" or more, is characterized by fineness and high fiber strength, and is used in high-value products such as sewing thread and expensive apparel. It is grown chiefly in west Texas, New Mexico and Arizona. ELS cotton and upland cotton are both eligible for nonrecourse loans. However, ELS cotton is not eligible for marketing loan repayment provisions or loan deficiency payments.

Ex-vitro. Conditions, reactions or responses of organisms removed from *in vitro* conditions. (In Latin *ex vitro* means 'from glass')

Eyespot. A fungal disease *(Cercosporella herpotrichoides)* which attacks all cereals, causing oval or eye-shaped areas of black dots on leaf sheaths as they emerge from the soil in spring, and causing straws to weaken and lodge.

F

F. f1 (First filial generation). The first generation of offspring resulting from an arranged cross between selected parents in breeding programmes.

f.c.s. An abbreviation for 'filled, carted and spread.' An agricultural valuers' expression for the labour required to transport manure from the farmyard to the fields. Now rarely used, the expression 'labour to farmyard manure' being more common.

F_2 (Second filial generation). The second generation of offspring resulting from crosses between individuals of the F1 generation

Face-packing. Packing fruits for market with the top layer of the container artistically arranged, while the fruits below are without arrangement and often inferior in quality

Facilitation. The process in which two individual plants or two populations of plants interact in such a way that at least one exerts a positive effect on the other. Double facilitation is equivalent to mutualism

Factice. Rubber-like substance which is prepared by the reaction of such oils as linseed, soyabean and cottonseed with sulphur or certain of its compounds; a vulcanized oil.

Facultative long and short night plants. Plants which do not require a night length longer or shorter than a given critical length for a response to occur, but which will respond faster if the dark period is longer or shorter, respectively, than a critical length

Facultative saprobe. Parasitic organism which is able to grow on dead organic matter.

Facultative. An organism which has power to live under a number of certain specific conditions or to adopt an alternative mode of nutrition or life style of environmental condition etc., e.g., a facultative parasite may be either parasitic or saprophytic.

Facutlative clestogamy. Plant that can possess either chasmogamic flowers or those that do not open during pollination.

FAD. Flavin Adenine Dinucleotide. A coenzyme that is responsible for transfer of hydrogen form one compound to another.

Failed acreage. Tracts of properly-planted and managed crops that did not grow or were destroyed due to a natural disaster. Failed acreage is eligible for indemnification if covered by the federal crop insurance program.

Fair information practices. A set of standards developed in the early 1970s to describe how information should be used and protected; usually stated as eight principles in this country and including openness, individual access, collection limitations, data quality, use limits, disclosure limits, security, and accountability. The agriculture community is starting to consider whether it needs such a set of standards, and whether this set is the best model.

Fairy ring. An are or ring of lush, usually darker coloured grass in a pasture caused by fungi which grow in expanding circles in the form of a perennial, Mycelium. Such rings may be many years old and many yards across. The ring of lush grass is due to nitrate production, which acts as a fertilizer.

Fall colors. Fall colors are due to the presence of other plant pigments besides chlorophyll. Reds and purples are caused by the presence of anthocyanins, produced after the chlorophyll is destroyed. Yellows are due to the presence of xanthophylls, a yellow to gold pigment present in the leaf throughout the growing season but masked by the chlorophyll. Oranges are the result of carotenes and xanthophylls. Browns are the presence of tannins in the leaf. Fall colors are influenced by several factors including the plant genetic makeup. Some plants naturally produce stunning fall colors: red maples, oaks, burning bushes, ashes. Environmental conditions also affect fall colors. Cool, bright, sunny fall days increase photosynthesis production and thus sugar accumulation in the leaves. In fall, leaves develop an abscision layer at the base of the petiole (leaf stem), preventing sugars from being transported to the roots or stems for storage; the increased sugar concentration produces more intense colors. The environmental conditions during the growing season can also cause differences in color intensity. Hot, dry weather usually reduces the chances of good fall color, as does excessive moisture in the early spring. Heavy autumn rainfalls also diminish falls colors. Frosts and freezes usually spell the end of the reds, yellows, and oranges.

Fall. (1) An American term for autumn. (2) The dropping of an animal from its mother's uterus to the ground during birth. Also the number of offspring born in a herd or flock, etc., in a season.

Fallen wool. A type of woolloddment purchased by the British Wool Marketing Board comprising wool which is either cliped from a dead sheep, or is badly shorn and is in the form of torn or broken pieces as distinct from an intact fleece.

Fallow (crop science). (1) Cropland left idle during the growing season. (2) A crop grown in well spaced out rows, e.g., potatoes or turnips, so that the land between the rows can be hoed and cultivated to control weeds. Also called a cleaning crop.

Fallow cropland. Cropland left idle during the growing season, sometimes called summer fallow. It may be tilled or sprayed to control weeds and conserve moisture in the soil. The amount of cultivated summer fallow has ranged between 22 and 32 million acres over the past 10 years, or 7 to 10% of the cropland used for crops.

Fallow. (1) A system in which cultivated land is left without a crop for long periods. (2) Land left unsown, usually for a season, during which it is ploughed and cultivated to kill perennial weeds by desiccation. The practice of fallowing is now less common.

Falls. The outer whorl of petals of an iris flower, often broader than the inner petals and often drooping or flexuous.

False fruit (Pseudocarp). A fruit that incorporates in addition to the ovary wall. other parts of the flower, such as the receptacle.

Fan branches. The main branches of a shrub or a tree appearing like the leaves of a fan.

Fan-and-pad cooling system. A cooling system, utilized in greenhouses, where cooling is brought about by evaporation of air drawn in through wet pads with the help of fans.

Fanged. A term for not caps having forked roots.

Fanner's lung. A distressing respiratory condition caused by the inhalation of dust particles, *e.g.*, from mouldy feed or litter, containing fungal spores, which results in an allergic reaction in the air sacs of the lungs and causes breathing difficulties. Cows and horses can also suffer.

Fanning mill. Machine which is operated by hand or electricity and equipped with a hopper, vibrating sieves, a fan etc. It is used for removing chaff, straw, weed seeds and dirt from grain.

FAPRI. Food and Agricultural Policy Research Institute.

Fardel-bound. Constipated. Especially used for cattle and sheep having food retained in the Omasum.

Farinograph. Instrument which is used to measured the protein strength of flour so as to judge its baking quality.

Farm acreage base. The total of the crop acreage bases (wheat, feed grains, cotton, and rice) for a farm for a year, the average acreage planted to soybeans and other non-program crops, and the average acreage devoted to conserving uses (excluding Acreage Reduction Program land). The FAIR Act of 1996 eliminated the need to calculate a farm acreage base.

Farm Credit System. A network of cooperatively owned lending institutions and related service organizations serving all 50 states and the Commonwealth of Puerto Rico. The FCS specializes in providing farm real estate and rural homeowner loans, operating credit, and related services to farmers, ranchers, and producers or harvesters of aquatic products. The FCS may also finance the processing and marketing activities of these borrowers, certain farm-related businesses, and agricultural, aquatic, and public utility cooperatives. It is chartered under authorities in the Farm Credit Act of 1971, as amended, but does not receive any direct government funding. The System provides about one-fourth of the total credit used by U.S. farmers, ranchers, and cooperatives. Historically, the FCS consisted of a Federal Land Bank (FLB), a Federal Intermediate Credit Bank (FICB) and a Bank for Cooperatives (BC) in each of the 12 districts across the nation. Within each district, Federal Land Bank Associations (FLBA) and Production Credit Associations (PCA) served as local lenders for the FCS providing farm real estate and operating credit, respectively. A severe financial crisis led to the enactment of the Agricultural Credit Act of 1987, which provided federal financial assistance to weak institutions in the FCS, but required the FCS to streamline its operations as a condition for assistance. As a result, each district was required to merge its FLB and FICB to form a Farm Credit Bank (FCB). For FLBAs and PCAs that share a similar geographical territory, stockholders were given the option of merging institutions to form an Agricultural Credit Association (ACA). The Central Bank for Cooperatives and 10 of the 12 BCs also agreed to a merger forming a National Bank for Cooperatives (CoBank).

Farm duty of water. Seasonal quantity of water which is delivered to individual farm units under an irrigation project. Such duty is usually expressed in terms of acrefeet per acre, total depth of water over a unit area, or number of acres served per unit of flow. Such duty is able to express the actual rate of use of water at the farm, after all canal and other losses are eliminated, but includes losses in the farm ditches and also waste. Farm Forestry. Refers to the practice of raising small woods on a farm in addition to normal cultivation for making the farm more self-sufficient in fuel, small timber, grazing facilities fodder and leaf manure, in addition to the other indirect benefits such as protection of crops against high winds and control over erosion. 10 to 15 per cent of farm area considered best are put under farm forests.

Farm equity. The net worth of the farm sector's assets (i.e., farmland, machinery, equipment, facilities, crop and livestock inventories) against which there is no debt. This represents all farm proprietors' residual claims to farm assets. Increases in farm equity in the late 1970s became increasingly important for most agricultural producers as a source of additional collateral against which to obtain credit for operating and expansion purposes. The level of farm equity ranges widely from one farm to another. The overall debt-asset ratio is a measure of the farm sectors financial condition.

Farm forestry. Practice of growing trees specifically for fuel, food and a variety of other resources on a farm

Farm income and balance sheet. The income statement measures the profitability of a farm business for a particular period of time, usually one year. The balance sheet measures the wealth or financial position of the business at a particular point in time by reporting the farm's assets, debt, and net worth. The Economic Research Service publishes the income statement and balance sheet of the Nation's farm sector, and the farm sector financial statement each state.

Farm income. Several measures are used to gauge the earnings of a farming operation over a given period of time: Gross cash income is the sum of all receipts from the sale of crops, livestock, and farm related goods and services as well as all forms of direct payments from the government. Gross farm income is the same as gross cash income with the addition of nonmoney income, such as the value of home consumption of self-produced food and the imputed gross rental value of farm dwellings. Net cash income is gross cash income less all cash expenses such as for feed, seed, fertilizer, property taxes, interest on debt, wages to hired labor, contract labor and rent to nonoperator landlords. Net farm income is gross farm income less cash expenses and noncash expenses, such as capital consumption, perquisites to hired labor, and farm household expenses. Net farm income is a longer term measure of the ability of the farm to survive as a viable income-earning business, while net cash income is a shorter term measure of cash flow.

Farm inputs. The resources that are used in farm production, such as chemicals, equipment, feed, seed, and energy. Most farm inputs are purchased (a change from the days when animals powered most operations), making production costs susceptible to nonfarm economic conditions. Over time, prices of farm inputs have increased relative to commodity prices, creating what farmers describe as a cost-price squeeze. The relationship between prices paid for inputs compared to prices received for output is quantified in the parity ratio.

Farm management. Branch of agricultural economics dealing with the business principles and practices of farming with an object of getting the maximum possible return from the farm as a unit under a sound farming programme.

Farm map. Map of a farm. It depicts clear picture of the situation of blocks, plot or sub-plots, farm buildings, wells, undergound pipe lines, drainage, roads and bunds, fencing etc. It facilitates in planning the cropping scheme.

Farm operator. A person who operates a farm, either by doing or supervising the work or by making the day-to-day management decisions. Nationally, farm operators own about 57% of their land and lease or rent the remainder.

Farm processing. Method or treatment which is used to prepare farm products for use or preservation. It is including the operation that maintains or improves the quality of farm products or changes their forms.

Farm program acreage. The number of harvested acres of feed grains, wheat, and cotton needed nationally to meet domestic and export use and to accomplish any desired increase or decrease in carryover levels. The acreage base for an individual farm was calculated as the producer's share of the national farm program acreage. The FAIR Act of 1996 eliminated the need to calculate a national program acreage.

Farm programs. This term is generally meant to include the commodity programs administered by the Farm Service Agency, as well as the other USDA programs that directly benefit farmers. Some examples of the other programs include farm loans, federal crop insurance, the noninsured assistance program (NAP), the Conservation Reserve Program (CRP), and conservation cost sharing.

Farm. (1) An area of land, usually with a house and necessary buildings, used for agricultural purposes. The term originally meant such land leased or rented by a landlord and worked by another person, the fanner.

(2) As defined for purposes of the Census of Agriculture since 1978, a farm is any place that has, or has the potential to produce, $1,000 or more in annual gross sales of farm products. According to the 1992 Census of Agriculture, there are about 1.925 million farms in the United States. This number includes all farm sizes and ownership structures, including corporate farms, partnerships, and family farms.

Farmed wetlands. Under the swampbuster program, these are wetlands that were partially drained or altered to improve crop production before swampbuster was enacted as part of the December 23, 1985, farm law. Farmed wetlands may be farmed as they were before the 1985 date, and the drainage that was in place before that date can be maintained, but no additional drainage is allowed.

Farmers market. A market where producers, generally farmers, sell their goods directly to consumers.

Farmers stock peanuts. Picked or threshed peanuts produced in the United States, which have not been changed (except for removal of foreign material and excess moisture) from the condition in which picked or threshed peanuts are customarily marketed by producers.

Farming system research. Research on farming systems, multi and inter-disciplinary in nature, to enhance and stabilize agricultural production.

Farming system. Appropriate combinations of cropping systems, livestock. fisheries, poultry and forestry, ensuring ecological and socio economic balances.

Farming. (1) To devote to agriculture (2) To manage and cultivate as a farm (3) to grow or cultivate in quantity.

Farmland Protection Program (FPP). A program established by the FAIR Act of 1996 to fund the purchase of conservation easements of 170,000-340,000 acres of land having prime or unique soil or other desirable production qualities that are threatened by urban development. Eligibility depends upon already having a pending offer from a state or local government to protect qualifying land by limiting nonagricultural use. USDA is authorized to use up to $35 million of funds from the CCC. During the first year of operation, USDA awarded almost $15 million to 18 states to protect an estimated 50,000 acres of farmland.

Farmland protection. Programs, operated mostly at state and local levels by government agencies or private entities such as land trusts, that are designed to limit conversion of agricultural land to other uses that otherwise might have been more financially attractive to the landowner. Every state has at least one such program; the most popular programs are tax relief through preferential or differential assessment, and right-to-farm laws. Easements are used in some states.

Farmland. Land used for agricultural purposes. The federal government recognizes prime farmland and unique farmland as the most important categories. According to USDA, the United States has had roughly 1 billion acres of farmland. Farmland consists of cropland, pastureland, and grazing land.

Farmstead. A farm house and the farm buildings associated with it.

Farmyard manure. The faeces and urine of farm animals mixed with litter, mainly straw, to absorb the urine. This mixture is also called 'Long Dung'. Its composition is variable dependent on the animals contributing the dung, their diet and the kind of litter used. Farmyard manure is usually stored in manure heaps where bacterial activity releases ammonia, fermentation occurs, and the material generally degrades into simpler compounds. It becomesstructureless and the liquids drain as Dung Liquor. In this condition it is called 'Short Dung.' This material is spread on farmland to improve soil structure (being rich in Organic Matter) and also as Fertilizer, its main constituents being compounds of Nitrogen, Potassium and Phosphorus. Also called muck.

Farrand test. A method for determining the Alpha Amylase content of Milling Wheat. Used as an alternative to the Hagberg Test. An extract of ground wheat is reacted with a solution of 'limit' dextrin (a derivative of wheat starch). After a measured time it is mixed with iodine solution, and the colour produced compared with that of a suitably diluted iodine 'limit' dextrin mixture. Loss of colour, due to Alpha Amylase in the wheat extract reacting with the 'limit' dextrin, indicates the activity of the enzyme, which is expressed as units. Sound wheat has an activity of 10 units or less. Wheat with an activity over 40 units is not normally accepted for milling. 40 Farrand Units is approximately equivalent to a Hagberg Falling Number of 160.

Farrow. (1) To give birth to a Litter of pigs. Also the litter itself. (2) A descriptive term for a cow or heifer not in-calf, or baren.

Farrowing crate, Farrowing rail. A steel crate used in a pig pen to contain a sow and prevent it from lying on and killing the piglets, whilst allowing them access to her teats. They are normally in the form of a long narrow rectangle, but are sometimes circular (the Ruakura pen). A farrowing rail serves the same purpose and is usually fitted a short distance form the pen walls and floor.

Farrowing Index. The number of Litters produced per year by a breeding sow.

Farrow-to-finish. Typically, a confinement operation where pigs are bred and raised to their slaughter weight, usually 200-250 pounds. Facilities that have 2,500 or more swine are considered by the Environmental Protection Agency to be a concentrated animal feeding operation (CAFO) subject to point source pollution permit requirements.

Fasciated. Abnormally much flattened, and seemingly several units fused together. Much flattened or cockscomb-like growth; seemingly several units fused together.

Fasciation. A malformation in plant shoots resulting in an enlarged and flattened shoot, as if several shoots are fused.

Fascicle. A tight bundle or cluster, as in the needle leaves of pines

Fascicled. Clustered, like the leaves on a spur of barberry.

Fastigiate. With upright branches; having a narrowly columnar or pillar-like growth form.

Fat content. Amount of butterfat in milk, usually stated in percent-age.

Fat free lean index. One of several measures of hog quality (in this case, leanness) that can -be used in determining value. The index was developed by the National Pork Producers Council, an industry trade group.

Fat soluble vitamins. These are usually found associated with the lipids of natural foods e.g., vitamin A, D, E and K.

Fat. (1) A term applied to animals reared for their meat (e.g., fat cattle, fat sheep) which have been fattened up and are in a ready condition for sale to a butcher at a market. (2) A dry measure of nine Bushels.

Fats. Storage materials found in living organisms, mainly in liquid form (oils) in plants, and in solid form (also called adipose tissue) in animals, comprising mixtures of glycerides (conden-sation products of glycerol and Fatty Acids). Most animals deposit fat in the body as an energy store. Some plants, mostly tropical, produce seeds rich in oil which is extracted mainly for cooking, the residue being used for cattle Cake.

Fatstock. Livestock fattened for sale in a market.

Fatty acids. Organic acids, thirteen of which occur in natural Fats. They are either (a) saturated (i.e. molecules to which no further atoms can be added), found mainly in solid form in Animals Fats, or (b) unsaturated, found mainly in liquid form (oils) in plants. The most common fatty acids in natural fats are palmitic, oleic and stearic. (Steapsin)

Fauna. (1) Pertaining to the animal life of a locality or region or that existing during a specific geological period or time. (2) The population of animal life in a place or area, or during a particular period.

Favism. Haemolytic anaemia caused by eating broad beans or by inhaling the pollen of its flower

FCIC. Federal Crop Insurance Corporation.

Feather. (1) One of the dermal outgrowths forming the plumage of a bird. (2) Long hairs found on the sides and on the back of the legs of some breeds of horse e.g. the Shire.

Feathered maiden. A Maiden tree which has grown side shoots.

Feathering. Spreading out roots before planting a potted plant.

Fecal coliform bacteria. Bacteria found in the intestinal tracts of mammals. Their presence in water or sludge is an indicator of pollution and possible contamination by pathogens.

Federal crop insurance. A voluntary risk management tool, available to farmers since the thirties, that protects them from the economics effects of unavoidable adverse natural events. Administrative costs are appropriated by the Congress and 30 percent of the insurance costs are federally subsidized.

Federal grain inspection program. The grain inspection program administered by the Grain Inspection, Packers and Stockyards Administration. The program establishes official U.S. standards for grain and certain other commodities such as rice, hops, and processed grain products. The program offers a user-financed nationwide inspection and weighing system to certify that grain meets approved standards. By law, all grain exported from the United States must be officially inspected.

Federal seed act. Originally enacted by Congress in 1939, the Act requires accurate labeling and purity standards for seeds in commerce, and prohibits the importation and movement of adulterated or misbranded seeds. The law works in conjunction with the Federal Noxious Weed Act to authorize the Animal and Plant Health Inspection Service to regulate the importation of field crop, pasture and forage, or vegetable seed that may contain noxious weed seeds.

Federal-state marketing improvement program. Sometimes referred to in budget documents as Payments to States and Territories, the program provides matching funds to states for research and innovative projects aimed at identifying new market opportunities for producers or at improving the efficiency of agricultural marketing systems. The program is administered by AMS and has been funded federally at just over $1 million per year recently.

Feed block. A block of foodstuffs left on pastures, particulary in hill areas, for sheep to lick at will as an aid to maintaining Condition. Normally they contain a Carbohydrate source (e.g., cereals, glucose, or molasses), Protein and Urea (sometimes excluded), Minerals, Vitamins and Trace elements.

Feed grain. Any of several grains most commonly used for livestock feed, including corn, grain sorghum, oats, rye, and barley. These grains and the farms producing them historically have received federal price and income support. They qualify as contract commodities and receive production flexibility contract payments.

Feed off. The practice of allowing a crop to be eaten by animals whilst still in the ground.

Feed ratio. The relationship of the cost of feeding animals to their market weight, expressed as a ratio to the sale price of animals, such as the hog/corn ratio. This serves as an indicator of the profit margin or lack of profit in feeding animals to market weight.

Feed ring. A railed circular container mainly used for hay and straw, from which livestock feed around its circumference.

Feeder cattle. (1) Cattle past the calf stage that have weight increased making them salable as feedlot replacements. (2) A steer or cow mature enough to be placed in a feedlot where it will be fattened prior to slaughter. Feeder calves are less than 1 year old; feeder yearlings are between 1 and 2 years old.

Feeder roots. Ultimate branches of a root system with a large absorbing area.

Feedingstuffs. The various foods available for farm animals, loosely classified into four main groups, viz, (a) Green Forages, (b) succulent foods, mainly roots and tubers, e.g., turnips, potatoes etc., (c) dry roughage or coarse fodder, e.g., hay derived from grasses, cereals and legumes, or Straw derived from cereals and the haulms of peas and beans, and (d) Concentrates. Also called feeds.

Feedlot. (1) A confined cattle feeding facility where feeder cattle (usually less than a year old) are put on higher protein rations to prepare them for slaughter as fed cattle at "good" or better grades. Commercial feedlots of 1,000 head or more are considered by the Environmental Protection Agency to be concentrated animal feeding operation (CAFO) and therefore subject to rules requiring permits setting effluent standards. (2) An area of land in which animals (very commonly beef cattle) are accommodated at very high density. The feedlot does not contribute at all to the production of animal feed, all of which has, therefore, to he brought to the animals from outside the feedlot.

Feer. To plough the first furrow when starting ploughing.

Fehling's Reagent. Refers to a solution of cupric sulfate, sodium potassium tartrate, and sodium hydroxide. It is used to test for the presence of reducing compounds such as sugars.

Fell. (1) The hide of skin of an animal, particularly when covered with hair. A fleece. (2) A term mainly used in Northern England for mountainous moorland and hill pastures. (3) To cut down a tree.

Fen. A lowlying, marshy area of land, with a Peat soil which may be alkaline to slightly acid, as distinct from the very acid soil of a Bog.

Fenland rotation. A three-course system of crop Rotation developed on the deep, rich, silts and Fens of East Anglia, in the sequence potatoes-sugarbeet-wheat. The short sequence increases the likelihood of attack by pests and disease, and other high value crops are frequently included, such as rootcrops, brassicas, bulbs, peas and celery.

Fermentation. (1) Anaerobic breakdown of organic substances by microorganisms, producing heat, wastage gases and a variety of end products. (2) Process after maceration (rolling) in the production of tea. (3) A step in cocoa processing to get rid of the pulp around the beans and to induce chemical changes for improving the quality. (4) A step in processing of parchment coffee where the mucilage is removed by natural fermentation. (5) The breakdown of organic substances by the action of Enzymes, usually secreted by living organism such as Bacteria and Yeast. Heat and gas (e.g., carbon dioxide) are usually evolved. The principle of fermentation is used in the making of Silage, wine, vinegar, etc.

Fernery (Conservatory). A greenhouse of tropics which protects plants from fierce sunlight and strong wind.

Ferret. An albino variety of the polecat, half-lamed and used to catch or unearth rabbits and rats.

Ferruginous soil. Soil having large amounts of iron minerals, espe-cially limonite and hamatite. It is characterized by the colour or shades of yellow and brown.

Fertile. (1) A descriptive term for plants or animals able to reproduce successfully, as distinct from being sterile. Also those able to produce offspring prolifically. Used also to describe eggs capable of producing a chick, or seeds capable of Germination. (2) A descriptive term for land or soil, rich in nutrients, and capable of producing good crops. (4) Capable of producing seed.

Fertiliser grade. An expression indicating the percentage of plant nutrients in a fertilizer e.g., a 12-5-5 grade of fertilizer means 12 per cent nitrogen, 5 per cent phosphoric acid (P_2O_5) and 5 per cent potash (K2,O).

Fertiliser. Substance which supplies mineral nutrients to plants in required quantity.

Fertility (soil). Means the ability of a soil to supply all the essential nutrients in optimum amount in a form that is readily available to plants.

Fertility gradient. Means the variation in nature fertility in any direction across an area of ground.

Fertilization. (1) The fusion of male and female gametes to produce zygote, containing chromosomes of both the parents. 2) The application of fertilisers to the soil to increase fertility. (3) The enrichment of soil by the application of Fertilizer. (4) The union of male and female gametes in reproduction to produce a zygote.

Fertilizer ratio. Referring to the relative proportion of three major plant nutrients, keeping the percentage of nitrogen as one in the ratio. Thus a 6-12-6 fertilizer mixture having 6 per cent nitrogen. 12 per cent available P2O5. and 6 per cent soluble potash has a nurtrient ratio 1:2:1.

Fertilizer requirement. The quantity of particular plant nutrients required, in addition to those already contained in the soil, to increase crop growth to a desired optimum.

Fertilizer. (1) Any organic or inorganic material, either natural or synthetic, used to supply elements (such as nitrogen, phosphate and potash) essential for plant growth. If used in excess or attached to eroding soil, fertilizers can become a source of water pollution. (2) In general terms anything added to the soil to increase the amount of plant nutrients available for crop growth. Normally applied to inorganic chemicals and non-bulky organic substances *Fertiliser Grade* (i.e., excluding Compost, Farmyard Manure, etc.,, which are commonly called Manures). Fertilizers basically provide the major plant nutrients *(i.e.,* Nitrogen Phosphorus and Potash), either individually (Straight Fertilizer) or as a combination of two or three (Compound Fertilizer)., sometimes together with Trace Elements: (a) The nitrogenous fertilizers include Ammonium Nitrate, Nitrate of Soda, Sulphate of Ammonia, Anhydrous Ammonia, Urea, Ammonium Phosphate and Liquid Nitrogen Fertilizers. (b) The phosphatic fertilizers may be quick acting (water sol-uble) or slow acting (water insoluble) and include Ammonium Phosphate, Superphosphates, Bone Flour, Bone Meal, Fish Guano and Reddzlaag. (c) The Potash fertilizers include Muri-ate of Potash, Sulphate of Potash, Kainit and Nitrate of Potash. Fertilizers applied to the soil usually quickly dissolve in the Soil Water and are taken up by plant roots or absorbed by soil colloids (Cation Exchange). Excess fertilizer may be leached through the soil into the Ground-Water. Fertilizers now-a-days are mostly produced in easily stored granulated form but pow-ders and crystals, and also less concentrated liquids are avail-able. Inorganic chemicals such as Limestone and Gypsum applied to correct such soil conditions as acidity and salinity may also be considered as fertilizers.

(3) Organic or inorganic plant food used to amend the soil in order to improve the quality or quantity of plant growth. (4) Those numbers are confusing: 5-5-10. 10-6-4. The first number is the percentage of nitrogen (N), the second phosphorus (P) and the third potassium (K). Different plants need different proportions of nutrients; your gardening center will guide you.

Fertilizers distributors. Machines which distribute Fertilizer on the land usually by broadcasting if in a solid form (i.e., crystals, granules or powders), or by spraying if in liquid form, and sometimes by injection into the soil (e.g., Anhydrous Ammonia). Broadcast types comprise a hopper which feeds the fertilizer to a spreading mechanism, the most popular being either a spinning disc or oscillating spout, which distribute fertilizer in swaths of variable width. These machanisms are less accurate than the 'full width' distributors such as the once popular 'plate and flicker' type or the more modern agitator or pneumatic type spreaders. Fertilizers can also be applied by a Combine Drill, Placement Drill, and by aircraft as top dressing, particularly to cereals.

Fertilizing. Hostas require little or no fertilizing. In fact, over fertilizing can be detrimental to some of the variegated types, causing the colors to fade. If plants are slow to develop and multiply, decomposed manure can be incorporated into the soil around clump. A teaspoon of 10-10-10, 12-12-12 or 15-15-15 fertilizer per clump will provide the necessary nutrients for yearly growth.

Fibre. A constituent of animal Feedingstuffs consisting mainly of Cellulose and Lignin, which aids digestion by stimulating mus-cular activity in the digestive tract. It also facilitates the diges-tion of concentrates by opening them up to the action of the digestive juices. All farm animals require a certain amount of fibre except when very young and on a liquid diet. Ruminants require the most fibre and derive energy from it by bacterial digestion in the rumen. Cows require adequate fibre to produce milk with a high percentage of Butterfat. Sheep and horses require less fibre than cattle but more than pigs. Excess fibre in the diet can prevent an animal eating sufficient food to derive adequate essential nutrients. Hay and and Straw are character-ised by a high fibre content.

Fibrous root system. Root system in which primary and lateral roots develop almost equally and have a limited quantity of cortex.

Fibrous. Having long narrow shreds or flakes.

Ficundity of insects. The average numbers of eggs laid by an insect has been from 100-200 but in some particular cases, especially the social insects, there may be more than this, e.g., a queen may lay 2,000 eggs a day; a queen termite may lay 60 eggs a minute with total of several millions.

Fiddle. A small portable device carried on the shoulders for broadcast sowing, especially grass and clover seeds in hilly areas. It comprises a hopper which slowly drops seed on to a ribbed disc which the operator causes to rotate and throw the seed by pushing and pulling an attached 'bow'. Very rarely used.

Fiddlehead. An unfurling fern frond that resembles the end of a violin.

Field bindweed. Perennial with very deep roots related to the morning glory. Located throughout the state. Has a showy trumpet flower. Takes over areas with low water use such as the edges of fields. A problem in central and northern orchards.

Field capacity zone. Buffer zone on which water has been held against gravity under free drainage. It is this zone which does not allow the moisture and consequently the salts from moving in a vertically upward direction.

Field capacity. The state of soil when all the soil moisture that is able to drain away freely has done so. The remaining moisture is held by the forces of surface tension around soil particles and in capillary pores (Capillary Water).

Field corn. Any variety of corn that is grown extensively in large fields primarily for livestock feed, as contrasted with the horticultural varieties, such as sweet corn or popcorn. Most field corn is of the dent variety.

Field crops. Herbaceous plants which are grown in cultivated field under more or less extensive system of culter.

Field duty of water. Quantity of water which is given to individual fields or orchards under an irrigation project. It has been ex-pressed in terms similar to those for farm duty of water.

Field efficiency (Plough). Refers to the ratio of effective field capacity, and theoretical field capacity. It is expressed in per-centage.

Field gleaning. The collection of remaining crops from farmers' fields that have already been mechanically harvested or from fields where it is not economically profitable for the farmer to harvest.

Field irrigation requirement of a crop. Means water which is required by crops exclusive of rainfall and contribution from soil profile.

Field moisture deficiency. Quantity of water needed to restore the soil moisture content to field moisture capacity.

Field moisture equivalent. Minimum moisture content which is expressed as a percentage of the weight of the oven-dried soil at which a drop of water kept on a smoothed surface of the soil will not immediately be absorbed by the soil but will spread out over the surface and give it a shiny appearance.

Field moisture. Refers to the per cent moisture content in a field sample of soil at any one time.

Field office technical guide. A manual placed in all Natural Resources Conservation Service district offices and field service centers that gives the technical specifications and guidelines for all approved conservation practices.

Field room. The time needed for a crop stooked (Stock) in a field to dry sufficiently for stacking.

Field service agency. Generally refers to any one of the following USDA agencies that administer programs and provide services to farmers and other rural residents through an extensive network of state and local offices: the Farm Service Agency, Risk Management Agency, Natural Resources Conservation Service, Rural Housing Service, Rural Business-Cooperative Service, and Rural Utilities Service. The Foreign Agricultural Service, because of its overseas offices, also is considered a field service agency under the Administrative Convergence plan being developed by USDA in 1998. Although other USDA agencies and mission areas also have field offices nationwide and overseas, they generally are not considered field service agencies by the Department.

Field test. An evaluation test whereby the field performance of experimental plants, their response to inputs and physical factors etc. are assessed in comparison to control.

Field waste. That portion of the irrigation water which has not been absorbed or evaporated as it passed over the irrigated land.

Field. An area of land, usually enclosed by a ditch, fence, hedge, etc.. for cultivation or grazing purposes. Fields are usually sown to a single crop at one time. In recent years the trend has been the removal of field divisions (*e.g.,* hedges) to facilitate cultivation and increased efficiency, and the larger fields created are sometimes termed management units. Such larger fields also allow cropping flexibility and sometimes several different crops are grown in the field at the same time.

Fieldman. A farmworker, the major part of whose work is in the fields, as distinct from around and in the farm buildings.

FIFRA. Federal Insecticide, Fungicide, and Rodenticide Act.

Filament. (1) The slender stalk or stem of the anther or pollen sac. (2) The stalk of a stamen or male flower part.

Filbert. The nut of the cultivated Hazel.

Files. A general term for winged insects, but particularly applied to the two-winged insects of the order Diptera, the mouth parts of which form a proboscis for piercing and sucking, especially blood. Many are important in agriculture, causing or transmit-ting disease to plants and animals. Some are parasitic (*e.g.,* Cleg, Gadfly, Ked, etc.) whilst the Larvae of others attack livestock and crops (*e.g.,* Blow Fly, Crane Fly, Frit Fly, Warble Fly, etc.). Clouds of flies, especially when biting, make cattle very restless.

Filiform. Thread-like, long and very slender.

Filler plants. typically have compact, upright growth and round out the top of the container.

Filler. A tree temporarily growing in between fruit trees in an orchard until they mature and require the space.

Film yeast. Type of yeast (microscopic fungs) growing on the surface of certain nutrient liquids and producing heavy folded films. Such yeasts are widely distributed in nature. They are generally found in fermentation process on the surface of sugar containing solutions, pickle brines, or other acid and sugar containing liquids. These yeasts generally grow in very high concentrations of salt, acid, and sugar.

Filter (bacterial). The term used for a special type of filter through which bacterial cells fail to pass. A filter is used for separating microorganism from a gas or liquid in which they become dispersed. Filtration is a convenient method to sterilise heat labile liquids and solutions which cannot be autoclaved. This is carried out merely by passing the virus tree liquid through a suitable filter.

Filter strip. An area of vegetation, generally narrow and long, that slows the rate of runoff, allowing sediments, organic matter, and other pollutants that are being conveyed by the water to be removed. Filter strips reduce erosion and the accompanying stream pollution, and can be a best management practice.

Filterable virus. Microorganism which is capable of passing the pores of microbiological filters (or a bacterial filter).

Fingers. Units with flat faces (known as ledgers projecting from and bolted to the cutter bar of a Binder, Mower or Combine Har-vester, across which the knives slide backwards and forwards and cut the grass blades or crop against the fingers.

Finish. An Open Furrow being the last one ploughed in a field.

Fire blight. A notifiable Disease caused by bacteria *(Erwinia amy-lovora)*, mainly of pears and apples, giving a scorched appear-ance to the fruit. Spread by insects.

Fire curing. The term used for the process of curing tobacco specially for jaffina tobacco used for chewing purposes. In this process the leaves are wilted for a few hours in the field tied into bundles and hung on laths in a smoke hut then smoked for 12 hours by burning.

Fireblight. A serious bacterial disease that kills members of the rose family.

Firebreak. An existing barrier which is natural or, otherwise, or one prepared before a fire takes place from which all or most of the inflammable materials have been removed. It is designed to stop light surface fires and to serve as a line from which to work and counter-fire if necessary.

Fire-calf Heifer. A term used in some areas for a Heifer after having borne its first calf but before it has borne a second calf

Fireplace. Affecting Flame Colors. Colors can be added to a burning fireplace with the addition of specially treated pine cones, blocks of wood or kindling, wood chips, or even one-inch thick rolls of tightly wrapped newspaper. Materials are soaked in the solution and then dried. The following chemicals will give off the color listed. Most can be purchased from chemical supply companies, drugstores and pharmacies.

First cross. The progeny from the mating of a male and female from separate pure breeds of animals or varieties of plants.

Fish and Wildlife (FWS). FWS, in the Department of the Interior, is the federal agency charged with managing and protecting the nation's wild plants and animals, including endangered and threatened species. It generally works closely with state agencies, which have management primacy for most species. (The federal government has assumed responsibility for marine mammals, migratory birds, and endangered and threatened species). It manages the National Wildlife Refuge System, and cooperates with private landowners in habitat conservation.

Fish farming. Usually, freshwater commercial aquaculture; catfish farms are an example. The breeding and rearing of fish in tanks, ponds, cages, etc., ultimately for harvesting and slaughter for the table.

Fish gauno. A phosphatic fertilizer made from unmarketable fish and fish waste products after the extraction of variable composition. The nitrogen content is usually 7-8% and the phosphoric acid content 4-8%. Usually only used for horticultural purposes. Also called fish meal.

Fish ladder. An inclinded flume which is carrying water from above to below a dam, at a velocity against which fish can easily swim.

Fish leaves. In tea, the first two leaves formed on a shoot as it grows from a bud.

Fish manure. The unprocessed offal from gutted fish and waste or unsold rotting fish, spread on fields as Manure.

Fish meal. An animal Feedingstuff consisting of dried and ground waste or unsold fish and fish filleting residues, by law containing not more than 6% oil and 4% salt. White fish meal is derived from non-oily 'white fish (*e.g.*, cod, haddock, etc.) and usually contains about 66% protein. Other fish meals are derived from oily fish (*e.g.*, herring, pilchards, etc.) after most of the oil has been extracted and are usually richer in protein than white fish meal. Fish meal is mainly fed to pigs and polutry, but is also included in rations for dairy cows, calves and other farm ani-mals. Fish Guano.

Fish pond fertilizer. Inorganic or organic matter which is applied to a fish pond for promoting the growth of plankton and aquatic plants on which the fish depend directly or indirectly for food.

Fish pond. (1) Pond having fish (2) Pond natural or artificial where fish life has been supported by the sewage discharged therein.

Fish screen. (1) Refers to a screen which is placed across the head of, or in, an intake canal or pipe line, to prevent fish from entering. (2) Barrier placed across the inlet or outlet of a pond for prevent-ing the passage of fish.

Fishway. Structure with pools and drops which is usd to facilitate the migration of fish around dams or other obstructions in streams. They are needed on many streams where fish periodically migrate upstream to the head waters to spawn, the young fish coming downstream again. One form is known as fish elevator, another form fish ladder.

Fissure. Bark that is furrowed and ridged, or splitting lengthwise

Fissured. Torn lengthwise, in vertical furrows, as applied to bark, or to pith, for which the more general term spongy is used.

Fives. A term for samples taken from hop Pockets (for assessing type) whose identity numbers end in 5 (*i.e.*, 5th, 15th, 25th, etc.). Samples are thus taken from one pocket in ten.

Fixation (photosynthesis). Process of producing carbohydrates by adding hydrogen to carbon and oxygen.

Fixation of nitrogen. Refers to the fixation of atmospheric nitrogen in the soil by symbiotic and non-symbiotic bacteria.

Fixation of phosphorus. Conversion of soluble phosphorus nutrient in the soil into less soluble and unavailable forms.

Fixative. Reagent which will bring about fixation.

Fixed ammonium. Ammonium which is held in a soil and is neither water soluble nor easily exchangeable.

Fixed equipment. In general terms anything on a farm, excluding growing crops, which cannot readily be moved.

Fixed ground water. Ground water which is held in saturated material with interstices so small that it gets permanently attached to the pore walls or moves to slowly that it is usually not available as a source of water for pumping.

Fixed moisture. Moisture which is held in the soil below the hygroscopic limit.

Fixed phosphorus. (1) *Phosphorus.* It is changed to less soluble form because of reaction with soil; moderately available phosphorus (2) *Applid Phosphorus.* Phosphorus which has not been taken up by plants during the first cropping year (3) *Soluble Phosphorus.* Phosphorus which gets attached to the solid phase of soil in forms highly unavailable to crops (4) *Unavailable Phosphorus.* Phosphorus other than readily or moderately available forms.

Fixed potassium. Potassium which is held by a soil and is neither water-soluble nor readily exchangeable.

Flae beetle. A small dark beetle *(Phyllotretra sp.}* which causes extensive damage to Brassica seedlings, eating small holes in the cotyledonous leaves (Cotyledon), especially during hot dry weather.

Flail hedger. A type of Hedge Cutter commonly used now-a-days for regular trimming. They are usually driven hydraulically or mechanically by p.t.o. from a tractor and consist of a high-speed cylinder bearing free swinging flails.

Flail. (1) A hand implement used to thresh corn, comprising a wooden beating bar or swingle, hinged to or loosely attached (often by leather) to a handle. Also called a dreshel or a Stick and a Half. (2) One or more free swinging arms on certain machinery *(e.g.,* Flail Forage Harvester, Flail Hedger) which, when rotated at speed, sever vegetation.

Flaked maize. A highly digestible, starch-rich feedingstuff derived from maize which has been steam treated, rolled and dried. It has a relatively low oil content and is often given to pigs.

Flash pasteurisation. Term includes all short time heat treatment methods which are not scientifically controlled. It was formerly used for milk destined for manufacturing purposes, but has now been almost entirely replaced by the accurately controlled HTST process.

Flat grafting. Grafting in which smooth transverse cut is made on both scion and stock (as in cactus) and the flat surfaces are held firmly in position by thread or raffia.

Flat rate feeding. A system of feeding by which cows are fed a standard fixed quantity of concentrate per day, from calving to turnout, thereafter depending on grass alone. A variation com-monly used with autumn calving cows is stepped feeding, in which the level of concentrate fed to groups of cows in the herd is reduced in one or two steps through the winter, with concen-trate feeding being discontinued on turning out to grass. The success of these systems depends on the cows having unre-stricted access to good quality forage, usually grass silage. If necessary, the silage intake can be reduced by supplementation with other suitable feedingstuffs, *e.g.,* Brewers' Grains, Dried Grass.

Flat. Shallow tray made of wood, plastic or metal with drainage holes for germinating seeds or rooting of cuttings.

Flax. Varieties of the plant *Linum usitatissimum,* selected and grown for their long stems rich in fibre, used in linen manufacture.

Fleam. A veterinary instrument for bleeding cattle.

Fleshy fruits. Class of fruits having soft and fleshy pericarp at maturity; includes berry, pepo, hesperidium, drupe and pome.

Fleshy root. A root that accumulates and stores a rich supply of reserve food for the plant.

Fleshy. For a stem, not hard and woody; for a fruit or bud scales, not dry. Succulent.

Flex acreage. The Omnibus Budget Reconciliation Act of 1990 mandated that deficiency payments not be made on 15% of a farm's crop acreage base, called normal flex acres.

The acreage could be planted to any program crop (called flexing), but not fruits and vegetables. An additional 10% of the farm's base acreage could be flexed at the option of the operator. Flexing did not diminish the crop acreage base of a farm. The FAIR Act of 1996 effectively provides total flexibility among all commodities, except for fruits and vegetables.

Flocculation. The aggregation of small particle to form larger ones. Particularly applied to soils and the formation of crumbs (De-flocculation).

Flood plains. Lowland and relatively flat areas adjoining inland and coastal waters, including floodprone areas of islands. This land includes, at a minimum,, those areas that are subject to a 1 percent or greater chance of flooding in any given year.

Flora. (1) The plant life of a locality or of a geological era. (2) Plants taken collectively. (3) A book in which all the plants from a particular area are described. (4) The plant population of an area or country. Also a botanical ordering of those plants together with descriptions for identifi-cation.

Floral foam. A water absorbing material (a substitute for pin holder or wire mesh) used in floral decoration to keep the cut flowers in place.

Floral preservative. Chemical used to prolong the vase life of cut flowers.

Floral primordium. Very early stage of flower bud.

Florescence. Refers to anthesis or flowering time, the state of being in bloom.

Floricane. A flowering and fruiting stem of a bramble, the season after it was produced (in the season of production, it is called primocane).

Floriculture. The art and science of growing and marketing flowering and foliage plants. The study of producing and marketing fresh flowers.

Florida. With abundant flowers.

Floriferous. Having many flowers compared to most cultivars or species.

Florigen. Hypothetical hormone like substance implicated in floral initiation in many plant species.

Florist. One who is involved in the procurement and selling of flowers.

Flotation. A method for separating a valuable material from a less valuable material by placing them in a liquid in which one will sink and the other will float.

Flower blasting. Flower bud abortion after the completion of flower differentiation.

Flower bud initiation. Formation of floral primordium.

Flower bud. An unopened flower.

Flower differentiation. Complete morphological differentiation of the floral organs following flower bud initiation.

Flower induction. An invisible developmental process that occurs prior to visible flower bud initiation.

Flower. The angiosperm reproductive structure which may include calyx, corolla, androecium and gynoecium.

Flowering Shrubs — As Cut Flowers. To enjoy flowering shrubs as cut specimens, prune in the early morning, remove all foliage beneath the water line. Lightly pound stems with a hammer or mallet to aid in water absorption. Place in a tall container and place in cool, dark location for longest life. A floral preservative can be used to prolong the cut limb's life.

Flue dust. A fertilizer containing 5-15% Potash derived from indus-try, mainly as waste products from iron and steel manufacturing processes.

Flue-cured tobacco. A type of cigarette tobacco, it and burley tobacco account for more than 90% of U.S. tobacco production. Flue-cured tobacco production is limited by national marketing quotas and acreage allotments, and is eligible for nonrecourse price support loans. Flue-cured production is centered in North Carolina.

Fluorosis. Also known as fluorine poisoning. Chronic and insidious disease of entile which occurs due to prolonged ingestion of fluorine. Characterized by stunted growth, lameness, mottled and irregular teeth, thickening of jaws and hones of limbs.

Flush. (1) A sudden, generally rapid growth of foliage, usually applied to grass. (2) To give an animal a richer diet prior to mating in order to improve its condition.

Flute budding. Refers to a method of budding where in the bark is removed from the root stock almost completely encircling it, leaving a narrow connection (about 1/8 of the circumference) between the upper and lower part of the slock. It is replaced by a similar bark having an active bud from a desired variety. The narrow connecting strip of bark on the stock keeps the top of the stock alive, if the bud patch does not unite.

Fly wool. Pieces of loose wool that fall from a fleece during shearing. (Wool Oddments).

Flying bent. A coarse tussock-forming grass found in wet habitats on Moorland, Heath, Fens and Marshes, with purple flowers eaten only by sheep. Also called Purple Moor Grass or Purple Melick-Grass.

Flying flock. A flock of sheep temporarily imported onto a farm, normally for less than a year, and then sold out.

Focus. The point or area where the eye is drawn first. Place the focus below the tallest point to achieve balance. Develop focus by using large, coarse, or bright colored plant material in that area. All plant material should radiate out from the focus.

Fodder beet. A root crop of the genus *Beta* which also includes sugar beet and mangel, usually grown following cereals and used to feed stock. It is intermediate between mangel, and sugar beet, having a higher Dry Matter and sugar content but smaller root than mangel, but a lower dry matter and sugar content and larger root than sugar beet. Fodder beet may be classified into medium dry matter (14-18%) and high dry matter (19-20%) varieties.

Fodder crop. A crop grown for use as animal feed, *e.g.*, Hay Kale, Rape, Lucerne, Trifolium, etc., and usually consumed in the green state. Also called green crop or forage crop.

Fodder. Food supplied to livestock, particularly dry roughages such as Hay and Straw. Sometimes used loosely to mean Forage.

Fog. Pasture grasses allowed to grow during late summer and autumn, providing winter grazing for sheep and cattle. Land left with such grass on it. Also to graze such grass in winter.

Foggage. Synonymous with Fog. Also used to describe rough coarse grasses of mountain and hill areas and Rough Grazing.

Fold, folding. An enclosure or pen for sheep, usually movable and made with hurdles or wire. Usually used to enclose sheep (and also pigs) in a field with a growing arable crop (*e.g.*, turnips, kale), moved each day to another part of the field or enlarged once the crop in that area has been consumed. This practice is termed folding.

Foliage plant. Any plant grown primarily for its foliage, utilized for interior decorations or landscaping.

Foliar diagnosis. Estimation of mineral nutrient deficiency or excess by analysing the chemical composition, colour and/or growth characteristics of the foliage.

Foliar embryos. Vegetative embryos developed at the edge of leaves, capable of growing into new plants

Foliar feeding. (1) Feeding plants with nutrient solution through the leaves. (2) Fertilizer applied in liquid form to the plants foliage in a fine spray.

Foliar nutrient. Any liquid substance applied directly to the foliage of a growing plant for the purpose of delivering an essential nutrient in an immediately available form.

Foliar. Of or concerning foliage or leaves

Follicle. (1) A small dry fruit opening down one edge. Ex. Columbine, milkweed. (2) Simple dry dehiscent fruit having one locule which splits along one suture.

Followers. (1) Young cows in a dairy herd not yet in milk and still maturing to replace their mothers. (2) Cattle put out to graze a pasture more thoroughly after a dairy herd has taken the best grass. (3) A banana sucker which has been allowed to grow for bunch production.

Food conversion ratio. The number of kgs of food consumed by an animal required to produce a liveweight gain of 1 kg. A measure of the efficiency of the animal in converting food into flesh, a small ratio indicating high efficiency. Also called feed conver-sion ratic.

Food grain. Cereal seeds most commonly used for human food, chiefly wheat and rice.

Food insecurity. Life Sciences Research Office (LSRO) defines food insecurity as when the availability of nutritionally adequate and safe foods in socially acceptable ways is limited or uncertain.

Food pantry. A service that collects and distributes unprepared food and grocery products to the needy.

Food poisoning. General term which is applicable to all stomach or intestinal diturbances due to food contamination with certain micro-organisms or their toxins.

Food power. The act of withholding or making available agricultural commodities for export or aid by an exporting nation or group of nations for the purpose of influencing the actions of another country or group of countries. Food power implies a foreign policy motivation rather than a financial or humanitarian motivation to export activities.

Food recovery. The collection of wholesome food for distribution to the poor and hungry. This includes field gleaning; perishable food rescue or salvage; the collection of perishable produce from wholesale and retail sources; food rescue, the collection of prepared foods from the food service industry; and nonperishable food collection, the collection of processed foods with long shelf lives.

Food security. Access by all people at all times to enough food for an active healthy life. Food security at a minimum includes the ready availability of nutritionally adequate and safe food, and an assured ability to acquire acceptable foods in socially acceptable ways, that is, without having to resort to emergency food supplies, scavenging, stealing, or other coping strategies. The World Food Summit, convened in Rome in November 1996 by the Food and Agriculture Organization of the United Nations, estimated that 800 million people worldwide do not have enough food to meet their basic nutritional needs. Representatives of the more than 180 nations attending the Summit pledged to work to reduce this number by half by no later than 2015. Causes of food insecurity may include poverty, civil conflict, governmental corruption, environmental degradation, and natural disasters. A U.S. position paper on international (world) food security, released in October 1997, argues that food security also requires "...social and economic conditions which empower individuals to gain access to food, either by producing food themselves or earning income to buy food."

Food unit. 1 kg of average barley, equivalent to 0.71 kg starch equivalent. Also called a fodder unit.

Foot rot. A chronic wound-infection which has been characterized by extreme lameness and frequently by distortion of the hoof. It occurs due to irritation by mud which dries and cracks the skin, by sand and by gravel between the claws. Animals badly affected may walk on 3 legs.

Foot-and-mouth disease (FMD). A major disease of cloven-footed animals (e.g., cattle and pigs) that does not exist in the United States. The Animal and Plant Health Inspection Service conducts a surveillance program to track the disease in foreign countries, regulates the importation of animal products from countries where FMD exists, and tests imported animals in quarantine.

Forage blower. A machine used to fill a Silo, particularly of the tower type, with finely chopped material for making Silage, by blow-ing it up a duct, often to a height of 15 m (50ft) into the silo. Some kinds of forage blower also chop the crop, these being called cutter blowers.

Forage box. A large mobile container, usually with wheels, similar to a Dump Box, used to transport Forage from a Silo to a Manager. The forage is discharged in a steady even flow, usually by a transverse moving floor mechanism or elevator belt, driven by p.t.o. from a tractor. Also sometimes used to transport forage from the field to managers for herds under Zero-Grazing.

Forage crop. Crops which are grown primarily for livestock feed, to be either harvested for hay, silage or green feed or harvested by grazing animals.

Forage harvester. A machine which cuts, chops and loads (by chute) green crops *(e.g.,* grass, lucerne, etc.) into an adjacent or towed trailer, for subsequent silage making. Sometimes the crop is pre-cut and allowed to wilt in windrows before being collected by a pick-up reel substituted for the cutter mechanism. There are three main types, viz., Double Chop Harvester, Flail Forage Harvester and Metered Chop Harvesters. They are mainly trailed or semi-mounted but self-propelled machines are available.

Forage Value Index (FVI). A derived index of the relative change in the previous year's average monthly rate per head for pasturing cattle on privately owned land in the West. Used in calculating federal grazing fees.

Forage. (1) Herbaceous plants or plant parts fed to domestic animals. (2) Vegetable matter, fresh or preserved, that is gathered and fed to animals as roughage includes alfalfa hay, corn silage, and other hay crops.

Forbs. In a prairie or meadow, the flowering plants that are not grasses and sedges.

Forcing. The manipulation of environment factors for producing a marketable pot plant or cut flower, especially during off season. The process of hastening a plants growth to maturity or bloom, stimulation of flowers or growth by controlling light and temperature.

Foremalin. An aqueous solution of formaldehyde used as a sterilant, mainly of glasshouse and potting soils and implements, and occasionally used as a disinfectant.

Forest health. A term used for a collection of concerns over the alleged deterioration in forest conditions, including both current problems (e.g., insect and disease infestations, wildfires, and related tree mortality) and risks of future problems (e.g., too many small-diameter trees (overstocking), excessive biomass, and an unnatural mix of tree species in mixed stands).

Forest plans. Land and resource management plans for units of the National Forest System under the National Forest Management Act. The Act specifies a detailed process and numerous requirements, including public participation and periodic revision, intended to achieve multiple use in the national forests.

Forest. A large area of uncultivated land covered by trees and underwood. Blocks of woodland in excess of 1000 acres are classed as forest by the forestry commission. The original meaning was unenclosed woodland or open, mainly treeless, areas reserved for hunting, usually belonging to the Crown, *e.g,* the New Forest. A plant community, predominantly of trees and other woody vegetation, usually with a closed canopy. A forest is an area with a high density of *trees* (or, historically, a wooded area set aside for *hunting*). These plant communities cover large areas of the globe and function as *carbon dioxide sinks*, animal *habitats*, *hydrologic flow modulators*, and *soil* conservers, constituting one of the most important aspects of the *Earth*'s *biosphere*. Forests can be found in all regions capable of sustaining tree growth, at altitudes up to the *tree-line*, except where natural fire frequency is too high, or where the tree growing environment has been impaired by natural processes or humans. Forests sometimes contain many tree species in a small area (e.g. tropical rain and temperate deciduous forests), but other forest types have relatively few species over large areas (e.g. taiga and arid montane coniferous forests). As a general rule, forests dominated by *angiosperms* are species-rich, while those dominated by *gymnosperms* are not so rich, although exceptions do exist (e.g., species-poor aspen and birch stands in northern latitudes). Forests are often home to many animal and plant species, and *biomass* per unit area is high compared to other vegetation types. Much of this biomass occurs below-ground in the root systems and as partially decomposed plant *detritus*. The woody component of forests contains *lignin*, which is relatively slow to *decompose* compared with other organic materials such as *cellulose* or carbohydrate. Forests are differentiated from *woodlands* by the extent of *canopy* coverage: in a forest the branches and foliage of separate trees often meet or interlock, although there can be gaps of varying sizes within an area referred to as forest. A woodland has a more continuously open canopy, with trees spaced further apart, which allows more sunlight to penetrate to the ground between them

Forestland. A classification of land use in the Natural Resources Inventory (NRI). It includes areas where trees cover at least 10% of the land and must be at least an acre in size. Forestland was found on 395 million acres, almost 30% of all private lands, in the 1992 NRI.

Forestry. (1)The science of developing, caring for, or cultivating forests; the management of growing timber. (2) Management of forest for maximum sustained yield of their resources and benefits. (3) The science or management of forests. (4) The science, art, and practice of managing and protecting tree and forest resources for human benefit.

Forfeiture penalty. A penalty paid to the Commodity Credit Corporation by a processor of sugar beets or sugarcane who, having taken out a nonrecourse loan, decides to hand over sugar pledged as collateral to the CCC rather than accept the then-market price (see loan forfeiture). The penalty is 1 cent/lb. on raw cane sugar, 1.072 cents/lb. on refined beet sugar. Some view this penalty as lowering the price support levels authorized by the FAIR Act of 1996 by the penalty amount (i.e., for raw cane sugar, from 18 cents/lb. to 17 cents/lb.).

Forkert budding. A modification of patch budding where the lower portion of the patch of the rootstock is not cut but retained as a flap. The bud patch is inserted in the portion and the flap is drawn over the inserted bud, followed by wrapping as usual.

Form, pl. formae. A minor variation within a species that is sufficiently distinctive to warrant recognition as a separate taxon. In cultivated plants, it is current practice to name these forms as cultivars.

Form. (1) A botanical variation of a variety differing in only one trait, such as color. A subdivision of species that occurs occasionally in the wild, seldom breeds true, and does not develop a natural population or distribution. (2) A member of a population that differs from the others to a degree not great enough that it can be called a cultivar. (3) A subdivision of species which occurs occasionally in the wild, seldom breeds true, and does not develop a natural population or distribution. (4) Vary the form of the plant material you choose; use tall linear species to add height; mounded species to add mass; and low growing, cascading species to fill in, add depth, and soften the edges of the container.

Formula funds. Federal dollars distributed to the land grant colleges of agriculture through formulas found in the Hatch Act, the Smith-Lever Act, the McIntire-Stennis Act, and the Evans-Allen Act for (1) agricultural research at the state agricultural experiment stations, (2) extension programs and (3) forestry research at the land grant colleges of agriculture, and (4) research at the 1890 institutions, respectively.

Formula pricing. An arrangement where a buyer and seller agree in advance on the price to be paid for a product delivered in the future, based upon a pre-determined calculation. For example, a packer might agree to pay a hog producer the average cash market price on the day the hogs will be delivered, plus a 2-cent per-pound premium. Such transactions have been used widely in agriculture, particularly for livestock. Users believe that formula pricing brings efficiency and predictability to market transactions. However, as the use of formula pricing expands, fewer animals are sold in cash markets, where prices are more widely reported and understood by producers. Some of these producers believe that formula pricing makes it harder to determine the true value of their animals in the marketplace, and creates greater opportunity for buyers to manipulate and pay lower prices.

Formula-based tariff reductions. A method of negotiating tariff reductions using an agreed-upon formula applied to tariff rates (with limited exceptions being granted for very sensitive items) by all contracting parties.

Forward. A descriptive term applied to both livestock and crops that are more advanced in their development at a particular time than normal.

Foul brood. A bacterial disease *(Bacillus sp.)* of bees which kills the larvae and causes their decomposition.

Foul-in-the-foot. A bacterial infection (usually *Actinomyces necro-phorus)* of the hooves, particularly of the hind feet, of cattle, normally around wounds or cuts. Swelling, heat, pus production and lameness occur. Muddy fields and gateways are particular sources of infection.

Foundation plant. A plant suitable for planting around houses and below windows.

Foundation planting. Planting of trees, shrubs or ground covers in front of a building foundation.

Four-firm ratio. It is common to express the degree of concentration within an industry, including agriculture (that is, the degree to which a few firms dominate sales or production) as a ratio, by stating the share (%) held by the top four firms.

Fowl cholera. A contagious bacterial disease *(Pasteurella apisep-ticus)* of poltry characterized by sudden high fever and profuse green diarrhoea.

Fowl pox. A viral disease of fowls causing warts on the comb, wattles and other parts of the head.

Foxy hops. Over-mature dried hops with a reddish-browncolour. The colour may also be caused by disease or decay.

Fragmented. Not continuous, as applied to bundle scars.

Frame harrow. A type of harrow with an articulated sectional frame and no wheels, providing flexibility to follow the contours of the land.

Frame work. The 'skeleton' of main branches of a tree or shrub.

Frass. Wood fragments mixed with excrement produced by insect larvae.

Free clones. Clones that can be purchased and resold without any agreement or prohibition.

Free market. A system in which the market forces of supply and demand determine prices and allocate available supplies, without government intervention. The concept of a free-market approach in agricultural policy, in its purest form, is no government price and income support programs, supply management programs, export subsidies, or barriers to international trade.

Free pollination. Pollination without control. The offsprings are of unknown parentage.

Free rider. In agricultural policy, the term generally refers to a firm or person who benefits from a collectively funded activity (such as a generic advertising and promotion, or check-off, program) without contributing to its costs.

Free stocks. Commodity stocks owned by farmers or others in the trade, rather than by those owned or controlled by the government. (Supplies in the Food Security Commodity Reserve are government-controlled and not considered free stocks.).

Free water. Water released by the cell when freezing occurs in intercellular spaces.

Freedom-to-farm. A phrase that was used in the congressional arena to characterize the production flexibility contract provisions of the FAIR Act of 1996.

Free-grazing. The most common method of grazing stock with few, if any, controls. Stocking rate is usually adjusted to the supply of herbage but over-and under-grazing often occurs. Fencing costs are kept to a minimum and intensive stock supervision is avoided. (Grazing systems).

Free-martin. A heifer calf born as a twin with a bull-calf, usually sterile.

Freeze-drying. A method of food processing where the material is initially frozen at super low temperature and then vacuum dried.

Freezes and Frosts. A freeze occurs when both the surface temperature around a plant and the air temperature are 32 degrees Fahrenheit or below. A frost occurs when the surface temperature around a plant is 32 degrees Fahrenheit or below. The air temperature is above 32 degrees F.

Freezing agents. Substances used for chilling foods.

Fresh weight (Wet weight). The weight of a plant or plant part including the water content.

Fringed. Ciliate with glands or scales rather than fine hairs.

Frit fly. A small shiny black fly *(Oscinella frit)* that lays its eggs on oat seedlings and tillers, which are then attacked by the larvae. The next generation of files lay eggs in the glumes and the larvae attack the developing grain. Wheat and maize may also suffer from frit fly attack.

Fritillaria. This group consists of about 100 deciduous, perennial bulbs native to the Northern Hemisphere. They are suitable for growing in borders, meadows, woodland gardens, and containers. Some species grow up to 4 feet tall, while others average 10 inches in height. In the spring, all Fritillaries produce beautiful, cup- or bell-shaped, nodding flowers, with a shiny yellow receptacle of nectar near the inner base of the petals. The delicate flowers are curiously blotched or checkered with various colors such as shades of brown, green, and white. F. imperialis (Crown Imperial) is the largest and most popular Fritillary. The bulbs produce a whorl of shiny leaves from which a 2- to 4-foot stem arises. A cluster of flowers, which may be red, orange, or yellow, top the stems. Each flower is crowned with a compact tuft of small leaves. The bulbs and flowers of the Crown Imperial have a skunk-like odor, which some people may find offensive. F. meleagris (Checkered Lily; Meadow Fritillary; Snake's Head; Guinea-Hen Flower) is a lovely species producing 1 to 3 interesting, checkered and veined flowers per each slender stem. The blossoms range from dark brown to rosy lilac, wine, or white. This unique species is suitable for growing in moist, woodland gardens, meadows, and borders. F. persica (Persian Fritillary) produces as many as 30 dark violet to reddish-purple flowers per each 2- to 3-foot stem. F. atropurpurea (Leopard Lily) grows up to 16 inches high with a few slender leaves, up to 3 inches in length. One to four flowers are produced per stem. They may be green edged in yellow with purple spots or purplish-brown spotted with white or yellow.

Frond. (1) A leaf, once applied only to the leaves of ferns but now also said of the leaves of palms too. (2) General term designating leaf of coconut, oilpalm, fern etc. (3) The term used to describe the branch and leaf structure of ferns or palms.

Frost damage. Damage caused by low temperature.

Frost day. Day with a minimum temperature less than 0°C.

Frost free dates. The average date of the last frost in Central Illinois is April 15. This doesn't mean it is a frost-free date. There is a 50 percent or less chance of frost on this date. The odds of a frost drop an additional 10 percent for each week past this date. In central Illinois, the FROST FREE date is May 12. Add approximately 1 week for each 100-120 miles north of Springfield and subtract 1 week for each 100-120 miles south of Springfield.

Frost injury (freezing injury). Injury or killing of plant tissues caused by exposure to low temperature and consequent formation of ice.

Frost pocket. A depression in the terrain into which cold air drains but from which it cannot escape, making the plants of the area prone to freezing injury.

Frost protection. The following methods can be used to limit frost injury. However, protection for temperatures below 25 degrees Fahrenheit may not be possible. Sprinkling with water. It takes more cold energy for water to freeze. Maintaining a sprinkler running in the early spring may limit cold injury to strawberry flowers and tender transplants. Water in the fall can also limit cold injury.

- Smudge pots. Pots generate localized heat. Small burning piles, where allowed, may provide similar effects. Check for dry grass and plant material nearby; even green plants may burn.
- Fans. Still air cools faster than moving air. Cold settles; circulate air near the ground.
- Covers. Old sheets, blankets and light rugs may limit cold injury to annual flowers and vegetables. Avoid plastic. Plastic will transfer the cold to the plant and provides little insulating capabilities. Leaving air spaces between plants and covers increases insulation. Avoid heavy materials which might damage plants or limbs.

Frost. Minute crystals of ice collected or deposited on ground or plants when the nights are cold and temperature falls below freezing point.

Fructification. (1) The act of producing fruit. (2) The production of a reproductive structure.

Fruit buds. Buds containing underdeveloped flowers; also referred to as flower buds.

Fruit confections. General term used to describe candies in which fruits are used.

Fruit drop. Any substantial dropping of immature fruits.

Fruit fly. (1) Small yellow-brown flies *(Drosophila sp.)* which feed on fruit (2) Large flies, *e.g.*, celery fly, gall flies, etc., whose larvae feed on fruits.

Fruit growing. The cultivation of edible fruits that are consumed either fresh or processed

Fruit juice concentrate. A fruit juice from which water has been mostly removed by heating or freezing.

Fruit juice powder. Fruit juice converted into a free flowing, highly hygroscopic powder by puff-drying, freeze-drying, vacuum drying, spray-drying or drum-drying.

Fruit preserve. Product made by cooking prepared fruit in sugar (sucrose) syrup until the concentration of sugar reaches 55 to 70 per cent. The fruit should be crisp and retain its form.

Fruit set. Persistence and development of an ovary and/or adjacent tissue following bloom.

Fruiting habit. The relative position of fruiting on the shoots of a plant (See Bearing habit).

Fruticose. Shrubby, in the sense of the stems being woody.

Fruticose/fruticosa/fruticosum/fruticosus. Shrubby or shrub-like, in the sense of the stems being woody. Often in a genus of smaller form.

FTE. Full-time equivalent (generally refers to agency staffing levels).

Fugacious. Falling off soon or fading away.

Full bloom. The plant at the period at which most of its flowers are blossoming

Full mouthed. A descriptive term for livestock in which a complete set of permanent teeth have grown, *e.g.*, after 3 years in sheep.

Full-cost water. An annual rate for water delivered from Bureau of Reclamation facilities, which includes project construction costs attributed to irrigation, as well as outstanding deficits on operation and maintenance charges, with interest on both accruing from October 12, 1982. The term is defined in Section 202 of the Reclamation Reform Act of 1982. The Bureau charges full-cost for water delivered to lands above the acreage limitation.

Fumigant. A vaporized pesticide used to control pests in soil, buildings and greenhouses, and chambers holding products such as fruits to be treated. Methyl bromide is an example.

Fumigate. To destroy bacteria, insects and other pests by exposing the place they are infecting *e.g.*, a farmbuilding, to poisonous gas or smoke (Disinfection).

Fungibility. The characteristic of interchangeability. Bulk commodities are generally described as fungible, whereas those with special characteristics may be marketed as identity preserved. Futures contracts for the same commodity and delivery month are fungible due to their standardized specifications for quality, quantity, delivery date and delivery locations.

Fungicide. (1) Any pesticide used to control, deter, or destroy fungi, which are forms of plant life (including molds and yeasts) that lack chlorophyll and are unable to make their own food (such as the plant pathogen, powdery mildew). (2) A chemical that kills fungi or prevents them from infecting healthy plant tissue. (3) A chemical used to destroy fungi and thus control fungal diseases. Any pesticide used to control, deter, or destroy fungi, which are forms of plant life (including molds and yeasts) that lack chlorophyll and are unable to make their own food (such as the plant pathogen, powdery mildew). (4) A chemical used to kill fungus, ills, diseases or viruses.

Fungus. One of a large group of plants including moulds, mushrooms and yeasts, characterized by lack of chlorophyll. They are either parasities, many of which cause crop disease (*e.g.*, Potato Blight, Mildew, Ergot, Rust) or Saprophytes which are impor-tant in the release of nutrients to the soil from dead plants and animals. They are mainly constructed of thread like hyphae and are classified into three groups, vis., Ascomycetes, Basidiomy-cetes and Phycomycetes.

Furrow irrigation. Small, shallow channels guide water across the surface of a leveled field. Crops are typically grown on a ridge or raised bed between the furrows. This is the major irrigation system that is based on gravity.

Furrow opener. Another name for a Coulter.

Furrow press. A heavy type of multi-ring roller usually trailed behind a plough, with wedge-shaped rims to some of the wheels (the press wheels) which compress the lower parts of the furrow slices. Sometimes used when preparing light land for wheat following a Ley. Also called press.

Furrow wheel. A wheel on a trailed plough which runs in the previously turned furrow.

Furrow. The groove or trench in the soil made by a plough as the mouldboard turns over the soil in a more or less continuous strip (the furrow slice). A completely ploughed field, before other cultivations are undertaken, is said to be 'in the furrow'.

Fusarium basal rot. Occurs in onions. Fungal disease that attacks bulbs and causes a watery rot.

Fusiform. Spindle-shaped; tapering to each end from a smaller mid-section.

Fym. An abbreviation for Farmyard Manure. Manure is *organic matter* used as *fertilizer* in *agriculture*. Manures contribute to the fertility of the soil by adding organic matter and *nutrients*, such as nitrogen that is trapped by *bacteria* in the soil. Higher organisms then feed on the *fungi* and bacteria in a chain of life that comprises the *soil food web*.

G

Galia or sand leaves. Lowermost 3 to 4 leaves of interior quality in tobacco plant. These constitute about 10 to 15 percent of the total bulk.

Gall. (1) An abnormal growth of plant tissue, usually in response to parasitic infection, mainly by a variety of Insects (e.g., certain wasps, saw-flies, midges, Aphids, and some moth and beetle Larvae) and Mites, the offspring of which are provided with food and shelter by the swollen tissue. Eelworms and Fungi can also cause galls to be produced. (2) A painful swelling, particu-larly on a horse, or a sore caused by rubbing. (3) A swelling on a leaf or stem caused by fungi or insects like wasps.

Gallon. A measure of volume or capacity, equivalent to 8 pints or 4.55 litres, used mainly for liquids but also as a dry measure for grain. The British Imperial gallon is equal to 277.3 cubic inches. The American gallon is equivalent to 0.83 British gallons.

Galls. A gall is an abnormal development or out growth of plant tissue resulting from an irritation produced by bacteria, fungi, or insects. They take on many shapes, sizes and colors. Abnormal warts, swellings, and knots on leaves, twigs and branches of trees and shrubs are usually galls. Most galls on leaves are caused by small insects or insect-like pests such as wasps, midges and mites. These creatures attack leaves as they are forming in the spring. Gall tissue is essentially tissue created by the plant in response to the injury. Removing the leaf galls in most cases will cause holes in leaves.

Gamma garden. Place where the plants or insect pests have to be treated with gamma rays which are emitted from the source.

Gammon. The thigh and adjacent parts of a side of bacon.

Garden agriculture. Cultivating crops like vegetables in land more or less adjacent to settlements, utilising organic wastes

Garden. A piece of land used for growing flowers, fruits. vegetables etc. Often used as a place for recreation.

Gardenia. Gardenias are popular shrubs and ornamentals in the southern part of the United States where severe cold won't damage the plants. Plants always seem to have more blooms at the time you receive them than any other time in their life due to less than favorable growing conditions in most homes. Gardenias thrive on 68-74 degrees F. day temperatures and 60 degrees F. evening temperatures. Plants prefer full sun indoors; if grown outdoors for the spring, summer and early fall, keep plants in partial shade. An east or covered west porch will be satisfactory. High humidity is essential. Avoid misting the foliage, though, as leaf spot fungal problems will develop. The soil should be kept uniformly moist, but don't overwater. A loose, well-drained organic soil is recommended. Fertilize monthly between April and November with an acid fertilizer. Check regularly for insects and other pests such as aphids, mealybugs, spider mites, thrips and scale. Follow recommended control practices if pest problems occur. The most irritating problem encountered is "bud drop" — flower buds will abort just before blooming. Common causes include low humidity, over watering, under watering, insufficient light, high temperatures, rapid temperature fluctuations, cold drafts, or change in plant locations. In other words, the plants are temperamental. Plants that don't set flower buds may be experiencing too much warmth.

Gardening. (1) Cultivation of plants in enclosed areas for ornamental purposes. (2) A plot of land used for the cultivation of flowers, vegetables, herbs, or fruit. (3) Gardens Grounds laid out with flowers, trees, and ornamental shrubs and used for recreation or display. Often used in the plural: *public gardens; a botanical garden.*

Gas chromatograph / Mass spectrometer. (1) Instrument which is used in gas chromatography to detect the presence of valatile compounds. (2) An analytical technique for identifying the molecular composition and concentrations of various chemicals in water and soil samples.

Gasohol. Registered trade name for a blend consisting of 90% unleaded gasoline and 10% fermentation ethanol. Gasohol emissions contain less carbon monoxide than those from gasoline.

Gathering. A method of systematic ploughing in which the tractor and plough are turned to the right each time they come out or go into work, so that already ploughed land is circled in a clockwise direction. Usually carried out alternatively with casting when ploughing marked lands in a field.

Generation. A single complete life cycle in the life of a plant or insect.

Generative nucleus. The nucleus of pollen grains which by division forms sperms.

Generic advertising and promotion. The promotion of a particular commodity without reference to a specific producer, brand name, or manufacturer. Because individual producers of nearly homogeneous agricultural commodities cannot easily convince consumers to choose one egg or orange or a single cut of beef over another, they join together in commodity promotion programs to use generic advertising to expand total demand for the commodity, thereby helping their own sales as well. Activities are intended to expand both domestic and export demand; examples include advertising, nutrition education, research to improve product quality and appeal, market research studies, and technical assistance. These activities are often self-funded through assessments on marketings called check-off programs.

Genetic breakdown. Loss of vigour and often the early death of F2 plants which lack the necessary adaptive complexes of either or both the original parents.

Genetic drift. A change in gene frequency from generation to generation as a result of chance events

Genetic effect of radiation. Inheritable changes which are chiefly mutations and produced by the absorption of ionizing radiations. On the basis of present knowledge these effects have been purely additive, and there occurs no recovery.

Genetic engineering. (1) A term applied to techniques, still being developed, which involve the separation of cells from collected 'early embryos', and nuclei transfer, by which multiple 'identi-cal' offspring can ultimately be produced. These techniques are combined with those of Embryo Transfer to implant the 'engi-neered' embryos in host mothers. The term also embraces the techniques, also being developed, by which DNA altertion and Gene manipulation can hopefully enhance the genetic quality of stock and crops. Genetic engineering provides a powerful alter-native to the time-consuming conventional methods of breeding (Cross-Brceding) and promises to revolutionise the agricultural industry by permitting scientists to raise greatly improved ani-mals and crop plants (*i.e.*, larger, greater disease-resistance, improved food-conversion ratio, improved nutritional value, etc.). (2) The use of recombinant DNA or other specific molecular gene transfer or exchange techniques to add desirable traits to plants, animals, or other organisms, or to enhance biological processes. Organisms modified by genetic engineering are sometimes referred to as transgenic, bioengineered, or genetically modified. The Agricultural Research Service does in-house research in this field, and the Animal and Plant Health Inspection Service regulates the release of genetically engineered organisms for field experiments. (3) Technology involving man made changes in the genetic constitution of cells (apart from selective breeding).

Genetic map. Arrangement of mutable sites on a chromosome which has been deduced from genetic recombination experiments.

Genetic purity. Trueness to type /varietal purity conforming to the characteristics of the variety as described by the breeders.

Genetic shift. Change in genetic make up of a variety due to growth for several consecutive generations in areas outside their adaptation.

Genetic sterility. Type of male sterility which has been conditioned by nuclear genes in contrast to cytoplasmic sterility. It may be transmitted by either the male or female parent.

Genetic stock collection. Stock having known genes, transloca-tions, inversions, additions or substitutions, and lines with resis-tance or tolerance to known races of pathogens, insects or weed pests.

Genetically Modified Organisms (GMO). A term, currently used most often in international trade discussions, that designates crops that carry new traits that have been inserted through advanced genetic engineering methods (e.g., Flavr Saver tomato, Roundup Ready soybeans, Bt cotton, Bt corn). GMO crops are meeting resistence from some trading partners, particularly the European Union, that are responding in turn to consumer concerns over public health and environmental safety aspects of GMOs. USDA also is being pressured to declare GMOs unacceptable in the proposed National Organic Program. The U.S. scientific community maintains that research shows GMOs to be safe and that the regulatory process for their commercial approval, which includes USDA, Food and Drug Administration, and the Environmental Protection Agency, is an adequate safeguard against any potential problems.

Genetics. Refers to the science of heredity, including the study of its chemical foundation, its development, expression and its bear-ings on variation, selection, adaptation, evolution, breeding and the activities of man.

Genome. All the genetic material in the chromosomes of a particular organism. USDA's research agencies have a Plant Genome Mapping Program to identify, characterize, and map the position of agriculturally important genes on the chromosomes of plants grown as crops or trees in order to better use these genes for improving the characteristics of the plant (resistance to disease, higher yields, etc.) through breeding.

Genotype. (1) The genetic make up of a nucleus or individual. (2) The true genetic makeup of a plant. (3) The genotype is the specific genetic makeup (the specific *genome*) of an individual, usually in the form of *DNA*. It *codes for* the *phenotype* of that individual. Typically, one refers to an individual's genotype with regard to a particular gene of interest and, in *polyploid* individuals, it refers to what combination of *alleles* the individual carries (see *homozygous, heterozygous*). Any given gene will usually cause an observable change in an organism, known as the phenotype. The terms genotype and phenotype are distinct for at least two reasons: (a) To distinguish the source of an observer's knowledge (one can know about genotype by observing DNA; one can know about phenotype by observing outward appearance of an organism). (b) Genotype and phenotype are not always directly correlated. Some genes only express a given phenotype in certain environmental conditions. Conversely, some phenotypes could be the result of multiple genotypes. Inspired by the *biological* concept and usefulness of genotypes, *computer science* employs simulated genotypes in *genetic programming* and *evolutionary algorithms.* Such techniques can help *evolve mathematical* solutions to certain types of otherwise difficult problems.

Genotypic selection. The selection of animals in stock breeding on account of their genotype, revealed by their performance fol-lowing close study of their breeding behaviour and the results of progeny testing.

Genus. (1) A group of closely related species (Genera is the plural). In a scientific name genus is the first of the two names given for an organism. (2) In biology, a genus (plural genera) is a grouping in the classification of living organisms having one or more related and morphologically similar species. In the common binomial nomenclature, the name of an organism is composed of two parts: its genus (always capitalized) and a species modifier. An example is Homo sapiens, the name for the human species which belongs to the genus Homo. See scientific classification for more details of this system.(3) Genus - singular; Genera - plural). A category of related organisms, usually containing several species; the first name of an organism in the binomial system of classification.

Geocarpy. (1) The development of a fruit beneath the ground which originates from a flower above the ground. (2) A prostrate southern Brazilian plant *(Arachis hypogaea)* widely cultivated in tropical and warm temperate regions, having yellow flowers on stalks that bend over so that the seed pods ripen underground.

Geographic Information System (GIS). Computerized systems used to compile, retrieve, analyze, and display spatially referenced data. Farming activities that utilize GIS typically include harvesting, fertilizing, pest control, seeding, and irrigation. Use of GIS is called precision farming.

Geotaxis. (1) Plant orientation with respect to gravity. (2) Movement of a motile organism using the earth's gravity for orientation.

Geotropism. (1) Growth in response to gravity. (2) The growth of a living organism in response to gravity, as the downward growth of plant roots.

Germ. (1) A micro-organism causing disease, particularly Bacteria. (2) A loose term for seed or the nucleus of a seed.

Germicide. (1) A chemical substance that kills germs. (2) Any compound that kills disease-causing microorganisms. Germicides must be registered by the Environmental Protection Agency as pesticides.

Germinate. The process of a seed sprouting.

Germination cabinet. Type of seed germinator which is most commonly used in the seed testing laboratories.

Germination test. In a germination test the seeds are kept under optimum environmental conditions of light and temperature for inducing germination. The conditions needed to meet legal standards are specified in the rules for testing the different kinds of seeds.

Germination. (1) Development of seeds into seedlings or leafy young plants; the first stage in the development of a plant from seed. (2) The initiation and development of vegetative growth of an embryo, spore, seed or other reproductive body, forming a new individual plant.

Germinative capacity. A term used in seed testing for the number of seeds in a sample which germinate in a given time.

Germinative energy. The rate of germination of a seed.

Germinative. Refers to the ability of the seed to sprout, grow and develop into normal plant.

Germplasm. (1) A collection of genotypes of a particular organism. (2) Refers to an entire array of cultivars in a crop species, related wild species in the genus, and hybrid between the wild and cultivated species.

Germtrap. Simple device which is used to prevent the spread of some diseases.

Gestation period. (1) The length of time between conception and birth, during which a developing young animal is carried in its mother's womb. For cows about 283 days, for sheep 144-150 days, and for pigs 116-120 days. (2) The period taken from germination / planting to commencing regular flowering and fruiting.

Gibbereilins. (1) Plant growth stimulating chemicals which are able to induce a number of effects on plant *e.g.*, rapid stem growth, overcoming of dormancy, production of seedless fruit and other responses. (2) A group of plant growth substances that influence cell elongation, cell division and other physiological processes in plants e.g. GA1, GA2 etc.

Fig.: Germinate

Gilchrist's disease. Refers to the chronic infection of animals and man. It is characterized by suppurative and granulomatous lesions in lungs and other affected organs. The disease is quite common in dogs and only once in a horse. It is caused by fungus *Blastomyces dermatitides.*

Ginning (Cotton). Process of separating lint from seed cotton in ginning factory.

Ginning Percentage (GP). Means weight of cotton lint obtained from seed cotton. It is usually expressed in terms of percentage of seed cotton or percentage of lint obtained from 100 unit of seed cotton by weight. It depends upon the cotton variety *eg.*, 40 *Ib* or kg of lint obtained from 100 *Ib* or kg of seed cotton by weight then GP is 40.

GIPSA. Grain Inspection, Packers & Stockyards Administration.

Girdle. (1) Encircling of plant roots, stems, trunks or branches resulting in a constriction of the plant part, or a reduction of water and nutrient flow through the girdled plant part. (2) To intentionally remove bark for grafting or to kill a plant. The choking of a branch by a wire or other material, most often in the stems of woody plants that have been tied to tightly to a stake or support.

Girdling root. Roots that strangle a plant by encircling it tightly.

Girdling. The removal of a narrow ring of bark from a shoot, cane, spur or trunk to favour fruit set, size, and / or ripening.

Glace fruit. Product obtained by coating candied fruit with a concentrated solution of sugar and confectioners' glucose syrup, followed by careful drying to give a transparent glaze to the surface.

Gladiolus. *Planting:* Gladiolus need to be planted after danger of frost. The soil should be warmed to a depth of 6 inches; oaks will be fully leafed out. Corms planted in cool soil are apt to rot. Make sure the corms or clean and healthy. Do not remove the tunic or outer papery layer. Gladiolus prefer full sun, but will respond in locations with some morning or late afternoon shade. Good air circulation is a must. Incorporate no more than 2-3 pounds of a 5-10-10 or 5- 10-5 fertilizer per 100 square feet. Underfertilizing is much better than overfertilizing. Work the soil to a depth of 8 inches. As a rule of thumb, plant gladiolus corms three times their diameter; for example, a 2 inch diameter corm should be planted 6 inches deep and 6 inches apart. Corms planted in shallow soil are more likely to fall over when blooming. Often, corms are covered with an inch or two of soil until they sprout. Sprouts, like asparagus plantings, are covered until the final soil line is achieved. Hilling corms is also a common practice, especially if early summer blooms are desired. A "hill" warms up faster than flat soil. Gladiolus generally will bloom 12 weeks after planting.

Staking: Gladiolus require little staking if corms are planted correctly. Plant the corms three times their diameter — a 2-inch diameter gladiolus would be planted 6 inches deep. However, some large gladiolus still require staking. Bamboo or plastic staking should occur at the time of corm planting to minimize damage to the developing plant and root system. Stakes should be at least three feet tall. Tie flower stalks to the stake using old nylon panty hose or torn sheets. Two ties should be sufficient to maintain a straight stem.

Winter Protection: Corms should be dug after foliage has matured and started turning brown. Lift corms carefully with a spade or spading fork, taking care not to cut into the corm. Cut the tops off 1 inch above the corm and dry for 2 to 4 weeks in a warm location (70-80 degrees F.) with good air circulation. Remove the old corm which is beneath the new corm. Discard any rotted or damaged corms. Cut stems back to within an 1/8 inch of the corm. Place the corms in an onion sack or old nylon panty hose. Hang from a wall or ceiling. Ideal storage temperatures are between 35 and 40 degrees F.

Gland(s). Secreting organ(s). Leaf teeth and stipules often end in minute glands.

Glandular-bristly. With stiff gland-tipped hairs.

Glandular-ciliate. Fringed with small glands.

Glandular-pubescent. With gland-tipped hairs.

Glass house crops. Much of the glasshouse sector of the horticulture industry has been re-equipped since the mid-1960s. Widespread use is made of units with automatic control of heating and ventilation, semi-automatic control of watering, and carbon dioxide enrichment of the atmosphere. Tomatoes are the most important glasshouse crop and, together with lettuce and cumcum-bers, represent some 95% of the total value of glasshouse vegetable output. The area under sweet peppers and celery has increased significantly.

Glasshouse. (1) A building constructed almost entirely of glass providing a sheltered environment, and warmer conditions than the open air, for the growing of flowers and vegetables. Artificial heating may also be provided. Also called a green-house. (2) A structure that has a transparent cover and an artificial heat source for growing plants, mainly in temperate regions (See Green house).

Glat. A gap in a hedge. Repairing such gaps is known as glatting.

Glauca/glaucum/glaucus. Covered with a white or bluish waxy bloom.

Glaucous. (1) Covered with a white or bluish bloom which can usually be rubbed off easily. (2) A surface with a waxy, white coating that can be rubbed off.

Glazed fruit. When candied fruit is dipped for a moment in a boiling syrup to impart a glossy finish to it, drained and dried it is called glazed fruit.

Glazing. Covering of fruit/vegetable with a thin transparent coating of sugar, which imparts them a glossy appearance

Glean. To gather corn left by a reaper or by those hand-harvesting a grain crop.

Gleaning. Collecting unharvested crops from fields or obtaining unused agricultural products from farmers, processors, or retailers, usually for distribution to food banks and charitable feeding organizations.

Global Positioning System (GPS). A network of satellites that can be used by ground-based units to precisely determine their location by latitude and longitude. GPS is part of the infrastructure required to operate geographic information systems that are used to practice precision farming.

Gloxinia. Gloxinia are relatives to African violets and prefer similar growing conditions. Day temperatures should be on the warm side with night temperatures at least 65 degrees F. for continual bloom. However, like the African violet, gloxinia will rest for a period of time before re-blooming. Soil should be loose and moist. Avoid getting the foliage wet. Place plants in a bright southern window during the winter; move the plants during the summer months to a bright east or west window to reduce the chance of burn. Fertilize monthly during the active growing season, though it is hard to determine just when that season is with gloxinia. If the plant isn't blooming, don't fertilize with a nitrogen fertilizer.

Glume. The pair of *bracts* at the base of a spikelet.

Gluten. A mixture of insoluble proteins present in wheat (and other cereal grains) giving wheat flour its distinctive character. It is a sticky substance remaining after the strach has been extracted.

Glycerinizing flowers. To glycerinize plants for long term preservation, combine 1 part glycerin (available at drug stores) with two parts warm water. Immerse stems stripped of foliage. Place plants in a cool, dark location. Add further solution if needed, until plants stop absorbing the liquid. Glycerined plant material will be more pliable. However, not all plants can be glycerined. Glycerinized plant material may mold readily during high humidity conditions. Store accordingly.

Goad. A pointed stick used to urge cattle to move faster.

Goaf. A term used in some parts for a stack of hay, straw, etc., in a barn.

Goatling. A female goat between one and two years of age.

Gobar Gas Plant. Also called bio gas plant. It is a simple plant for carrying out anaerobic digestion of crowdung and liter to pro-duce a combustible gas which is useful for lighting and cooking puiposes.

Gobbler. An adult male turkey. Also called stag turkey or turkey cock.

Go-down. A cutting in the bank of a stream or river allowing animals to get to the water to drink.

Goottee. Synonym for air layering, marcottage, circumposition, and Chinese layerage

Gore. A triangular piece of land.

Gossypoi. Phenolic pigment in cottonseed that is toxic to some animals.

Gourmandisers. Strong shoots arising from the stem as in cocoa, coffee etc.

Gout fly. A small yellow and black fly related to the Frit Fly, the Larvae of which feed, in consecutive generations, on the shoots and the developing ears of cereals, particularly barley. The leaf sheaths become swollen (or 'gouty') and twisted. Also called the Ribbon-tooted Corn Fly.

Grades and standards. The segregation, or classification, of agricultural commodities into groupings that share common characteristics. Grades provide a common "trading language," or common reference, so that buyers and sellers can more easily determine the quality (and therefore value) of those commodities. Two USDA agencies—the Agricultural Marketing Service and Grain Inspection, Packers, and Stockyards Administration—serve as objective sources for this information. These agencies develop common grades and standards and conduct inspection and grading services for most food and farm products, and industry pays for most of the cost through user fees.

Grading certificates. A formal document setting forth the quality of a commodity as determined by authorized inspectors or graders.

Grading up. The establishing of a pedigree herd of cattle by mating pedigree bulls with a core of non-pedigree cows, and subse-quently mating each generation of females with further pedigree bulls of the same breed, thereby increasing the percentage of pedigree 'blood' with each generation.

Grading. The process of sorting into different lots conforming to certain pre-determined standards.

Graft chimeras. Chimeras produced artificially by grafting or budding.

Graft incompatibility. (1) Refers to the inability of parts of two different plants, when grafted together, to produce a successful union and of the resulting single plant to develop satisfactorily. (2) The inability of two different plant parts when grafted together to have a successful, long lived graft union.

Graft union. The region where rootstock and scion come together; there can be slightly deformed growth at the union that is noticeable, but does not effect the function of the tree

Graft. (1) A shoot of a desired plant variety (a scion) bearing a bud, joined on to the stem of another variety (the stock) with a weaker or stronger rooting system, as required, so as to develop into a whole plant, usually a tree or shrub, of the same variety as the scion. Grafting is a common method of propagation (Budding). Cherry varieties are often grafted on to wild cherry stocks. (2) A ditching spade with a long, narrow, concave blade. (3) A spade's depth. Also called spit. (5) Inserting a bud or stem into another plant for purposes of propagation. (5) Planting material consisting of a scion and a rootstock. (6)To unite a stem or bud of one plant to stem or root of another plant.

Graftage. (1) The term used for the method of plant propagation in which, the limbs of (wo different plants are jointed together to form a new plant. There are many methods of grafting e.g., splice, cleft, saddle, side, and topworking etc. (2) Fusing or uniting one part of a plant (rootstock) with that of another (scion), for producing a graft.

Grafting. The process of joining a stem or bud of one plant on to the stem or root stock of a different plant.

Grain drier. A machine used to dry grain after it has been harvested while too moist for safe storage with the possibility of fermen-tation occurring. There are three main types. *Batch Driers.* The grain is held in batches in special containers and dried using a hot air flow, followed by cooling with a cold air flow.

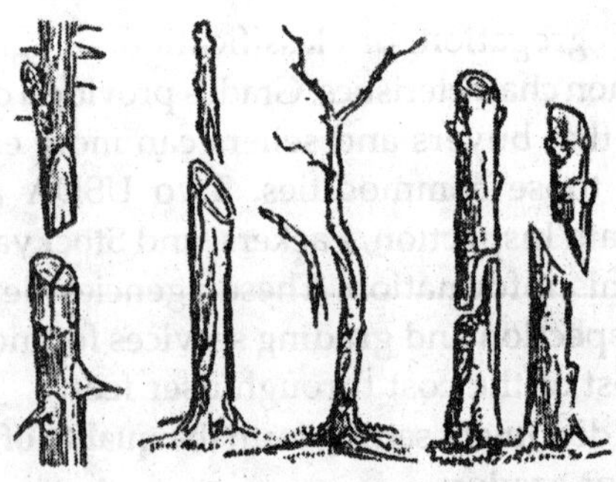

Fig.: Grafting

Grain elevators. Grain elevators in which bulk storage of grain is provided for the public for a fee. Grain of the same grade but owned by different persons is usually mixed or commingled as opposed to storing it "identity preserved." Some elevators are approved by exchanges as "regular" for delivery on futures contracts.

Grain lifters. Projecting devices fitted to the cutter-bar fingers of a combine harvester to lift cereals which have been flattened or laid-down by bad weather, so that the crop can be cut and gathered by the combine's reel.

Grain moisture content. The percentage moisture in harvested cereal grain. For long-term safe storage grain needs to be dried to 14% moisture content. Above this level there is a possibility that the grain will ferment and heat up, totally spoiling the grain if not well ventilated and occasionally turned.

Grain reserve. A phrase that might refer to the food security commodity reserve, food security wheat reserve, the farmer-owned grain reserve, or the strategic grain reserve.

Grain. (1) Cereal seed, either in hulk after harvesting or and individual seed. Also a growing cereal crop. (2) The smallest British Unit of weight, 1/7000 of a pound, 0.648 gm, equivalent to the average weight of a seed of com.

Graminae. The botanical term for the grass family, flowering plants with long, narrow leaves and tubular stems, including pasture grasses, cereals, reeds (but not sedges), bamboo and sugarcane.

Granary. (1) A farm building for storing grain and other feeding stuffs (e.g. Cake Meal) and for processing them (i.e., grinding and mixing) before feeding to livestock. Also a grange. (2) A figura-tive term for a rich grain-growing area or region.

Grandiflora/grandiflorus/grandiflorum. Epithet meaning larger flowers than normal for a species; a class of roses with large and clustering blooms.

Grange. (1) A farmhouse with accompanying buildings and stables. Also a country house. (2) A Granary. (3) The Order of the Patrons of Husbandry.

Grass blends. A turfgrass blend is a combination of two or more different cultivars of the same grass species, such as two or three different types of Bluegrass such as Monopoly, Touchdown and Adelphi. A blend increases the disease resistance of the turfgrass stand.

Grassed waterway. A generally broad and shallow depression planted with erosion-resistant grasses, which is used to convey surface waters off of or across cropland.

Grasshoppers. Many varieties throughout the state. Problem on New Mexico rangeland because the insects eat grass.

Grassland. Areas used for grazing stock, consisting mainly of grasses and clovers where cultivated, and also of mosses, lichens, heather, etc., where uncultivated or natural.

Gravel culture. Modification of hydroponics where granites of irregular shape and size are used as the medium and nutrient solutions are applied.

Grazing district. An administrative unit of BLM-managed rangelands established by the Secretary of the Interior under Section 3 of the Taylor Grazing Act of 1934.

Grazing fee. A charge, usually on a monthly basis, for grazing a specific kind of livestock. For federal lands, the grazing fee is based on a formula found in the Public Rangelands Improvement Act (PRIA). The federal grazing fee is equal to a base fee of [$1.23 x the

Forage Value Index (FVI)] + [the Beef Price Index (BPI)] - [the Prices Paid Index (PPI)] [100] and is charged per animal unit month.

Grazing Lands Conservation Initiative (GLCI). A program started by USDA under its discretionary authority in 1991 and then specifically authorized by the FAIR Act of 1996 to provide increased technical and educational assistance to conserve and enhance private grazing lands. More than 60% of these grazing lands are considered to have serious environmental problems that could lessen their productive capacity if corrective actions are not taken.

Grazing permit/license/lease. Official written permission to graze a specific number, kind, and class of livestock for a specified time period on defined federal rangeland.

Grazing preference. The status of qualified holders of grazing permits acquired by grant, prior use, or purchase, that entitles them to special consideration over applicants who have not acquired preference.

Grazing privilege. The benefit or advantage enjoyed by a person or company beyond the common advantage of other citizens to graze livestock on federal lands. Privilege may be created by permit, license, lease, or agreement.

Grease. An inflammation of the skin of a horse's heels resulting in swelling, itching and the discharge of greasy pus.

Green box policies. Domestic or trade policies that are deemed to be minimally trade-distorting and that are excluded from reduction commitments in the Uruguay Round Agreement on Agriculture. Examples are domestic policies dealing with research, extension, inspection and grading, environmental and conservation programs, disaster relief, crop insurance, domestic food assistance, food security stocks, structural adjustment programs, and direct payments not linked to production. Trade measures or policies such as export market promotion (but not export subsidies or foreign food aid) are also exempt. See blue box policies.

Green budding. A method of budding where seedlings of 2-6 months are used as stock while the bud wood is only 6-8 weeks.

Green feed. The term used for the mechanically harvested forage feu to animals while fresh and succulent. They constitute an excel-lent source of vitamin-A as they are rich in carotene.

Green fertilizer/manure. A fast-growing crop used to add organic matter when plowed under.

Green forages. A group of feedings tuff comprising the grasses and young cereals, some legumes (e.g., Clover, Sainfoin; Lucerne, Vetches, etc) and certain Brassicas (e.g., Kale, Cabbage, Rape, etc.) green Maize, Beet lops, and Silage, etc. They characteris-tically have a higher protein content, particularly when very young, than succulent foods such as roots and tubers.

Green house. A structure which protects plants from wind, precipitation, excessive radiation, extremes of temperature etc.

Green leaf manuring. The practice of collecting and incorporating green leaves and tender green twigs into the soil

Green manure crop. Crop grown and incorporated when succulent, to improve soil, especially of organic matter

Green manure trampler. An implement which is used to trample and press the green manure crop in the fields.

Green manure. (1) A crop (e.g., Italian Ryegrass, Mustard, etc.) specifi-cally grown for subsequent ploughing-in, in order to provide humus to the soil. Green manure crops are often planted in the autumn after a fallow period when no main rotation crop is to be planted, to prevent accumulated nitrates from being leached out (Leach) during the winter. (2) A crop planted with the intention of turning it under for use as organic matter. (3) Any crop that is grown expressly to be plowed or dug under so as to improve the soil.

Green manuring. The practice of growing green manure crops and incorporating into the soil.

Green pruning. Pruning of actively growing rose plants without the benefit of a dormancy period.

Greenhouse effect. (1) The hypothesized warming of the Earth's atmosphere as a result of increasing atmospheric levels of carbon dioxide and other gases that trap infrared radiation emitted from the earth's surface. While the increase in such gases is well documented, the effect on climate remains debatable. Estimates of the temperature effect range from zero to an increase of several degrees average global temperature by 2050; changes in temperature would affect rainfall patterns. Significant climate change would inevitably affect agricultural practices. (2) The quality of the atmosphere to permit most of the insolation and prevent back radiation. [Solar radiation is predominantly of short wavelengths which passes through the atmosphere. Terrestrial radiation is of much longer wavelength and is trapped or reflected by the atmosphere. The selective transmission of radiant energy by the atmosphere is similar to that of the glass in a greenhouse.

Greenhouse gardening. Growing of unusual and out of season vegetables, fruits and flowers using greenhouse, especially under temperate climatic conditions.

Gregarious. Plants of one kind which occur naturally in close proximity.

Grey leaf. A deficiency disease of cereals, mainly of oats, but sometimes wheat and barley, due to inadequate manganese in the soil. Characterised by grey streaks on the leaves, causing many young plants to die and resulting in poor grain yield. Also called grey speck.

Greyface. A crossbred sheep derived from a Border Leicester ram and a Blackface ewe, produced mainly in Scotland. Greyface ewes are crossed predominantly with Suffolk rams, producing lambs with a high quality carcase.

Grip. (1) A drain, small ditch or furrow used to drain surface water off grassland. (2) A narrow opening made in the soil, using a spade, into which young plants or cuttings are placed. (3) A Dunging Channel in a Cowhouse.

Gripping machine. A machine, normally drawn by a Crawler Trac-tor, similar to a large plough, used to cut surface drains or grips in boggy conditions in hill areas or river valleys, and usually fitted with wide tracks to prevent sinking. The machine nor-mally has an arched frame supporting the cutting mechanism, which removes a continuous strip of soil leaving a U-shaped furrow.

Grooming. (1) The cleaning, brushing, combine and general lending of the coat, tail, legs and feet of certain animals, particularly horses and cattle, especially when kept indoors. (2) The operation of cleaning out lichens and mosses which develop on rubber trees during monsoons.

Gross energy value. The energy released as heat by a food when a given weight is completely oxidised, usually in a bomb calo-rimeter. Also known as energy of combustion.

Gross farm income. The monetary and non-monetary income received by farm operators. Its main components include cash receipts from the sale of farm products, government payments, other farm income (such as income from custom work), value of food and fuel produced and consumed on the same farm, rental value of farm dwellings, and change in value of year-end inventories of crops and livestock.

Gross Processing Margin (GPM). This refers to the difference between the cost of a commodity and the combined sales income of the finished products that result from processing the commodity. Various industries have formulas to express the relationship of raw material costs to sales income from finished products.

Ground cover. (1) A group of plants usually used to cover bare earth and create a uniform appearance; a spreading short plant used to cover soil; a living mulch. (2) Plants grown for their low, spreading habit to prevent the growth of weeds and for protecting soil.

Ground frost. A temperature at ground level of less than 0°Cor 32°F. Most plant tissues are not destroyed until the temperature falls well below the freezing point of water.

Ground.limestone. Any of the natural forms of limestone ground to produce a powder for use in liming. The powder acts to rectify soil acidity more quickly than limestone applied in lumps.

Groundwater. (1) Water within the earth that supplies wells and springs. (2) The water from wells and underground aquifers. An estimated 95% of the drinking water used in rural areas is from groundwater. Because of its use as drinking water, there is concern over contamination from leaching agricultural and industrial pollutants or leaking underground storage tanks. (3) Rain water which has percolated through to under-ground rock strata from the surface as distinct from draining to streams or rivers as runoff. Such water is the source of well and spring supplies.

Grove. (1) A small group of trees (2) Orchard: an area where many trees are commercially grown, e.g. for their fruit an orange grove.

Grow together container. Transplanting an assortment of plugs into 10-inch, 12-inch, or larger container and growing them on to a marketable stage.

Growing point. Region at the tips of both stems and roots where cell division occurs and new cells are formed

Growing season. (1) The number of days between the average date of the last killing frost in spring and the first killing frost in fall (2) The time period, usually measured in days, between the last freeze in the spring and the first frost in the fall. Growing seasons vary depending on local climate and geography. It can also vary by crop, as different plants have different freezing thresholds. It also is an important component in defining wetland areas.

Growing-off. A condition of hops in which the bines easily detach themselves from the rootstocks, due to Hop Canker.

Growth habit. (1) General appearance or mode of growth. (2) Refers to the genetic tendency of a plant to grow in a certain shape and to attain a certain mature height and spread.

Growth promoter. A substance included in an animal feed (Feeding-stuffs) which has the effect of increasing the efficiency of feed conversion (Food Conversion Ratio) and/or leads to improved daily liveweight gains. Examples included antibiotics such as flavomycin, Virginiamycin and zinc bacitracin, and mineral copper sulphate.

Growth ragulator. A chemical applied to a plant, which regulates or alters its rate of growth, either by stimulation or retardation. Often applied to control weeds in crops (*e.g.*, Hormone Weedkill-ers, Translocated Herbicides) when growth is severely retarded and death usually results. These chemicals act in a similar manner to auxins' the natural plant growth regulating sub-stances or hormones. See Plant growth regulator. A chemical used to increase growth or shorter stems.

Growth retardant. A chemical (such as chloro choline chloride or CCC) that selectively interferes with normal hormonal promotion of growth, but without appreciable toxic effects.

Growth ring (Annual ring). A distinct layer of growth as seen in the cross section of a woody stem.

Growth. An irreversible process in which there is an increase in size, dry weight or volume of an organism.

Grub out, Grup up. To remove plants from the ground, particularly digging out the roots so that the land is cleaned. Especially applied to Orchards.

Grub. A term applied in the general sense to insect larvae such as caterpillars and legless maggots, but sometimes used more specifically for the legged larvae of beetles.

Grubber. (1) A hand implement for grubbing out weeds, etc., or for clearing and stirring up the soil, consisting of a long stick with a forked tip behind which is fitted a curled metal pivot. (2) A heavy cultivator.

Guano. (1) A fertilizer rich in nitrogen and phosphorus, derived from the accumulated droppings of certain sea birds, found particularly on islands off the Peruvian coast. (Fish Guano). (2) Bat or bird dropping used as a fertilizer high with about 13% nitrogen.

Gully erosion. Also called ephemeral gully erosion, this process occurs when water flows in narrow channels during or immediately after heavy rains or melting snow. A gully is sufficiently deep that it would not be routinely destroyed by tillage operations whereas rill erosion is smoothed by ordinary farm tillage. The narrow channels, or gullies, may be of considerable depth, ranging from 1 to 2 feet to as much as 75 to 100 feet. Gully erosion is not accounted for in the universal soil loss equation. In a few states gully erosion is substantial, but in most areas more soil is lost through sheet erosion and rill erosion.

Guttation. Exudation of water by uninjured plants usually on the edges or tips of leaves from hydathodes

H

Habit. (1) The entire aboveground shape characteristic of a plant (such as a vase-shaped or columnar tree or a mounded shrub) and aspects of its pattern of growth that might influence the three-dimensional shape of it in a landscape (whether it forms colonies, for instance, or loses its lower branches with age, or tends to be flattened by a hard rain). (2) Growth form or overall plant shape. (4) Same as growth habit.

Habitat. (1) Natural home or dwelling place of an organism. (2) The place where a population (e.g., human, animal, plant, microorganism) lives, characterized by physical features (e.g., desert) and/or dominant plants (e.g., deciduous forest). (3) Type of region in which a plant is native.

Habituation. The acquired ability of cultures to grow and develop without depending on external source of plant growth substances.

Hag. A firm spot or higher spot in a marsh or bog. Also a place from which peat has been dug, often a hole filled with water.

Hagberg test. A test used mainly by millers to determine the alpha amylase content of milling wheat. A suspension of 7 gm of ground wheat in 25 ml water, in a special tube, is stood in a boiling water bath and stirred. The suspension thickness and subsequently thins if the enzyme is present. After 60 second the stirrer is allowed to fall under its own weight and its time of fall is measured. A good milling wheat has a falling time of over 200 sees. Whereas with falling times below 100 sees, are not nor-mally accepted for milling. Highly active wheats show falling time of less than 5 sees. Falling numbers are sometimes quoted. These include the total time in the water bath. Thus a falling time of 100 sees is equivalent to a falling number of f6() sees. (Farrand Test).

Hail forage garvester. A type of Forage Harvester consisting mainly of one or more swinging arms or flails, each bearing a number of knives, and attached to a horizontal high-speed rotor. The flails are covered by a hood from which a chute delivers the cut forage to a trailer. Flail forage harvesters may be trailed or semi-mounted on a tractor, and offset versions of both types exist which avoid tractor wheel damage to the crop.

Hairy vetch. Hairy vetch is used as a winter cover crop to improve the soil's structure and nitrogen content when tilled in or turned over in the spring.The plant is an annual and will die out. It is considered extremely winter hardy. Hairy vetch is often sown as seed in late July through August. Plants will germinate and develop shoots and root through the fall. Vetch will go dormant during the winter months; a severe winter with below 0 degrees F. may kill the plants, limiting the need for burn down in the spring. Plants surviving the winter can be rotortilled or plowed in the spring. Dense populations can be knocked down with an application of 2,4-D; hairy vetch is quite sensitive to herbicide applications. Vetch may flower and produce seed that will germinate. Cultivation or spot treatment with 2,4-D can control the problem. Make sure to read and follow all labeled directions when using a pesticide. Hairy vetch is NOT the same as crown vetch, though both are legumes and will improve the soil fertility. *Seeding Rates:* Sow seeds at the rate of 1/2 to 3/4 pounds per 1000 square feet.

Hake bar. The pan of a trailed plough which is coupled to the draught linkage behind the tractor.

Half- or sub-shrub. A plant with stems that are woody at the base, usually dying back to the woody stems or even back to the ground after severe winters.

Half-standard. A fruit tree, the lowest branches of which are about 1.2 cm (4 ft.) above the ground. (Standard). (2) In floriculture, trees or shrubs grown with about one metre of clear stem.

Halt-Sib. Term used in population improvement. A Half-Sib family comprises progeny from *a* cross of bulk pollen into a selected parent—that is, one parent in the cross is identified (selected).

Hammer nill. A machine usually housed in a barn, which grinds Cereals and other Feed ings tuffs. It consists of steel hammers which rotate at high speed in a casing and are surrounded by a perforated screen. When the grain is sufficiently reduced in size by the impact of the hammers it passes through the holes in the screen. Various mill sizes are available with differing screen mesh sizes, and throughout rates vary between 3 tonnes per hour for large powered mills and about 50 kg per hour for small automatic mills.

Handler. Generally, the first buyer of a farmer's commodity destined for fresh market use (in contrast to processing). Under marketing orders, handlers are defined as anyone who receives the commodity from producers, grades and packs it, and sells it to someone else for further marketing. Usually, the requirements spelled out in marketing orders technically apply to handlers, although producers absorb their effects.

Hanging drop technique. Technique for observing the presence of motile microorganisms in a drop of fluid.

Hanging. The practice of hanging animal and bird carcases, particu-larly game, on hooks at room temperature for a period so that the meat condition improves.

Hank (cotton). One hank is equal to 840 yards length of yarn. American cotton on an average is having 38 counts. It means 38 hank can be obtained from one proud of lint.

Haptachlor. An insecticide of the chlorinated hydrocarbon type, similarto and sometimes incorporated inchlorane, used for seed dressing.

Hard pan. An impervious layer in a soil which restricts root penetration as well as movement of air and water.

Hard seed. (1) Seeds that have a seed coat impervious to water or oxygen required for germination. (2) Clover seeds with very tough seed coats preventing Germination and which require scratching in order to do so.

Hard water. Water which contains certain minerals, usually calcium and magnesium sulphates, chlorides or carbonates in solution, to the extent of causing a curd or precipitation rather than a lather, when soap is added.

Hard wood cutting. Cuttings from hardwood for propagation as in the case of pomegranate, grape, fig, plum etc

Harden off. (1) To allow a nursery plant grown in a Glasshouse to acclimatise to the more rigorous outside conditions by transfer-ring first to a cloche or frame, and subsequently to the open air. (2) To gradually reduce the temperature in a brooder over a period of time until young chicks no longer need artificial heat to survive. Now-a-days temperature is usually reduced by 5.5°C (10°F) per week over the first four weeks of life, from about 32°C to about 15.5°C (90°F to 60°F).

Hardened. A plant condition created by various factors enabling it to withstand environmental stresses.

Hardening off. (1) Gradually adapting seedlings and cuttings to the outdoors. Increase the time you leave them outdoors each day until you transplant them. (2) The process of gradually acclimatizing greenhouse or indoor grown plants to outdoor growing conditions. (3) The process of plant adjustment to cold temperatures.

Hardening. (1) The process of gradually exposing tissue culture plants to higher temperature and light intensity and lower relative humidity to make them establish *ex vitro.*

(2) The treatments given in the nursery to enable seedlings /planting materials to withstand the transplanting shock and/or the adverse conditions in the main field.

Hardiness. (1) Quality of plants to resist injury from unfavourable temperatures. (2) Refers to a plant's tolerance to winter climatic conditions. Factors that influence hardiness include minimum temperature, recent temperature patterns, water supply, wind and sun exposure, genetic makeup, and carbohydrate reserves. (3) The ability of a plant to withstand cold temperatures. (4) The ability of a plant to withstand low temperatures or frost, without artificial protection.

Hardscape. Constructed elements of a landscape such as decks, walls fences and driveways. Structures and features of a landscape constructed from materials other than plants, such as walks, walls, and trellises made of wood, stone, or other materials, pathways, pools and ponds.

Hardwood cutting. (1) Cuttings which are made from the past season's growth or olderwood that has shed its leaves and has got lignified e.g. grape, fig. etc. (2) A cutting taking from mature or lignified stems.

Hardwood. (1) A deciduous tree. (2) A term describing broadleaf trees, usually deciduous, such as oaks, maples, cottonwood, ashes, and elms.

Hardy. (1) A term applied to plants and animals capable of surviving the cold during winter. (2) Used in the horticultural sense, enduring winter conditions.

Harmonization. The process of establishment, recognition, and application of internationally recognized measures or standards. Used most often in reference to tariffs (as in the Harmonized Tariff Schedule of the United States (HTSUS)), technical standards, or sanitary and phytosanitary measures applied to imported food products.

Harmonized system. The international classification system for goods, implemented by most countries on January 1, 1998, which is used for tariff classification, trade statistics, and ultimately, transport documentation. Officially known as the Harmonized Commodity Description and Coding System, conversion was begun by the Customs Cooperation

Council in 1970 as a replacement for the Customs Cooperation Council Nomenclature also known as the Brussels Tariff Nomenclature.

Harrow single action disc. Harrow with two gangs placed end, which throw the soil in opposite directions. The discs are arranged in such a way that right side gang throws the soil towards right and left side gang throws the soil towards left.

Harrow spike tooth. Harrow with peg shaped teeth of diamond cross-section attached to a rectangular frame. It is used to break the clods, stir the soil, uproot the weeds, level the ground, break the soil crust, and cover the seeds. Its principal use is So smoothen and level the soil directly after ploughing.

Harrow tandem disc. A disc harrow, comprising of four gangs or more in which each gang can be angled in opposite direction.

Harrow, acme. Special type of harrow having curve knives.

Harrow, spring tooth. Harrow with tough, flexible teeth, suitable to work in hard and stony soils. Spring tooth harrow is fitted with springs, having loops of eliptical shape. It gives a springing action in working condition. It is best suited for hard and stony ground. It is used in soil where obstruction like stones, roots and weeds are hidden below the ground surface. The type pulverises the soil and helps in killing weeds.

Harrow, triangular. Spike tooth harrow with a triangular frame. The frame is made of wood and pointed spikes are fitted in the frame. The teeth of the spike are fixed and not adjustable.

Harrow, zigzag. Spike tooth harrow with a zigzag frame and teeth attached at the junction of the frame members.

Harrow. Secondary tillage implement which is used to cut the soil to a shallow depth for smoothening and pulverising the soil as well as to cut the weeds and to mix materials with the soil.

Harrowing. (1) Working the soil with the help of a wooden implement harrow. (2) Secondary tillage operation which pulverises and smoothens the soil.

Harvest index. The ratio of grain weight to total plant weight in a cereal crop. The effect on crop performance of variation in spike number per plant within the crop, is poorly understood. The harvest index (HI) of plants with different spike number was measured from two nitrogen experiments, one of which was irrigated from anthesis. There were few significant differences in whole-plant HI between plants of different spike number or nitrogen treatment, and few significant differences in plant mass between nitrogen treatments. Spikes from tillers had lower HI than main spikes, but on plants with more tiller spikes, tiller spike HI increased at high nitrogen to compensate for their greater number. This is likely to be because early tillers on plants with more tillers were more like the main stem, than on plants with few tillers. If these results, from plants differing in spike number because of combined effects of plant density and nitrogen, can be extended to changes in plant density, increasing HI by reducing the proportion of tiller spikes, is unlikely to be successful. A useful in-season crop simulation model would complement the sophisticated grain marketing techniques being adopted by farmers. Many crops display large within-crop variation in the number of spikes per plant, and a better understanding of the impact of this variation on crop yield may aid the development of such models. The issue of plant spike number is also relevant to debates about sowing rates and reduced tillering wheats. In this paper, the effect of number of spikes per plant on HI is investigated in crops supplied with various rates of nitrogen in two Wimmera environments. Yield of the plant parts of economic interest (economic yield) as percentage of total biological yield in terms of dry matter.

Harvest time (apples). Below is a list of apples and the time of ripening based on when Red Delicious ripen. For example, Gala with a listing of "-3" means that it will ripen three weeks earlier than Red Delicious. Granny Smith with a "+6" means it ripens 6 weeks later. There is no sure-fire method of determining a ripe apple other than taste. Ripe apples will twist easily from a branch, and the ground color (the undercolor or color next to the stem and blossom end) will change from green to yellow or red. When apples start falling from a tree, they're also probably ripe.

Harvest year. The first, second, third, etc., year following the Seeding Year of a Ley.

Harvest. The time when ripe or mature crops are cut (eg., cereals), lifted (e.g., root crops) or picked (e.g., fruit, hops) and gathered in, the whole process being called harvesting (sometimes called ingathering). Sometimes applied loosely to the yield of a par-ticular crop. Most arable crops now-a-days are harvested by machine, e.g., Combine Harvester, Potato Lifter Sugar Beet Harvester.

Harvested acres. The cropland actually harvested for a particular crop, usually somewhat smaller at the national level than planted acres due to weather damage or abandonment because of low market prices.

Hatchery. A term applied (a) to either a cabinet or a walk-in type of incubator, or (b) to premises specialising in the hatching of eggs and production of Day-Old-Chicks, as distinct from the breed-ing or rearing poultry.

Hatching. The act of a young bird breaking out of an egg, aided by its egg tooth. A hen will normally sit on her eggs and incubate them to bring them to the point of hatching. Artificial incubation is used in commercial poultry rearing. A hatch is also a brood of chicks.

Haugh. A riverside Meadow or flat alluvial (Alluvium) land in a river valley, particularly in Scotland.

Haulm killer. A chemical applied to a potato crop to kill the above ground foliage to make harvesting easier.

Haulm. (1) Stems of herbaceous plants as distinct from hollow culms of grasses. (2) The stems and leaves of corn, peas, beans, potatoes, etc, especially after harvesting. Also called halm.

Haustoria. {Latin - haurire, to drink} Haustorium, singular; Haustoria, plural. An outgrowth of stem, root, or hyphae of certain parasitic plants which serves to draw food from the host plant.

Hay bale stack. Bales of Hay built into a stack for drying and storage, often under a Dutch Barn, and usually on a base of rough timbers to raise the bales off the ground. Airspaces are left between the bales to allow moisture to escape and to allow lower bales to expand under the weight of higher ones. When chemical addi-tives which prevent mould growth are introduced during baling, and hay is fully field-dried, bales may be more compactly stacked. Small temporary stacks are usually left in the field for a period to ensure full drying before barn stacking. Field stacks are occasionally thatched with hay or reeds, but more commonly a plastic sheet is used to keep out rain.

Hay rack. A metal or wooden open frame in which hay is placed and from which livestock may take what they require.

Hay stack. A storage stack of loose Hay common before the introduc-tion of the Baler and bale storage. Now usually only seen in hill areas and on small farms where topography precludes the use of a baler. Plastic sheets, as protection against rain, have now replaced thatch roofs on such stacks. Some stacks of baled hay are referred to as haystacks. Also called hayrick.

Hay sweep. An implement, either horse-propelled or fitted to the front of a tractor (and sometimes to an old car), once used commonly to pick up hay from Swaths or Haycocks and to transport it to a stack. It consisted of a series of wooden prongs about 3 in (10 ft.) long and spaced at 30 cm (1 ft.) intervals in a frame. Often used in conjunction with an elevator or stationary baler. Nowadays most hay is collected using a pick-up baler.

Hay. A term mainly applied to grasses, but also to legumes, some herbs, and occasionally Cereals, which have been cut and dried and conserved for Fodder. Hay may be classified into two main types, viz., (a) seeds hay (sometimes called clover hay), usually derived from a 1 to 2 year old Ley, at one time ma inly from a Red Clover-Ryegrass mixture, and (b) meadow hay, derived from permanent Grassland or a long ley. Normally stored in Bale form in a bam nowadays.

Haycock. A cone-shaped heap of loose hay drying in a field. Now usually only found on hill farms and smallholdings where the use of balers is not practical. Also called hatchel or pike.

Haying and grazing rules. Under previous commodity support law, farmers were permitted, for limited time periods (usually during droughts) and under specific circumstances, to harvest hay or graze cattle on land idled under acreage reduction programs. These rules were eliminated by the FAIR Act of 1996.

Hayioader. A trailed implement similar to a small elevator used to pick up loose Hay from a Swath and deposit it onto a trailer for carting to Haystack. Now seldom used except on some hill or small farms where the use of a baler is not practical.

Hayknife. A large broad-bladed knife with a cross-set handle at one end, used to cut Hay from a Haystack, or Silage from a clamp.

Haylage. A foddrer Crop wilted in the field, after cutting, to between 50-60% moisture content and then chopped and blown into an airtight Silo. Carbon dioxide released by cell Respiration inhib-its bacterial activity and rotting. Drilage is almost identical except that wilting is only allowed to between 60% and 70% moisture content.

Haymaking, The conserving of Hay for use as Fodder, involving drying and the maintenance of nutritive content. (Hay Quality). The moisture content of hay must be reduced from about 80% when freshly cut (Mower), to below 20% to prevent moulding rotting, heating, and possible combustion.

Hay-rake. A hand-rake with a long handle and normally with wooden prongs, used to gather hay.

Haywire. Twine, normally polypropylene nowadays, used in a baler to tie bales. Now usually called baler twine.

Hazel. A common shrub coppiced in oakwoods and scrub, and in certain areas used for making hurdles. The hazels are a genus of about ten species of *deciduous trees* and large *shrubs* native to the temperate northern hemisphere. The botanical name is *Corylus,* and is placed in the family *Corylaceae,* though some botanists include this family within the *Betulaceae.* They have simple, rounded *leaves* with double-serrate margins. The *flowers* are produced very early in spring before the leaves, and are *monoecious,* with single-sex *catkins,* the male pale yellow and 5-12 cm long, the female very small and largely concealed in the buds, with only the bright red 1-3 mm long *styles* visible. The *seeds* are *nuts* 1-2.5 cm long and 1-2 cm diameter, surrounded by an involucre (husk) which partly to fully encloses the nut; the shape and structure of the involucre are important in the identification of the different species of hazel.

H-Building. Refers to a method of plant propagation by as sexual means in which two longitudinal parallel cuts of equal length (3-4 cm long) are made in the bark of stock. A cross-cut is then made about midway between the ends of the two parallel longitudinal cuts. The two flaps of bark has been lifted and a bud removed from the scion with bark but without any wood has been inserted on the sock, covered by the flaps and tied.

H-class. Nursery plots of one or two rows 2 or 3 m long and 40 cm apart used for testing new introduction, selections and hybrids under normal growing conditions.

Head month. A month's use and occupancy of rangeland by a single animal or equivalent. A full head month's fee is charged for each month of grazing by adult animals if the grazing animal (1) is weaned, (2) is 6 months old or older when entering National Forest System land, or (3) will become 12 months old during the period of use. For fee purposes, a head month is equivalent to one animal unit month.

Head-house. A building in close proximity to or attached to a greenhouse, used as a workshop, storage area, pesticide room, potting area, eating area etc.

Heading back. (1) Also called stubbing, dehorning, or lopping. Main branches are cut to stubs with little regard for their location. Regrowth from below the cuts is dense, vigorous, and upright. The new shoots create a dense head and shade and are weakly attached. While this pruning method is often used to reduce the height of large trees, it is not recommended. (2) Cutting an older branch or stem back to a stub or twig.

Heading-back (Heading). A type of pruning where terminal portion of the shoot is removed, causing side buds to grow below the cut

Headland, headrig. The strip along the border of a field where a plough is turned during ploughing, and which is itself ploughed when the rest of the field is completed. Also called for racre.

Headwaters. The source, or upper part, of a stream. Often used in discussing water rights related to wilderness or other federal land designations.

Heart rot. (1) The decay of the heart wood of a free as a result of fungal disease. (2) A disease of beet due to boron deficiency, resulting in the death and browning of the root centre and death of young leaves. Similar to Brown Heart.

Heart wood. (1) The darker, denser part in the central portion of a woody stem, consisting of dead elements which lost their conducting ability. (2) The nonliving inner core of wood, usually darker than sapwood.

Heart. A term applied to fertile soil capable of producing good crops (in good heart) or infertile soil in poor condition (in poor heart).

Heath. An area of open, uncultivated, barren country, with poor acid soil, often sandy or gravelly, and with a characteristic vegetation cover of low shrubs, dominated usually by heather and other ericaeous plants.

Heaving. (1) Lifting of soil and plants due to frost and ice. (2) Occurs during alternate periods of freezing and thawing, where roots become exposed and frequently die.

Heavy land, heavy soil. (1) Land or soil with a high clay content, a high drawbar puli, and harder to cultivate than light soil, and thus sometimes called man's land. (2) A soil predominantly composed of clay.

Hectare (ha). (1) A metric measure of area equal to 10,000 square meters. One hectare=2.47 acres. One acre=0.405 hectares. See acre. (2) A metric unit of land measure. 100 acres of 10,000 sq. metres. Equivalent to 2.4711 Acres.

Hedge cutter, hedge trimmer. A tractor-mounted machine used to trim hedges. There are three main types, viz. *(a)* flail type (Flail Hedger), *(b)* circular-saw type, with a heavy saw blade mounted on an articulated arm driven hydraulically by p.t.o., and *(c)* reciprocating cutter-bar type, also usually operated by p.t.o. There are also various hand-held machines for light or awkward work.

Hedge planting (Single hedge planting). The method of planting where the spacing between plants within the row is closer than that between rows.

Hedge. (1) A close row of shrubs, bushes or small trees forming a fence or field boundary, and often planted or maintained so as to be stock-proof. Hawthorn is commonly used. Hedges last much longer than wire fences but are much more expensive. (2) To plant, maintain, trim or layer a hedge. (3) A linear planting of plants as a barrier. (4) Plants grown close together, forming a row of massed foliage.

Hedging. Taking a position in a futures market opposite to a position held in the cash market to minimize the risk of financial loss from an adverse price change; a purchase or sale of futures contract as a temporary substitute for a cash transaction that will occur later (i.e., long hedge and short hedge). Hedgers use the futures markets to protect their business from adverse price changes.

Heel cutting. Cutting, with a heel (small slice of older wood) at the base for maximising rooting.

Heel in. To store young plants temporarily prior to planting by keeping them in a trench and covering the root or rooting portions with soil.

Heel of cutting. The expanded base of a side shoot after removing it away from the main stem of a plant. *Heeling-in* Placing plants in the ground for temporary storage (when planting cannot be done immediately) keeping the roots in contact with moist soil.

Heel. Back part of the food which is composed of one single bone, the calcaneus (as calcis). The heel pad is made of fat and skin which bears the weight.

Heeling in. Temporarily setting a plant into a shallow trench and covering the roots with soil to provide protection until it is ready to be permanently planted.

Height (tree). Means the vertical distance between the ground level and the extreme top of a tree. On slopes it is generally measured on the upper side of the tree.

Heirloom. Any open-pollinated plant that has a history of domesticity and that has been grown for at least 50 years.

Helophyte. Marsh plant growing in muddy, non-saline sites.

Herb. (1) Plants grown for their flavour, aroma or medicinal value. (2) Any vascular plant which is not woody (2) A plant dying to the ground at the end of the season; one whose aerial stems are soft and succulent without appreciable parenchymatous xylem tissue, a plant not woody in texture. (3) A vascular plant which is distinguished from a tree or shurbs by having a non-woody stem. It is often used in medicine or for providing scent or flavouring. (4) Any fleshy plant. Used mostly to denote edible aromatic ones. A plant dying to the ground at the end of the season; one whose aerial stems are soft and succulent. (5) Name for any plant that is used medicinally or as a spice and for the useful product of such a plant. Herbs as condiments and seasonings are still important in culinary art; the use of medicinal herbs, however, has waned since the advent of prescription and synthetic medicines, although plants remain a major source of drugs. The term *herb* is also applied to all herbaceous plants as distinguished from woody plants.

Herbaceous cutting. Cutting made for vegetative propagation from succulent herbaceous plants.

Herbaceous. (1) Having no persistent woody stem above ground. (2) Relating to or characteristic of an herb as distinguished from a woody plant. (3) Green and leaflike in appearance or texture. (4) Not woody; having no persistent woody stems above ground. (5) Plants that are mainly soft and succulent, forming little or no woody tissue. (6) Plants that die down in winter and sprout forth the following spring; a plant with soft or succulent tissue.

Herbage. (1) Herbaceous vegetation. It is generally applied to grassland species. (2) Leaves, stems and other succulent parts of forage plants used as animal feed.

Herbarium. (1) A collection of pressed and dry specimens for research purposes. (2) Collection of dried plants, systematically arranged and labelled.

Herbicide. (1) Any pesticide used to destroy or inhibit plant growth; a weed killer. (2) A pesticide which kills weeds. The action of a herbicide may be either selective, killing the weeds only and leaving the crop unharmed, or non-selective (or total), in which all vegetation is killed. The latter type are usually applied to fields before planting. Herbicides are normally applied either as foliar sprays direct to the soil where they have a residual effect. Applications may be before crop sowing (pre-sowing), after sowing but before crop emergence (pre-emergence), or after crop emergence (post-emergence). (3) A chemical that kills or retards plant growth. Herbicides may kill the entire plant; or they may kill only the aboveground plant parts, leaving the roots alive. (4) A chemical used to kill specific plants or all plants in its path. (5) Substance used for killing or inhibiting the growth of weeds.

Herd improvement test. Kind of production lest which is carried out for dairy cows in which the whole herd, rather than any individual animal, is kept under test for the whole year.

Hereford. A heavy and hardy breed of cattle, characteristically deep red with white markings on the head, hack, chest, legs and tail tassel. A relatively early maturing breed, with short, strong legs, and a short, broad head. Sometimes polled now-a-days. A dominant breed throughout the world, known for it grazing ability.

Hesperidin. Crystalline substance which is found in orange and other fruits. Chemically it is a rhammo-glucoside of heoperetin, a flavanone closely related to flavone.

Hesperidium. (1) A fleshy berry-like fruit with a hard rind and definite longitudinal partitions. Ex. Orange (2) A syncarpous, polycarpellary berry with a separable leathery rind, characteristic of orange

Heterograft (Xenograft). (1) An interspecific graft. (2) A type of tissue graft in which the donor and recipient are of different species. Also called *xenograft.*

Heterophytes antibody. All the non-green plants which are not unable to prepare their own carbohydrate food but draw it from different sources. They are either parasites or saprophytes.

Heterosis. (1) Lite rally means a condition which is different from parents, Genetically it is the increased vigour growth, yield, or function of a hybrid over the parents that has resulted from the crossing of genetically unlike organisms. Literally means a condition which is different from parents, Genetically it is the increased vigour growth, yield, or function of a hybrid over the parents that has resulted from the crossing of genetically unlike organisms. (2) Superiority of the F1 over the parents (Synonym. Hybridvigour).

Heterotroph. An organism which cannot synthesize its own food, and hence depends on previously synthesized foodstuffs

Heterotrophic cells. Cells that need complex nutrient molecules like amino acids, glucose etc. from which to obtain energy and to build their own macromolecules.

Heterozygous. (1) A genetical term for a plant or animal having a dominant gene for a particular characteristic (e.g., hair colour) derived from one parent and a recessive gene from the other, as distinct from two identical genes (when termed Homczygous). (2) Condition where one or more genes on a chromosome differ from that on the other member of the chromosome pair.

Hexagonal system of planting. Refers to planting fruit trees in an orchard at each comer of an equilateral triangle, by which 6 trees form a hexagon, the 7th tree being planted in the centre. This system has been found to be useful where land is expensive and very fertile with a good supply of water.

Hidden hunger. (1) A condition when a plant shows no obvious symptoms of nutrient deficiency, yet the nutrient content is not sufficient enough to give the expected yield. (2) Expression which is used to designate a condition in plants and animals brought about through the lack of certain essential elements in feed or in the soil and is not easily detected by outward appearance.

High fanning. Fanning to a high standard with good management and producing good yields, as distinct from low tanning in which yields are low, due to poor management or infertile land.

High moisture feed grains. Corn and grain sorghum must have moisture content below CCC standards in order to qualify for marketing assistance loans. However, the FAIR

Act of 1996 makes recourse loans available to producers of corn and grain sorghum that have higher moisture content.

High pressure high volume sprayer. Hydraulic sprayer which is used for spraying tall trees and dense crops, similar to multipur-pose which is multi-cylinder, plunger type, larger tank-size, can cover 10 to 20 acres a day.

High Value Products (HVP). Agricultural products that are high in value, often but not necessarily due to processing. HVPs can be divided into three groups: 1) semi-processed products, such as fresh and frozen meats, flour, vegetable oils, roasted coffee, refined sugar; 2) highly processed products that are ready for the consumer, such as milk, cheese, wine, breakfast cereals; and 3) high-value unprocessed products that are also often consumer-ready, such as fresh and dried fruits and vegetables, eggs, and nuts. In recent years HVPs have accounted for a greater percentage than bulk commodities in total value of U.S. agricultural exports.

High Fructose Corn Syrup (HFCS). Produced from converting to fructose a portion of naturally occurring glucose in starch produced from corn. A natural sweetener, HFCS production expanded during the 1980s as a substitute for higher-cost sugar used in soft drinks. HFCS-55 (55% fructose), which is as sweet as sugar, has almost completely replaced liquid sugar in beverages. HFCS-42 (42% fructose) is roughly 90% as sweet as sugar, and is mainly used in cereal, baking, dairy, and processed foods. HFCS and two other corn-derived sweeteners (glucose syrup and dextrose) accounted for approximately 55% of total U.S. natural (caloric) sweetener use in recent years.

Highly Erodible Land (HEL). Land that is very susceptible to erosion, including fields that have at least 1/3 or 50 acres of soils with a natural erosion potential of at least 8 times their T value. More than 140 million acres are classified as HEL. Farms cropping highly erodible land and under production flexibility contracts must be in compliance with a conservation plan that protects this cropland.

Hill dropping. The method is used for sowing in which seeds are dropped at fixed spacing and not in a continuous stream. Thus the spacing between plant to plant in a raw remains constant.

Hill placement. Application of fertilizer either in bands or in local areas near the plants like cotton, cabbage, tomato etc.

Hinge. A term applied to the uncut soil left by the share of wing of a plough which does not cut a full furrow width.

Hirsel. A group of heft of sheep. Also the area over which such a heft is grazing, often under the charge of a shepherd. Sometimes called a herding.

Hitch. The mechanism by which trailers and trailed implements are connected to a tractor for towing. The two main types are the Drawbar and the Pick-Up Hitch.

Hoe. A trailed implement used to cultivate the soil between row crops to control weeds, consisting of horizontal blades drawn just below the soil surface. Various types exist including Tractor Hoes, Steerage Hoes, and front-or mid-mounted hoes which are rigidly connected to the tractor and are raised or lowered hydraulically, and can be operated by the tractor driver. Hand does comprising a thin blade on a long handle are used by gardeners to loosen soil and to chop and removed weeds.

Hog cholera. Acute, septicalmic, highly contagious and fatal disease or pigs. It is characterized by high fever, anorexia; severe leucopenia, haemorrhages in different parts of the body and it is after accompanied with digestive, nervous and pulmonary symptoms. It gets caused by a filterable virus with a particle size of about 35mu. The affected animals eliminate the virus in large quantities in all their excretions and secretions.

Hog flu. Acute,highly contagious viral disease of pigs which is characterized by sudden onset with fever, respiratory symptoms and marked prostration and is caused by a virus related to the human influenza virus.

Hog/corn ratio (corn-hog ratio). Number of bushels of corn equal in value to 100 pounds of live hogs (feed ratio). Put another way, the price of hogs, per hundredweight, divided by the price of corn per bushel. Since corn is a major input cost to hog producers, the higher the price of hogs relative to corn, the more profit there is in feeding hogs.

Holding pond. A pond or reservoir, usually made of earth, built to store polluted runoff.

Holiday Cacti. Holiday cacti are distinguished from each other by their leaf margins. Christmas cactus are identified from other blooming holiday cactus by their rounded or scalloped leaf margins. Thanksgiving and Easter cacti have pointed margins. Under normal conditions, the holiday cacti will bloom close to the holiday suggested in their name. Flowering cacti should be kept in bright, indirect light. Too much light can cause flower color to fade. Day temperature of 70 degrees F and evening temperatures of 60- 65 degrees F are considered ideal. Avoid overwatering during flowering. Do not fertilize when plants are flowering. Once flowers fall, continue to grow the plant as a houseplant. Soil should be loose; a sandy soil will limit overwatering problems. Fertilize monthly between April and October. Holiday cacti are short day plants, meaning they bloom when nights are long. Plants will also flower if exposed to prolonged cool temperatures between 50 - 55 degrees F.

Hollow trees. Hollow trees usually occur due to rotting organisms penetrating the tree and attacking the heartwood. However, since heartwood doesn't provide any water or food movement (it might, on some trees, be used as food storage), there is little problem of a tree dying. However, hollow trees MAY run the risk of wind damage and falling as the tree ages.

Holstein. Friesian cattle first imported into Canada in 1881 from Holland and bred there to produce a variety with dairy-like characters (i.e.,less flash) and giving high milk and butterfat yields (Milk Yields). Characteristically black and white in colour. Also called the Holstein-Friesian or Canadian Holstein.

Homeoclimactic: Regions that may be miles or continents apart but share similar climates. *Lauren Springer* writes, for example, "My northern Colorado garden mirrors the extreme climate of the steppes and dry mountains of Central Asia and the eastern Mediterranean."

Homestead cultivation. Cultivation in the surrounding land of a house, by the householders.

Homestead protection. When a USDA farm loan borrower lacks the financial resources to make payments on a delinquent loan, is ineligible for a restructured loan, and is unable to buy out the loan at the net recovery value of the collateral property, the borrower can convey the property to USDA in lieu of loan payments. Until eliminated by the FAIR Act of 1996, the borrower may have been eligible for homestead protection, whereby the borrower could lease and/or purchase the residence and up to 10 acres of adjoining land.

Homogenization. The process of physically reducing the particle size of fat in milk, thus enabling even distribution of fat throughout the milk.

Homozygous. (1) A condition where all the genes on one chromosome are similar to all those of the other chromosome pair. (2) A genetical term for a plant or animals having two identical genes for a particular characterisitic (e.g., hair colour), one genes derived from each parent. The genes may be both dominant or both recessive.

Honey dew. The sugary deposit forming on plant parts from the droppings of certain insects, such as aphids, mealy bugs, whiteflies and scale insects.

Honey. A thick, sweet fluid prepared by worker bees in a Beehive from nectar collected from flower and stored in the honeycomb (Comb) for feeding to the larvae. Different honey with charac-teristic flavours are produced from nectar derived from different plant species.

Honeydew honey. Honey produced wholly or mainly from secre-tions of, or found on living parts of plants other than the flowers, and which is light brown, greenish brown, black or of any intermediate colour.

Honeydew. The sticky secretion produced by sucking insects such as aphids.

Hoof and Horn Meal. A fertilizer, rich in protein, consisting of dried and ground animal hooves or horns (or a combination of both). It has a nitrogen content of 12-14% which is rapidly released when appplied to warm soils, normally before planting or sowing. Also called Hoof Meal, and sometimes Horn Meal.

Hoover. A type of potato digger with a chain elevator.

Hop mildew. A fungal disease *(Sphaerotheca humuli)* of hops, causing white spots on leaves, and attacking the female flowers and preventing cone formation. Sometimes late attacks occur causing cones to become reddish coloured (red mould).

Hop. A climbing plant *(Humulus lupulus)* of the mulberry family which is characterized by a long, rough twining stem and rough vine-like leaves (hop bine) and is bearing clusters of bitter, catkin-like fruits of cones (the hops) used for flavouring beer.

Hop-acre. A unit of hop production consisting of 1200 individual hop plants (Hills), and sometimes 1000 plants, depending on the planting distance.

Hop-dog. A hand tool with a wooden handle and serrated iron jaws used to remove poles (Hop) from the ground.

Hoppus food. A unit of measurement of round timber equal to 1.273 cubic ft. The extra 0.273 ft compensates for surplus wood assumed lost during sawing to produce squared planks.

Horizontalis/horizontale. Having habit or branches very horizontal; branches spreading outward rather than being erect.

Hormone weedkiller. A synthetic hormone applied to growing crops which acts selectively by causing distored growth and eventual death to weeds whilst leaving the crops unaffected, *e.g.*, 2,4-D, M.C.P.A. Usually called a growth regulator.

Hormone. (1) A *hormone* is any chemical produced in one part of the body that has a target elsewhere in the body. Plants have five classes of hormones. Animals, especially chordates, have a much larger number. Hormones and *enzymes* serve as control chemicals in multicellular organisms. One important aspect of this is the obtaining of food and/or nutrients. (2) Organic substances produced by plants and animals in minute quantities. Plant hormones (*e.g.*, Auxins) are involved in growth regulation. Animal hormones (*e.g.*, adrenalin) are usually secreted by various endocrine glands into the blood stream, and affect behaviour and a variety of body functions.

Horsefly Repellent. Horseflies in small dirty herds could be repelled by employing following mixture.

Content	*Parts w/w*
Water	2
Tar oil	1
Fish oil	1
Kerosene	1
Soap shavings	1
Pulverized Rosin	2
Water	0.5

The soap and rosin are boiled in water until dissolved. The remaining ingredients are then added and the whole mixture is boiled for 15-20 minutes. The mixture should be applied every 2-3 days.

Horticultural fruit. A fruit which is edible on ripening. The current trend toward increasing popularity of organically or naturally produced foods is relatively new having its beginning in the early sixties. Until recently, consumption of organically produced foods was considered a life style choice of a small group of people who had health and environmental issues with conventionally produced foods (4). Similarly, organic producers were individuals who selected organic production more so as a life-style choice rather than as a framing technique. According to Brummond they were characterized as individuals seeking safer foods and a better environment; were usually more observant and more patient than conventional growers, and had a better understanding and interpretation of biological systems, and, often incorporated their personal beliefs into their production systems. During the decade of the nineties, the interest in organics began to creep into the mainstream consumer purchases. Currently, there appears to be an influx of business oriented producers into the organic production field. These individuals view organics more of a marketing strategy than a life style choice or belief. Consequently, large-scale agriculture has begun to take notice. As a result, organically grown has increased 20% in each of the past five years. The increasing popularity of organic foods is, in a large part, due to the belief that food produced with this culture is free of pesticides and has greater nutritive value than conventionally produced foods. Organically produced food, now representing seven billion dollars in annual sales is beginning to assume the posture of "Big Business ". Since the seven billion-dollar figure represents only 1 % of the total US food supply, organic production has an abundant potential for market expansion. Organically grown is still in its infancy in Texas with a reported 1500 A of total production. However, based on the interest of conventional producers to capitalize on this increasing popularity, acreage is expected to increase steadily in the near future. Therefore, the intent of this publication is to serve as a guide to help producers make an informed decision as to whether or not organic production is a vital option for their operation.

Horticultural specialty crops. The Census of Agriculture includes as "horticultural specialties" bedding plants, florists' greens, flower and vegetable seeds, flowers, foliage, fruit stocks, nursery and ornamental plants, shrubbery, sod, mushrooms, and vegetables grown under cover (e.g., in greenhouses).

Horticulture therapy (Hort therapy). (1) Treatment of unhappy and mentally retarded people, by diverting their attention to ornamental gardening, flower decoration and growing vegetables, fruits and other horticultur-al crops. (2) There are four elements that are essential for an activity to qualify as horticultural therapy if it is to be considered a profession eligible for the same status as other caring professions: (a) a defined treatment procedure that focuses on horticultural or gardening activities, (b) a client with a diagnosed problem who is in treatment for that problem, (c) a treatment goal that can be measured and evaluated, (d) a qualified professional to deliver the treatment.

Horticulture. Latin, *Hortus*, "garden"; *cultura*, "cultivation" Science and art of gardening and of cultivating fruits, vegetables, flowers, and ornamental plants. Horticulture generally refers to small-scale gardening, and agriculture to the growing of field crops, usually on a large scale, although the distinction is not always precise (for example, *market gardening* could be classed either way). A horticultural variety of a plant is one produced under cultivation, as distinguished from the botanical species or varieties, which occur in nature. Although many horticultural practices are very ancient, comparatively recent knowledge of genetics, plant physiology, biochemistry, ecology, plant pathology, entomology, molecular biology, and soils, and the systematic application of such knowledge to practical use (e.g., in plant *breeding*), has expanded horticulture into an extremely complex science. Agencies such as the various bureaus of the Dept. of Agriculture, the state experimental stations, and the many agricultural colleges; organizations such as the American Horticultural Society and the various state horticultural societies and local granges and garden clubs; and the commercial flower-growing and experimental nurseries—all engage in developing, analyzing, systematizing, and disseminating improved horticultural practices for the benefit of both amateur and professional gardeners.

The Latin words *hortus* (garden plant) and *cultura* (culture) together form *horticulture*, classically defined as the culture or growing of garden plants. Horticulture is, however, much more. Horticulturists work in plant propagation, crop production, plant breeding and genetic engineering, plant biochemistry, plant physiology, and the storage, processing, and transportation of fruits, berries, nuts, vegetables, flowers, trees, shrubs, and turf. They improve crop yield, quality, nutritional value, and resistance to insects, diseases, and environmental stresses. Genetics is also used as a valuable tool in the development of plants that can synthesize chemicals for fighting disease (including cancers).

Horticulture involves five areas of study. These areas are *floriculture* (includes production and marketing of floral crops), landscape horticulture (includes production, marketing and maintenance of landscape plants), olericulture (includes production and marketing of vegetables), pomology (includes production and marketing of fruits), and *postharvest physiology* (involves maintaining quality and preventing spoilage of horticultural crops). In strict terms the scientific cultivation of fruit, vege-tables, flowers and shrubs. Also used to describe the commercial production of such crops, some on general farms, but mostly on specialised holdings where soil, climate, skilled labour and irrigation can produce maximum yields of high-quality crops and where access to markets or good roads enable their sale at economical prices. Most horticultural enterprises have increas-ing output per unit area with the help of cultivation and environ-mental control, and the widespread use of machinery. On some farms the shortage and high cost of labour have led to the introduction of 'pick your own harvesting for direct sales to consumers.

A noun referring to a specialist in horticulture. The terms ornamentals, landscape horticulture, and environmental horticulture are common terms used to identify the sub-grouping of horticulture dealing with the landscape setting.

Hose-in-hose. A double flower having floral envelopes (whorls of petals or petal-like structures) nested one within the other. The term relates only to appearance, not to the morphology of the structures.

Host indexing. Refers to a procedure which is used to determine whether a given organism (Plant or animal or microbe) is a carrier of a vims disease. Material is taken from that organism an transferred to another that will develop characteristic symp-toms if affected by the virus disease in question.

Host range. Various kinds of plants or animals which could be affected by a given pathogen.

Hosta. *Flowers:* Hostas aren't really grown for flowers. However, some plants have attractive flowers and should be allowed to bloom. Remove blooms stalks soon after plants have finished flowering to prevent seed development. *Planting:* Most hostas thrive in full shade, though a few can tolerate more sun, especially those with yellowish foliage. Blue and blue-green hostas should be planted in full shade. Variegated plants can tolerate partial shade. A well-drained, rich organic soil is ideal for growing plants. Work the soil to a depth of six to eight inches. *Ross Method of Propagation:* Some hostas are slow to divide, especially some of the large, thick crown types. The Ross Method, developed in the United States should help encourage additional plants to form. In the spring or early summer, carefully remove the soil around the hosta's stem, exposing the white basal plate. This is the area right above the roots. An extremely sharp, and thin, knife is inserted into the basal plate and cut down through the roots. Another cut at a 90 degree angle to the first cut can be made. The knife is removed, and soil replaced around the crown. While it isn't necessary, some growers will insert a toothpick into to the wound to encourage callus tissue formation. Callus tissue will form on the basal plate where the cuts were made. A growth bud usually will also form, giving rise to another plant soon or in the following growing season. Plants may slightly yellow or look anemic for a week or two. However, plants in a loose, high organic soil are quick to recover. Many hosta growers/hobbyist are using the Ross Method before plant large types to encourage side plant productions as soon as possible. Make sure knives are clean and sterilized before cutting.

Hot bed. (1) Propagating structure, in temperate regions, where temperature is enhanced for germinating seeds, rooting cutting and growing nursery stock (See Cold frame). (2) A heated soil bed used for propagation as with cuttings in winter.

Hot pack (Hot fill). Filling of pasteurised or sterilised food, while hot, into clean (not necessarily sterile) containers under clean (not necessarily aseptic) conditions.

Houghing rate. This can be approximately calculated by the follow-ing formulae which assume a field efficiency factor of 70% to take account of headland turning and non-productive time.

House plant. A plant adapted to the micro-environmental conditions of houses and other buildings.

Household (foodstamp). A food stamp household is composed of all those who purchase food and prepare meals in common. All related co-residents must apply as a single food stamp household, no matter how they purchase and prepare food. except for elderly persons who are medically certified as unable to purchase and prepare meals separately. Other co-residents may apply separately if they purchase and prepare food separately, and residents in certain eligible institutional settings (e.g., shelters for battered women, residential drug treatment programs) may apply as separate households no matter how they purchase and prepare food.

Huckleberry. Huckleberries are native to higher elevation forests and thrive in an acid soil. The plant is a close relative of the blueberry with fruit color ranging from midnight blue/ black to maroon. Grow huckleberries in the garden in a manner similar to blueberries. Don't expect similar yields to blueberries.

Huller. A mobile machine which operates like a threshing Machine and is used to separate the seed from the head of clovers, etc.

Humectant. Substance that is able to absorb moisture. It is used for maintaining the water content of products like tobacco, glue, films, dentifixices, cosmetics, baking products, leather, soaps and textiles.

Humid tropics. The tropical areas with excessive moisture and relative humidity, characterized by isothermal conditions which facilitate year round crop production.

Humidity — Increasing. A high humidity is essential for most houseplants during winter months. To increase humidity level:

- Place plants in typically humid rooms such as the bathroom, kitchen or laundry rooms.
- Group plants together. As water transpires through the leaves, a miniature humidity chamber is temporarily created.
- Use a cool water vaporizer.
- Use a humidifier. Turn the thermostat down to 65 degrees F.
- Set plants on a tray filled with pea gravel. Fill the tray with water, just below the bottom of the pot. As the water evaporates, the plants will be surrounded by a temporary humidity change.

Misting: Misting with a hand spritzer does LITTLE BENEFIT for houseplants. Water evaporates quickly into the air. Studies indicate a person would need to mist the air near a plant every one to five minutes in order to have an effect on the plant.

Humidity mixing ratio. Defined as the amount of water vapour in grams mixed with one kilogram of dry air. Common values are in the order of 4-30 gm per kg. For each temperature there is a corresponding saturated humidity mixing ratio.

Humification. (1) Process of organic decomposition. (2) The term used for the process of decomposition of organic matter leading to the formation of humus.

Humus. (1) An organic soil material consisting of partially or completely decomposed plant and animal matter. (2) Decomposed and partly decomposed Organic Matter in the soil derived from plant and animal remains as a result mainly of bacterial action, giving a dark colour to the upper Soil Horizons. Humus has colloidal properties and plays an important role in Cation Exchange, and assists in giving Crumb structure to soil. (3) The rich, brown or black organic part of the soil resulting from the partial decay of leaves and other matter.

Humus-nucleus. A stable complex of Lignin and Protein derived from decaying plant remains, present in the soil.

Hundredweight. One hundred pounds (abbreviated as cwt.). A standard unit of measure for milk, rice, and some meat livestock.

Hungry soil. One requiring plentiful supplies of fertilizer to produce good crops, due to lack of organic matter and relative infertility. (Sandy Soils).

Hurdle. A mobile wooden or metal frame used for temporary fencing, particulary for sheep.

Husbandry. The economic management of a farm.

Husk. (1) The dry, thin, outer covering of certain fruits and seeds, *e.g.*, the Glumes of cereal seeds. (2) A disease of cattle and sheep due to parasitic nematode worms, and typified by a husky cough and, when severe, difficult breathing. (3) Outer covering of a fruit or seed.

Hybrid sterility. The inability of some hybrids to produce viable gametes due to absence of homologous chromosomes.

Hybrid vigour. (1) Qualities in a Hybrid not present in either parent, *e.g.*, increased hardines, improved growth rate. (2) The increased vigour over the parental types exhibited by hybrids.

Hybrid. (1) A plant resulting from a cross between two or more other plants which are more or less alike. (2) An individual resulting from the union of gametes differing in one or more genes. (3) The offspring of parents of different species, varieties or breeds of plants or animals. They may be fertile or sterile. The likelihood of sterility is increased the greater the difference between the Genotypes of the parents, due to the increase chance of imperfect Chromosome pairing. (4) The offspring of two plants of different species or varieties of plants. Hybrids are created when the pollen from one kind of plant is used to pollinate and entirely different variety, resulting in a new plant altogether. Hybrid seed will not be true therefore it isn't suitable for seed saving. (5) Usually a cross between different clones within a species (intraspecific hybrid) or between clones from different species

Hydrangea. The color of some hydrangea can be changed by modifying the soil pH. Hortensia hydrangea (Hydrangea macrophylla) will turn blue if the soil is on the acid side. Fertilize yearly with an acid fertilizer according to direction; modify the soil when planting with two (2) pounds of sulfur and peat moss. A pH between 4.5 and 5.5 is desirable. To turn flower pink to light red, the soil should be neutral to alkaline. Use ground limestone ONLY around the plant if the color isn't pink. Pink colors occur between 6.5 and 7.5. At pH levels higher than 7.5, chlorotic leaves and poor growth results.

Hydric soil. Soil that, in its undrained state, is flooded long enough during a growing season to develop anaerobic conditions that support the growth and regeneration of hydrophytic vegetation (plants specialized to grow in water or in soil too waterlogged for most plants to survive). This term is part of the legal definition of a wetland included in the Food Security Act of 1985. The Natural Resources Conservation Service maintains a national list of hydric soils.

Hydroponics (Soil-less culture). The technique of growing plants without using soil, either in circulated, aerated nutrient solutions or in inert media such as gravel, sand, peat, vermiculite or saw dust, to which a nutrient solution is added

Hydroponics. (1) The growing of plants (*e.g.*, tomatoes) in nutrient solution without soil. (2) The growing of plants without soil by using an inert medium such as sand, peat, or vermiculite and adding a nutrient solution containing all the essential elements needed by the plant for its normal growth and development. Water culture, when plant roots are suspended in a liquid medium containing the nutrient solution while their crowns are supported in a thin layer of inert medium, is true hydroponics. Often called soilless culture, it also includes aeroponics where plant roots are suspended in a dark chamber and sprayed with the nutrient solution. (3) The science of growing plants in mineral solutions or liquid, instead of soil.

Hyena disease. An irreversible wasting condition of the hind-quar-ters in dairy heifers. The disease has not yet been recognized in Britain but is well known in France and the Low Countries.

Hypobaric (Sub-atmospheric pressure) storage. Storage of agricultural produce in low atmospheric pressure where the concentration of oxygen and ethylene are reduced, resulting in low rate of respiration

Hypocotyl. The part of the embryo or seedling below the cotyledonary node and above the radicle or root; the transition region connecting the stem and root

Hypogeal. (1) Plants in which the cotyledons remain below the surface of the soil (See Epigeal). (2) Subterranean or occurring beneath the surface of the soil or underground.

Hypogeous germination. Germination in which the cotyledons remain below the surface of the soil.

Hypoxia. A low oxygen condition in the water that may occur where a nutrient-laden free-flowing body of water (like a river) enters a lake or ocean. The high nutrient content promotes rapid growth of plankton/phytoplankton that subsequently die and, in the process, consume large amounts of oxygen (see biochemical oxygen demand). While fish and shrimp can migrate away from a hypoxic area, less mobile bottom-dwelling organisms are unable to escape. A Mississippi River/Gulf of Mexico Watershed Nutrient Talk Force under the leadership of the Environmental Protection Agency (comprised of representatives from the scientific, economic, ecological and agricultural communities) is investigating the recurring and increasingly large hypoxia problem at the mouth of the Mississippi River. Currently, scientists believe nitrogen making its way into tributaries that flow into the Mississippi River, and eventually the Gulf of Mexico, causes the hypoxia condition.

Hysteranthous. (1) Developing leaves after the appearance of flowers, as the almond. (2) Having the leaves expand after the flowers have opened.

I

Ikebana. (1) Japanese style of flower arrangement which has a spiritual and religious background. (2) The Japanese art of formal flower arrangement with special regard shown to balance, harmony, and form.

Illegitimate clonal seeds. Seeds collected from the budded area of known-rubber clones; but produced by natural cross pollination.

Imbricate. (1) Said of scales which overlap like shingles; the opposite of valvate in which the scales meet along a line without overlapping. (2) Scales that overlap like shingles; the opposite of valvate in which the scales meet along a line without overlapping.

Immunity. Natural or acquired resistance of a plant to active infection by the pathogenic microorganisms or to the adverse effects of a particular parasite.

Imperfect flower. A flower that lacks either stamens (pollen producing structures) or pistils (seed producing structures).

Imperfect stage. A stage in the life cycle of a fungus in which spores are produced without a previous sexual fusion; also called imperfect stage.

Impermeable seed. Seed which does not allow passage of water or gases through the seed coat (See Hard seed).

In burr, A stage in the growth of hop bines when the female flowers are fully developed and the stigmas are proturding.

In vitro. Latin for 'in glass'. Reactions, responses, or experiments in an artificial environment in isolation from the whole organism.

In vivo. (1) Latin for 'in living'. Biological processes that occur within the whole, living organism. (2) Within the living organism; applied to laboratory testing of agents within living organism.

Inarching. (1) A grafting method where rootstock is taken to the scion donor plant for grafting. Stock and scion are independent for their nourishment till the completion of the graft union (Synonym: Approach grafting). (2) To graft by uniting, as a scion, to a stock, without separating either from its root before the union is complete; — also called to *graft by approach. P. Miler.* (3) A type of grafting that attaches a parent plant using an arched stem. The grafting of a growing branch to a stock, without separating it from its parent stem. (Graft).

Inbred/inbreeding. (1) The progeny of either a single cross pollinated plant obtained by selfing or two closely related plants obtained by inbreeding. (2) An undesirable plant made by crossing two related plants. (3) The crossing of closely related plants, which leads to increase in homozygous gene pairs as well as genetically linked undesirable characters. (4) The mating of closely related animals (or plants), *e.g.*, parent with offspring, brother with sister, etc., to increase the istics. The practice tends to increase the number of Homozygous gene pairs, but genetically linked undesirable characteristics are also accumulated and progeny with such characteristics are usually culled. (See Breed in, Cross-Breeding, Line Breeding, Out-Crossing).

Incentive payments. Direct payments made to producers of wool and mohair, which were similar to deficiency payments made to producers of grains and cotton. The incentive payment rate was the percentage needed to bring the national average return to producers (the market price plus the incentive payment) up to the annually set national support price. Each producer's direct payment was the payment rate times the market receipts. Producers with higher market receipts got larger support payments. This created an incentive to increase output and to improve quality. The wool and mohair commodity programs ended after the 1995 marketing year as required by P.L. 103-130.

Incised. (1) Intermediate between toothed (dentate or serrate) and lobed, being a sharply inward cut leaf (the inward cuts are called incisions). (2) Cut by sharp and irregular incisions more or less deeply, but intermediate between toothed and lobed.

Included bark. This problem occurs at branch attachments when bark becomes embedded between the branch and stem. This included (embedded) bark creates a weak branch attachment, one which is susceptible to breakage.

Incompatibility. (1) Condition preventing pollen grains from functioning on a pistil. (2) Inability to form a successful, long lived graft union.

Incomplete flower. (1) Flower lacking one or more of the four sets of floral parts. (2) One which lacks any one or more of these parts; calyx, corolla, stamens, and pistils. (3) Incomplete flowers are missing one or more of the four main flower parts — male (stamen), female (pistil), petals or sepals.

Incubation period. (1) The period of time required fora newly laid egg to develop to the stage when the young bird is ready to hatch out. For chicken eggs the period is 21 days. (2) The period between an animal becoming infected by disease-causing germs (*e.g.*, Bacteria) and the symptoms developing, during which germ number increase dramatically.

Incubator. An apparatus used to hatch eggs by providing artificial heat, usually maintained at 37.2°C (99°F) and reduced for the last 2 days of the Incubation Period to about 36.10C (97°F). Humidity is maintained at about 60%. Types vary from small flat ones holding 50-100 eggs, used for a small breeding flock, to large cabinet types holding up to 40,000 eggs in trays or specially designed walk-in-rooms holding up to 80,000 eggs, normally used at a Hatchery. Trays are frequently tilted to turn the eggs before they are transferred for the final 2 days of incubation to a hatching section.

Indeterminate. Said of those kinds of inflorescence whose terminal flowers open last, hence the growth or elongation of the main axis is not arrested by the opening of the first flowers.

Indexing. Determining the presence of disease or verifying the assumption of freedom from disease or virus

Indicator plant. (1) A sensitive plant that alerts to some condition such as drought. (2) Plants which reflect deficiency of plant nutrients, soil moisture stress etc. indicating the suitability for growing crops.

Indigenous. A plant that is native to a particular country, habitat or climate, as distinct from an introduced plant

Indumentum. A wooly or hairy covering of leaves or young shoots. Can be on both top and bottom surfaces of new or young leaves, but is usually confined to the underside of mature leaves.

Industrial crops. Crops that primarily have industrial applications in contrast to food or livestock feed uses. Industrial uses account for a relatively small but a growing and potentially much larger share of the market for farm commodities. The USDA devotes a significant research effort to identifying and developing new industrial uses for crops; this effort is encouraged by the Alternative Agricultural Research and Commercialization Corporation (AARC). Some of the industrial and experimental crops include: castor beans (lubricants, nylon, cosmetics); chia (cosmetics); crotalaria (fibers); cuphea (soap, surfactants); guar (food stiffeners, drilling muds, paper manufacturing); guayule (natural rubber and hypoallergenic latex products); hesperaloe (specialty pulp paper); kenaf (twine, fiberboard, carpet padding, newsprint); lesquerella (lubricants, cosmetics); meadowfoam (cosmetics, lubricants, water repellents); milkweed (insulated clothing, filler for comforters, nonwoven textiles) and plantago ovato (high fiber additive to laxatives). While corn is the primary feedstock for ethanol, it is not considered an industrial crop because nearly 95% of production goes to feed uses.

Inert ingredient. Pesticide components such as solvents, carriers, dispersants, and surfactants that are not active against target pests. Inert ingredients may be toxic and may be subject to testing under the Federal Insecticide, Fungicide, and Rodenticide Act.

Infant formula cost-containment. Refers to statutory provisions in the Child Nutrition Act of 1966 that require state WIC agencies to solicit bids to infant formula companies for the sale of infant formula used in WIC food packages.

Infectious bovine rhinotrachoitis. An acutely contagious viral dis-ease affecting mainly the upper respiratory tract of cattle. Symptoms include runny eyes and nose, salivation, high tem-perature and loss of appetite together with a drop in milk yield in dairy cows. Fatalities are not common. The disease is more severe in store cattle.

Infectious bronchitis. Acute, highly contagious viral disease of chicken which is characterized by respiratory symptoms like gasping and drastic reduction in egg production. It is caused by filterable virus having a particle size of 70-106 mu.

Infectious disease. Those diseases capable of transmission from animals suffering from the disease-free animals.

Infectious kquine encephalomyelitides. Characterized by symp-toms of incoordination of movement and paraplegia. It has been caused by at least four distinct viruscs which differ from each other in antigenic characters, mode of spread, verulence etc.

Infectious laryngotruchitis. Acute, highly contagious viral disease of chickens, characterized by gasping respiration rates, sneez-ing and coughing which may be accompanied by expulsion of blood stained mucus from the nostrils. It has been caused by a filterable virus which can pass through a pore diameter of 0.80u.

Infectious necrotic hepatitis. Acute infectious toxaemic disease which affects affecting sheep and cattle. It is characterized by short course, blackish discolouration of the skin caused by subcutaneous uenous congestion and necrotic foci in liver. It is caused by localized bacterial infection of liver by *Clostridium oedematiens.*

Infertile soil. A soil which is deficient in plant nutrients and requires plentiful applications of fertilizer to produce good crops.

Infield. An old Scottish term for a field near the farm buildings under continuous arable cultivation and constantly manured. (Out-field).

Infilling (supplying)*:* The operation of replacing vacant sites in a plantation with new plants.

Infiltration. The downward entry of water into soil. Also called percolation. A high rate of infiltration means that soil moisture for crops will be higher. Many conservation practices, such as conservation tillage, reduce rates of runoff and increase infiltration rates.

Inflorescence. (1) A grouping of flowers attached to a specialized stem (peduncle). (2) A collective term for flowers when several are borne on one main axis, as panicle, raceme, umbel, etc

Inhibitory polyculture. Polyculture of crop species where there is undesirable interaction resulting in a net negative effect on all species.

Injurious weeds. Certain harmful weeds *(i.e.,* spear thistle, creeping thistle, curled dock, broad leaved dock, and ragwort) which the Ministry of Agriculture, Fisheries and Food may require an occupier to prevent from spreading, by written notice issued under the Weeds Act 1959. Also called Scheduled Weeds.

Inlay approach graft. Type of approach grafting to be used where the bark of the stock plant is considerably thicker than that of the scion plant.

Inoculum. The pathogen or its parts that can cause infection. That portion of individual pathogens that are brought into contact with the host.

Inorganic insecticide. Insecticide of inorganic origin like arsenical or fluorine compounds; lead or calcium arsenate and paris green examples.

Inorganic. Something that was never alive; of human manufacture.

Insect(s). A class of Arthropod with bodies divided into a head, thorax and abdomen, the head bearing a pair of feelers or antennae, the thorax three pairs of legs and wings. Adulthood is preceded by three stages: egg, larva and Pupa. Insects include Flies, fleas, Beetles, Aphids, Bees, etc.

Insectary. Also known as insectarium. A laboratory in which insects are bred in large numbers.

Insecticide solvent. Substances in which insecticide in the form of crystal or powder is dissolved so that it might be used as spray; water commonly used; acetone, ethyl alcohol, benzene, xylene, linseed oil, groundnut oil, cotton seed oil, turpentine, kerosene, petroleum and other solvents; methylnaphthalenes superior solvents.

Insecticide. A pesticide which kills Insects. Most of those in farm use today are synthetic organic compounds and include Chlorinated Hydrocarbons, organo-phosphorous compounds and Carbamates. They are available in granular, liquid and powder forms and may be applied in various ways.

Insecticide. A substance used to kill, control, repel, or mitigate any pest. Insecticides, fungicides, rodenticides, herbicides, and germicides are all pesticides. Environmental Protection Agency regulates pesticides under authority of the Federal Insecticide, Fungicide, and Rodenticide Act (FIFRA). In addition, under FIFRA, a substance used as a plant regulator, defoliant, or desiccant is defined as a pesticide and regulated accordingly. All pesticides must be registered and carry a label approved by EPA.

Any substance that kills insects. An insecticide is a *pesticide* whose purpose is to kill or to prevent the multiplication of *insects*. Insecticides are very widely used in *agriculture*, as well as in people's dwellings and workplaces. The use of insecticides is one of the major causal factors behind the increase in agricultural productivity in the 20th century. Various plants have been used as folk insectides for centuries, including *tobacco* and *pyrethrum*. Some insecticides have been banned due to their adverse effects on animals or humans. *Misuse* of insecticides is a major factor in *pollinator decline*. *DDT* is an example of a heavily used and misused pesticide. One of the better known impacts of DDT is to reduce the thickness of the egg shells on predatory birds. The shells sometimes become too thin to be viable, causing reductions in bird populations. This occurs with DDT and a number of related compounds due to the process of *bioaccumulation*, wherein the chemical, due to its stability and fat solubility, accumulates to progressively higher concentrations in the body fat of animals farther up the *food chain*. The near-worldwide ban on agricultural use of DDT and related chemicals has allowed some of these birds—such as the *peregrine falcon*—to recover in recent years. Despite these positive results of the reduction of DDT utilization, opponents of *environmentalism* often erroneously cite it as an example of environmentalism going too far and interfering with *malaria* eradication, going so far as to estimate the number of human lives resulting from the DDT ban; for instance, *Michael Crichton* states in his bestselling book, *State of Fear*: "Since the ban, two million people a year have died unnecessarily from malaria, mostly children. The ban has caused more than fifty million needless deaths. Banning DDT killed more people than Hitler." The reasoning behind this accusation is demonstrably erroneous, however, because there is in fact no ban whatsoever on use of DDT for eradication of malaria or any other insect borne disease. Groups fighting malaria have praised the ban on agricultural use of DDT, since it reduces the rate with which mosquitos become resistant to DDT, which is the real reason it is not used more often to fight malaria:

"The outcome of the treaty is arguably better than the statu--s quo going into the negotiations over two years ago. For the firs-t time-, there is now an insecticide which is restricted to vector co-ntrol onl-y, meaning that the selection of resistant mosquitoes wi-ll be slower th-an before." In fact, according to *Agricultural production and malaria resurgence in Central America and India*, Chapin, Georgeanne & Robert Wasserstrom, *Nature*, Vol. 293, 1981, page 183), the lives actually saved due to banning agricultural use of DDT can be estimated: "Correlating the use of DDT in El Salvador with renewed malaria transmission, it can be estimated that at current rates each kilo of insecticide added to the environment will generate 105 new cases of malaria." Some of the newer insecticides are more specific in their actions and are designed to break down into non-toxic components within a few days of application.

Inseminate. (1) To sow seed. (2) To artificially transfer Pollen to the Stigma of a plant.

Instar. The period or stage between molts in a larva. These are usually numbered such as 2nd instar.

Instream use. Water use taking place within the stream channel. Examples are hydroelectric power generation, navigation, fish propagation and use, and recreational activities. Often used in discussions concerning water allocation and/or water rights.

Integrated Pest Management (IPM). System of pest management blending cultural, biological and chemical controls along with host plant resistance or tolerance. A new approach to insect control that combines different methods, starting with the least harmful to the environment. A pest control strategy based on the determination of an economic threshold that indicates when a pest population is approaching the level at which control measures are necessary to prevent a decline in net returns. In principle, IPM is an ecologically based strategy that relies on natural mortality factors, such as natural enemies, weather, and crop management, and seeks control tactics that disrupt these factors as little as possible. Also, a USDA/Environmental Protection Agency program that aims to decrease pesticide applications by teaching farmers to use a variety of alternative control techniques to minimize pesticide use. These techniques include biological controls, genetic resistance, tillage, pruning, and others.

Intensive cropping. Intensive land utilization adopting multiple cropping, intercropping, frequent succession of crops etc

Intensive fanning. A method of fanning in which the aim is to produce the maximum number of crops per year, of high yield, from the amount of land available and to maintain a high Stocking Rate for livestock. (Extensive Fanning).

Intensive protection. Protection of crops from pest damage to the extent that such damage is negligible and does not impede precise measurement of research parameters.

Inter crops. The crops which are raised in an orchard or other widely spaced crops for increasing the income from the same piece of land by short duration crops like vegetables, pulses, papaya etc.

Intercropping. (1) Growing short term crops or dissimilar crops in the interspaces of the main crop. (2) Growing two or more crops simultaneously on the same field. (3) The growing of two or more totally different species together in the same field. A common practice in developing countries which provides an insurance if one crop fails.

Intermediate agricultural products. Generally refers to agricultural products that have a higher per-unit value than bulk commodities; they are often partly processed but not necessarily ready for the consumers. Examples might include soybean meal, wheat flour, vegetable oils, feeds and fodders, animal fats, hides and skins, live animals, and sweeteners such as sugars. Applied to trade policy, intermediate products are one of three categories of agricultural products used by the Foreign Agricultural Service to report export and import data under its BICO system (the others are bulk and consumer-oriented agricultural products). The agricultural trade title (Title II) of the FAIR Act of 1996 permits the Secretary of Agriculture to make available up to $100 million annually of Export Enhancement Program funds for the sale of intermediate agricultural products.

Intermediate stock (Interstock). Stem or root piece present between the (desirable) scion and original rootstock that provides the root system. Double grafting is involved.

International commodity agreement. An undertaking by a group of countries to stabilize trade, supplies, and prices of a commodity for the benefit of participating countries. An agreement usually involves a consensus on quantities traded, prices, and stock management. For example, the United States was a party to the International Natural Rubber Agreement.

International Grains Agreement (IGA). Replaced the International Wheat Agreement in 1995. The IGA comprises a Grains Trade Convention (GTC) and a Food Aid Convention (FAC). The IGA is administered by the International Grains Council (IGC), an intergovernmental forum for cooperation on wheat and coarse grain matters. The Grains Trade Convention provides for information-sharing, analysis and consultations on grain market and policy developments. Under the Food Aid Convention, donor countries pledge to provide annually specified amounts of food aid to developing countries in the form of grain suitable for human consumption, or cash to buy suitable grains in recipient countries. The International Grains Agreement does not contain any mechanisms for stabilizing supplies, prices, or trade.

International Grains Council (IGC). An intergovernmental forum responsible for administering the International Grains Agreement (IGA).

International Monetary Fund (IMF). A multilateral financial institution established in 1945 to help member countries with international payments problems and to maintain orderly exchange rate policies. U.S. agricultural exports benefit indirectly from activities of the IMF that maintain the global trade in commodities and food.

International Trade Commission (ITC). An independent, quasi-judicial federal agency that provides objective trade expertise to both the legislative and executive branches of government and determines the impact of imports on U.S. industries. It makes recommendations concerning countervailing duty and antidumping petitions submitted by U.S. industries seeking relief from imports that benefit from unfair trade practices. The agency also updates and publishes the Harmonized Tariff Schedule of the United States. Established by Congress in 1916 as the U.S. Tariff Commission, the Trade Act of 1974 changed its name to the U.S. International Trade Commission.

International Wheat Agreement (IWA). Replaced in 1995 by the International Grains Agreement.

Internode. (1) The portion of a stem between two nodes. (2) The length of space on a plant stem between two adjacent nodes or stem joints. (3) The part of a stem between two nodes. (4) The section of stem between two nodes. (5) The stem space between leaf or flower nodes.

Interspecific competition. Competition amongst plants belonging to different species.

Introduced. A plant that is exotic or brought from another region; a plant brought and sold in the nursery trade for the first time.

Inulin. Starch found in tubers and roots of artichokes, dahlias and dendelions. It is a fructosan as it is hydrolyxible to fructose. It readily dissolves in warm water and does not give any coloura-tion with iodine.

Invasive. An introduced (non-native) plant that disrupts the local ecosystem and isn't checked by climate, grazing, or other natural means. Extreme examples include kudzu vine (*Pueraria lobata*), which can grow 35 feet in one season and which the American

Horticultural Society's *A-Z Encyclopedia of Garden Plants* calls "the vine that ate the South".

IR-4. Inter-regional Project 4, also know as the Minor Crop Pest Management Program, is funded by CSREES to generate data to register pesticides and biological pest control agents for minor crops where there is no economic incentive to do so on the part of the pesticide manufacturing industry IR-4 provides coordination, funding, and scientific guidance for both field and laboratory research to develop data for the registration by the Environmental Protection Agency of pest control products on a wide variety of commodities. The program has been responsible for data to support over 2,074 food use clearances (1,127 of which were obtained during the past ten years), 3,602 ornamental registrations and, has supported research on 26 biopesticides which resulted in 18 minor use registrations. Each year, about 40% of all EPA newly registered pesticides are the result of IR-4 activities.

Iron Bacteria. Bacteria in the soil which secrete hydrated ferric oxide in badly drained soils whose water contains ferrous bicarbonate in solution a condition made possible only if calcium carbonate is absent.

Iron. (Fe). ATrace Element which, if deficient in the soil (particularly calcareous soils), can cause Deficiency Disease (Chlororis) especially in fruit crops. Also essential for Heamoglobin forma-tion (Deficiency causes anaemia in animals) and for a variety of metabolic (Metabolism) needs in both plants and animals. (2) Iron is a catalyst throughout the plant, involved in energy and enzyme actions.

Signs of Deficiency: Iron deficiency shows up as a chlorosis — a yellowing of the leaves with bright green veins.

Irradiation breeding. Creation of mutations by subjecting the plants or seeds to radiations and their utilization for crop improvement.

Irradiation. The process of exposing food or other items to radiation of various wavelengths in order to destroy contamination from undesirable organisms, achieve insect disinfestation or delay maturation. It is approved for most produce and some meat products. Recently the Food and Drug Administration approved its use and USDA proposed rules for its use in red meat products. While it has been used for produce and chicken in the Southeast, its expanded use is dependent on the construction of facilities on a wide-spread basis. There are various types of irradiation treatments (gamma, x-ray and ultraviolet) with various characteristics and limitations to consider in building irradiation facilities.

Irrigation efficiency. The ratio of the amount of water used by irrigated crops to the amount actually supplied.

Irrigation return flow. Part of artificially applied water that is not consumed by plants or evaporation, and that eventually "returns" to an aquifer or surface water body, such as a lake or stream. Commonly used when discussing water conservation techniques and measurement.

Irrigation water management. Managing irrigation applications based on the water-holding capacity of the soil and the need of the crop. The water is applied at a rate and in such a manner that the crop can use it efficiently and resource losses are minimized. Irrigation efficiency is the ratio of the amount of water stored in the crop root zone compared to the amount of water applied. Water conservation has become more important as costs have risen and demands have grown for wildlife and urban uses.

Irrigation water. Water which is artificially applied in the process of irrigation. It does not include precipitation.

Irrigation. (1) Applying water (or wastewater) to land areas to supply the water (and sometimes nutrient) needs of plants. Techniques for irrigating include furrow irrigation, sprinkler irrigation, trickle (or drip) irrigation, and flooding. About 51 million acres of land are irrigated in the United States. More acres of corn are irrigated than any other crop, but only about 15% of the harvested acres. In contrast, irrigation is used for 100% of rice, 81% of orchards, 64% of vegetables, and 36% of cotton. About 40% of freshwater withdrawals in the nation are for irrigation, making agriculture the single largest user of water. Nearly 90% of all irrigation water withdrawals are in the western states, where in some areas competition for available supplies among uses, including base stream flow, has become controversial. Consumptive use as a percent of withdrawals is about 61% for irrigation. (2) The application of water to soil to provide an adequate supply for crop needs, to increase crop yields or to aid their establishment. Mainly used for grass, vegetables, potatoes and sugar-beet in India. Water may be supplied by the mains, by abstraction under licence (Abstraction Licence), from a water-course, or from on-farm reservoirs. Application is measured in inches per acre (Acre-Inch) or milimetres per hectare, the amount and frequency being related to various considerations including Soil Moisture Deficit, soil type (Light Soils contain less Readily Available Moisture than Heavy Soils), rainfall pattern and Evapotranspiration. (3) Method of supplying water by artificial means.

Isolated seed garden. An area planted with selected clones / varieties and completely isolated from other clones / varieties by specific isolation distance all round

Isotopic tracer. An isotope of an element, (radioactive or stable) a small amount of which may be incorporated into a sample material (the carrier) so as to follow the course of that element through a chemical, biological, or physical process, and also follow the larger sample.

J

J. Jack. (1) Jack tree *(Artocarpus heterophyllus).* (2) Japanese tool for bending trunks and branches (in bonsai culture)

Jam. (1) Concentrated fruit pulp having a fairly heavy body form rich in natural fruit flavour. Pectin in the fruit imparts it a good set and high sugar concentration (more than 68%) and facilitates its preservation. It is prepared by boiling the fruit pulp and juice with sufficient quantity of sugar to a reasonable consistency to hold the fruit tissues in position. A good jam must be having bright colour, rich flavour typical of a fruit, should not be syrupy or stiff. It should be free from crystallization of sugar. (2) The product prepared by boiling fruit pulp with sugar (sucrose) to a moderately thick consistency

Japonica/japonicus/japonicus. Epithet meaning from Japan; such plants are widely Asian.

Jelly. (1) The product prepared by boiling fruit with or without water, expressing and straining the juice, adding sugar (sucrose) and concentrating to such consistency that gelatinisation takes place on cooling. A perfect jelly is clear sparkling, transparent and of attractive colour. (2) Semi-solid product which is prepared by cooking essentially a clear fruit extract and sugar. In jelly making pectin forms the most essential constituent. A perfect jelly should be transparent, attractive in colour, give strong flavour of the fruit and firm enough to retain a sharp edge when cut but quivers when pressed. The fruits rich in pectin and acid like sour apple, grapes, guava, lemon, orange, loquat etc. are best for making jelly.

Johnsongrass. Perennial grass that is a problem in many crops. Introduced and tested as a forage species. Didn't produce well as forage plant but became a nuisance.

Jointed. Having nodes or points of real or apparent articulation.

Jorquette. The point at which fan branches emerge from the vertical chupon stem of cocoa

Jungle. (1) Low or thin forest (2) Any impenetrable thick or tangled mass of vegetation (3) Any dense intermingled growth.

Juvenile. Chronologically speaking, it is an organ or tissue that is not fully developed; developmentally speaking, a plant that is unable to produce flowers (reproduce); contrast with mature which is able to reproduce

Juvenile budling. A budling in which the scion is derived from a young plant, usually less than two years old

Juvenile form. Young plant that is having leaves and other features different from those of a mature plant of the same species.

Juvenile hormone. Hormone which is secreted in insects by small cervical glands called

Corpora Allata. Its presence in the body of the insects results in the conservation of larval character and the inhibition of metamorphosis.

Juvenile period (Juvenile phase/condition). Early period of plant growth when flowering is not observed or cannot be induced. Also characterised by easiness in rooting of cuttings in difficult to root species

Juvenile plant. A young plant that has not yet gained the ability to flower.

Juvenile (1) Stage Special stage in the life history of some algae from which the ordinary plant is developed as an outgrowth. (2) An early non-reproductive phase of plant growth, usually characterized by different leaf shapes or by non-flowering, vigorous increase in size, and often thorniness.

Juvenility. In some plants, the differences in the internal stages of young seedlings from that of mature plant have been manifested by striking differences in leaf shape or greater thorniness in younger plants. This immature condition is referred to as juve-nility. It is important in several aspects of plant propagation.

K

Kahn test. Flocculation test which is used for diagnosis of syphilis in which cardiolipin (a non-treponemal) antigen is used. Such tests are also called standard serological tests (SSL).

Kainit. A low-grade fertilizer containing 12.30% potash and about 50% common salt used mainly on sugar beet. Sometimes also containing about 10% magnesium.

Kalanchoe. There is some argument over the correct pronunciation of the kalanchoe. Some will argue that it should be ka-LAN-cho while others say kal-an-CO-ee. Either is acceptable, though the first is preferred. Kalanchoes are succulent, and as such, can tolerate warm, dry conditions commonly found indoors during the winter. Flower colors range from red to orange to yellow. Plants prefer a bright, sunny location. Cool evening conditions during bloom will make flowers last longer. When done blooming, cut the plants back by one-fourth to one-half and re-pot in a slightly larger pot. Soil should be extremely well-drained; the addition of sand, vermiculite or perlite in equal parts to a packaged houseplant soil mix is recommended. Plants can be placed outdoors in partial shade. Allow the soil to dry before watering. Fertilize every two to three weeks. Pinch plants to keep compact. Like chrysanthemums and poinsettias, kalanchoes are short-day plants and bloom when nights are long. Avoid placing plants under artificial lights during the evening. Plants seldom do well after their second blooming. Rooted cuttings can be taken to encourage new plant production.

Kale. A crop of the Cabbage group characterised by a strong tall stem and open curied leaves, grown as a Fodder Crop. Fed either in the field or after cutting, or made into Silage. There are several types, viz, Hungry Gap Kale, Marrowstem Kale, Rape Cale, and Thousand-Headed Kale.

Karnal bunt. A fungus disease of wheat that reduces yields and causes an unpalatable but harmless flavor in flour milled from infected kernels. Appearance of the disease in the United States in early 1996 resulted in the Animal and Plant Health Inspection Service implementing an emergency quarantine, inspection, and certification program for wheat moving out of the infested areas, along with regulations on sanitizing machinery and storage facilities. Many foreign countries have a zero tolerance for karnal bunt in import shipments.

Ked. A brown, flat, wingless, hairy, blood-sucking fly *(Melopoagus ovinus)*, parasitic on sheep, and causing constant irritation. Also called cade, kade, kaid, seep louse and sheep tick.

Keel. (1) A ridge on the back of a leaf or bud scale; fused boat-shaped petals in members of the pea family. (2) Grass or other Fodder Crops on which livestock are grazed.

Kelps. The largest of the brown algae which usually consist of expanded leaf-like blades connected by a stalk to a holdfast.

Kemp. Coarse, brittle, white fibres, shed during the growth of a Fleece, which are difficult to dye.

Keratin. A tough, sulphur-containing, fibrous protein. The major constituent of the outer layer of skin and horn.

Ketchup. Strained concentrated extract of particular fruit vegetable which is having spiced vinegar.

Kharif. Crop sown just before the south- west monsoon and usually harvested in August-September.

Kibbled. Broken into coarse fragments, *e.g.*, kibbled maize, kibbled lime.

Kieserite. ($MgSO_4.H_2O$). Magnesium sulphate, a greyish white crys-talline powder containing about 16% magnesium, used as a fertilizer. Regarded as a concentrated form of Epsom Salts, having less water for crystallisation.

Kind. One or more related species or sub-species of crop plants, each individually or collectively known by one common name e.g. mango, banana etc.

Kitchen gardening. Practice of growing vegetables in home gardens or compound gardens on a small scale.

Kline Test. Microscopic flocculation test for the diagnosis of syphilis.

Knee roots. A root arising above ground level as a result of flooding or poor soil drainage.

Knees. (1) Pointed or dome-like outgrowths from cypress roots rising above the water. (2) Bent projection of trunk; pointed or dome-like outgrowths such as those at the base of a bald cypress from cypress roots rising above the water.

Knet. Portion of the branch which is embedded in the wood by the natural growth of the trees. The knot is 'loose' or 'light' depending on whether the branch was dead or living at the time it was embedded.

Kniffen. System of training grapevine, named after William Kniffen. The arms of the vine are tied to horizontal wire on the same level above the ground

Knives. Reciprocating sections which slide to and fro across the flat faces of the Fingers on the Cutter Bar of a Mower. Binder or Combine Harvester.

Knock down. Percentage of immediate kill achieved by the applica-tion of an insecticide.

Knotter. A mechanism on a Baler or Binder which knots the twine binding the Sheaves. Also called knotter bill.

Kochia. Same family as tumbleweed and has the same characteristics.

Kohl rabi. A fodder crop of the Cabbage group with a much Swolien, turnip-shaped, scarred stem, from which grow long-stemmed leaves.

Ksloes. A mainly black, shaggy haired, small breed of cattle once found on the Hebridean islands and which contributed, with mainland cattle, to the development of Highland cattle. Some-times used to mean Highland cattle.

L

Labelled compound. A compound having labelled molecules. By observations of radioactivity or isoiopic composition, this compound or its fragments may be followed through physical, chemical, or biological processes.

Lac Insect (Laccifer Lacca Kerr). Insect belonging to group of occids (scale insect and mealy bugs). It is having a large number of small lac glands all over the body producing the lac resin; quantity produced small; needs secretion of 1,50,000 insect to make up one *lb* of shellac. It thrives on sap of host plants.

Lacciferidae. Family of scale insects including the Lac Insect *Lac-cifer lacca.* They secrete a resinous substance from which shellac is prepared.

Lacquering. Process of coating the inside of tin cans with some material (lacquer) which prevents discoloration but does not affect the flavour or wholesomeness of the contents (food inside).

Laid crop. One, particularly a cereal crop, flattened in the field by the weather conditions, *e.g.*, wind or rain, or because the stem is not strong enough to support the grain vertically. Also called a lodged crop.

Laminarin. (1) Polysaccharide storage product which is found in brown algae (2) Carbohydrate food reserve in brown algae.

Lammert's cycle. Daily cycle of temperature changes taking place in beehives during winter. The temperature decreases to a minimum of 13°C. This stimulates the bees to greater muscular activity and the temperature increases rapidly to about 25°C. At this temperature the bees stop their activity and the temperature again decreases gradually to 13°C. In an average hive about 20 gms of sugar are used up during this cycle.

Lanceolate. Shaped like a lance-head, as applied to leaves.

Land capability (classification). The quality of soil resources for agricultural use is commonly expressed as land capability classes and subclasses, which show, in a general way, the suitability of soils for most kinds of field crops. Soils are grouped according to their limitations when they are used to grow field crops, the risk of damage when they are used, and the way they respond to treatment. Capability classes, the broadest groups, are designated by Roman numerals I through VIII, with I being the best soils and VIII being the poorest.

Land evaluation and site assessment system (LESA). A technique that can be used at the local level to determine the quality of land or agricultural uses and to assess sites or areas of land for their agricultural viability. It was first used in the early 1980s.

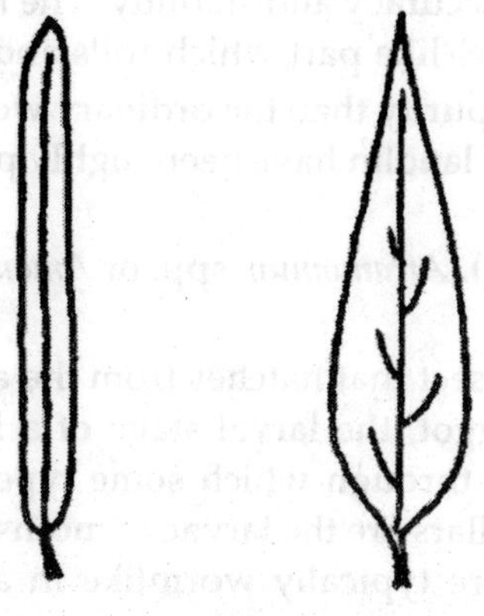

Fig.: Lanceolate

Land grant colleges of agriculture. The Morrill Act of 1862 granted federal land to states to sell, and instructed each state to use the proceeds to endow a college to teach "agriculture and the mechanical arts." States not having any federal land within their borders were given "land in scrip," permitting them to sell federal land located in other (usually western) states in order to establish an agricultural college. The original schools are called the 1862 Institutions. Subsequently, the Morrill Act of 1890 created the black colleges of agriculture, called the 1890 Institutions. The Elementary and Secondary Education Reauthorization Act of 1994 gave land grant status to 29 Native American colleges, called the 1994 Institutions.

Land trust. A private nonprofit organization, under § 501 (c)(3) of the federal tax code, that may receive donations of money, property or development rights, and may use its assets to purchase property or development rights.

Land use and development controls. Ordinances, resolutions, and controls enacted by local government under the authority of state enabling legislation to protect public health, safety, or welfare. Many controls can affect agricultural enterprises; how they are affected depends on the design of the control.

Land use plan. A coordinated collection of data, programs, and activities related to existing and potential uses of land and resources within a defined area. Commonly associated with local units of government trying to anticipate and organize uses of space so as to meet defined goals. For producers, conservation plans are a type of land use plan.

Lands. Strips or divisions marked out in a field due to be ploughed by dividing ridges or rigs, each subsequently ploughed systemati-cally (Casting, Gathering). The width of each land is determined according to tractor and plough size, *e.g.*, 40-50 m wide for a 4-furrow plough.

Landscaping (Landscape gardening). Planting and planning of outdoor space to have harmony between land-forms, architecture and plants to meet human needs for function and beauty.

Landside. A long piece of metal fitted to the frog of a plough which bears against the furrow wall on the unploughed side and resists the force of the Mouldboard as it turns the furrow, thus providing lateral accuracy and stability. The landside sometimes includes a secondary spiring-loaded, wheel-like part which rolls and reduces wear.

Lanolin. Wool wax having greater purity than the ordinary wool grease of common degrees; the USP and cosmetic grades of lanolin have been highly purified, refined, bleached and deo-dorised.

Large cardamom (False cardamom). *Aframomum* spp. or *Amomum* spp., important cash crop of Sikkim and Darjeeling.

Larva, Larvae (plural). A young insect that hatches from the egg and differs fundamentally in form from the adult. A maggot, the larval stage of a fly, is a good example of this difference. An immature stage through which some types of insects must pass before developing into adults. Caterpillars are the larvae of moths and butterflies, and grubs are the larvae of beetles. Larvae are typically wormlike in appearance. Immature insects such as caterpillars.

Late blight. Occurs in potatoes. This was the causal agent of the Irish Potato Famine. A soil-borne foliar disease that can also infect potato tubers. Infected plants die.

Lateral. Said of buds which appear along the sides of the twig; at or along the side. Side shoots which develop from lateral buds on the branches or stems of trees or shrubs, as distinct from terminal shoots which develop from terminal Buds.

Lath. An overhead structure of evenly spaced slats of wood or other materials used to create shade.

Lauren's Half-Shade Polarimetre. Instrument which is used for finding the optical rotation of certain solution. When it is used for finding the optical rotation of sugar it is called a sacchari-metre. If the specific rotation of sugar is known, the concentra-tion of the sugar solution can be found out.

Lauritsen Electroscope. Rugged yet sensitive electroscope which is using a metalized quartz fiber as the sensitive element.

Lawn mover. (1) Mechanical device which is used for cutting lawn grass to an even height. (2) Hand operated or power-driven machine used for cutting the grass in the lawn.

Lawn. (1) Open ground about a house or other building which is grown to the grasses and maintained in good turf, especially for its aesthetic value.. (2) An area covered by green carpet of grass and kept clipped by regular mowing.

Laxatives. Mildest and least active cathartic which is used to over-come constipation e.g., white mineral oil, sulphur etc.

Layerage. (1) Method of plant propagation by asexual means in which a portion of stem, shoot or branch has been covered with soil or some other medium in which root can develop, after which the rooted portion gets detached from the parent plant. (2) Method of asexual propagation in which rooting is induced from stem, branches or roots.

Layering. (1) Part of the process of hedging in which partly cut stems are being over and woven with others to produce a new hedge. Also called plashing or pleaching. (2) The pegging in the ground of a branch or half-cut branch from the siem of a plant so that it produces new roots and gives rise to a new plant. A common method of propagation. (3) A method of propagation, by which a branch of a plant is rooted while still attached to the plant by securing it to the soil with a piece of wire or other means; propagation method to root attached portions of side shoots.

Laying period. The period of season during which polulry are in lay normally of about 50 weeks duration, and usually commencing at 22-24 weeks of ege for egg-producing strains, and 26-28 weeks for Broiller strains. Most egg-producing fowl are slaugh-tered after the first laying period as fewer eggs are laid in the second period.

Layout (Layout plan). A planting plan with respect to land, space, time and planting materials.

LD50. (Lethal Dose 50%) Dose of a given lethal agent (say number *of* pathogenic microorganisms) which on given to each of the test animals, will be able to kill 50 percent animals in a test series.

Lea plough. A type of plough with a long pointed share, keeping the cutting edge well in front of the slightly twisted convex mould-board, thus producing a continuous smooth furrow slice, as distinct from the rough broken slice of the Digger Plough. This type only ploughs to about 15 cm (6 in.) depth and is rarely used now.

Lea. Open country such as meadow, pasture, or arable land left Fallow or under grass. Another term for ley.

Leaching. (1) The process by which chemicals are dissolved and transported through the soil by percolating water. Pesticides and nutrients from fertilizers or manures may leach from fields, areas of spills, or feedlots and thereby enter surface water, groundwater, or soil. Leaching from concentrated sources such as waste sites and loading areas vulnerable to spills can be prevented by paving or containment with a liner of relatively impermeable material designed to keep leachate inside a treatment pond, landfill, or a tailings disposal area. Liner materials include plastic and dense clay. (2) The removal or loss of excess salts or nutrients from soil. The soil around over fertilized plants can be leached clean by large quantities of fresh water. Areas of extremely high rainfall sometimes lose the nutrients from the soil by natural leaching. (3) The removal of salts and soluble minerals from the soil by flushing the soil with water.

Leader. (1) The major supporting stem or trunk of a tree or shrub. (2) A shoot selected to form or extend a branch or trunk. (3) The primary or terminal shoot; the trunk of a tree.

Leaf. (1) The foliage appendages of the stem, though not always serving as foliage; sometimes metamorphosed into a spine (barberry), or tendril (clematis), or reduced to a scale (juniper). Leaves originate at and mark the node or joints of the stem. Buds normally occur in the angles or axils above leaves and are correspondingly alternate, opposite or whorled on the stem. (2) A vegetative organ which, when complete, consists of a flat blade, a petiole or stalk, and (*usually two*) small leafy appendages at the base of the petiole.

Leaf bud cutting. Cutting consisting of a single leaf with the axillary bud and a small portion of stem.

Leaf roll. A virus disease of potatoes transmitted by Aphids causing the leaves to curl upwards and inwards, to thicken and turn yellow, and the plants to remain stunted. Yields are reduced by up to 90%.

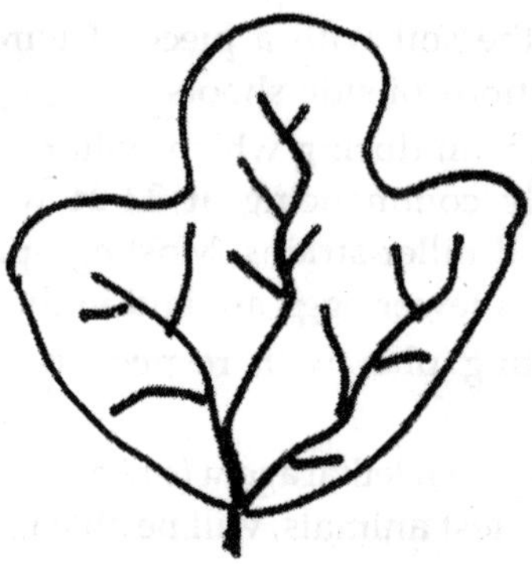

Fig.: Leaf bud

Leaf scar. (1) The healed structure which was once a point of leaf attachment. (2) Mark on a twig or stem left where a leaf was detached. (3) Scars from which leaves have fallen. They usually occur characteristically either singly (alternate) or paired (opposite) or in groups of more than 2 (whorled) at each node. Leaf scars differ greatly in size and shape, and offer some of the best winter characteristics. The points where woody strands of vascular tissue passed up into the leaf are usually evident, and are called bundle scars or bundle traces. Typical leaf scars are essentially at the level of the stem; but they are raised o a pronounced base or leaf cushion in some cases, or the buds are covered by an articular membrane in other.

Leaflet. (1) One of the divisions of a compound leaf. (2) Flat division of a leaf; one of the divisions of a compound leaf.

Leafspot. One of many types of fungal or bacteria diseases causing round marks.

Leafy spurge. Perennial with very deep roots. Contains a milky sap that is irritating and poisonous to some animals. Crowds out native plants in grasslands, meadows, and stream banks. Potentially a big problem since it reduces plant diversity in native areas.

Leam. An open drain, particularly used in Land Drainage systems in the Fens.

Lean to greenhouse. A greenhouse built against the side of another structure such that it has only one sloping roof

Leavening agents. Substance used in yeast foods to promote the growth of baker's yeast.

Leggy. (1) Condition where a shrub or multi-stemmed small tree does not have significant foliage in its lower portion. (2) Abnormally tall and spindly growth; a lot of stem space between leaves.

Legitimate clonal seed. A clonal seed of known parentage.

Legume. (1) A plant of the pea family (Leguminosae) which produces seed in pods and is characterised by five-pettalled flowers and by root nodules capable of Nitrogen Fixation. Some are culti-vated for their protein-rich seeds (*e.g.*, Peas and Beans) for human consumption and for feeding to stock. Others are sown with grasses to produce mixed leys (*e.g.*, Clovers, Lucerne, etc.) Legumes are often grown as a break between cereal crops to enrich the nitrogen in the soil. (2) A family of plants, including many valuable food, forage and cover species, such as peas, beans, soybeans, peanuts, clovers, alfalfas, sweet clovers, lespedezas, vetches, and kudzu. Sometimes referred to as nitrogen-fixing plants, they can convert nitrogen from the air to build up nitrogen in the soil. Legumes are an important rotation crop because of their nitrogen-fixing property. (3) A pod; the characteristic fruit of the pea family.

Lemma. (1) The bract of a spikelet above the pair of glumes. (2) The lower of the two bracts which enclose the flower of a grass.

Lenticel(s). (1) A dot or corky spot on stems or branches, usually of different color than the stem parallel sides, long and very narrow (2) Very small wart-like structures, breaking through the bark of most young twigs. Corky in texture and made of loosely packed cells, providing gaseous exchange between the inner tissues of the stem and the atmosphere. (3) Lenticles are small corky spots on the surface of stems and roots made of loosely packed cells, providing gaseous exchange between the inner tissues and the atmosphere.

Lepidote. Having scales. Tiny scales typically cover the undersides of the leaves. Characteristic used to separate the genus rhododendron into two major groups.

Leptospirosis. A bacterial disease *(Leptospira hardjo)* of cattle which causes abortion, results in serious masttis and a dramatic loss of milk. It is transmitted via the urine and can also affect man. It is estimated that 30% of cattle herds in southern Britain are affected, and up to 90% in bad areas like South Wales. Another strain of the bacteria, *Leptospira australis,* can cause infertility in pigs.

Level. In general terms a flat, usually low-lying area of land. Specifi-cally applied to a Drainage District or lowland area in a river valley Fen with more or less uniform drainage characteristics and usually managed as one unit, *e.g.,* Pevensey Levels, Romney Marsh.

Lewing. A screen of material, *e.g.,* coir mesh, plastics-coated netting or plastic strips, erected on the exposed boundaries of hop gardens and Orchards where natural windbreaks do not exist.

Ley Corn. A cereal crop which follows a Ley in a Rotation.

Ley Farming. A system of farming in which several fields (and sometimes an entire farm) are cropped as Leys with regular reseeding, or, alternatively any rotation or cropping system which includes leys. Also called Alternate Husbandry.

Ley. Land temporarily sown to grass (Grassland Species) and ploughed after 1 -3 years (short duration ley) or after a longer period, up to 10 years in some cases (long duration ley). Seed mixtures are available to suit particular conditions and the type of ley re-quired. Leys may be established by (a) Undersowing, either by broadcasting (Broadcast) or preferably by drilling (Drill), (b) Direct Seeding, (c) Direct Reseeding, sometimes with a Com-panion Crop, to encourage early grazing, and (d) direct drilling into old grassland previously treated with a herbicide. Short duration leys are an important element in crop Rotation.

Lichen. A lowly type of plant comprising an alga and a fungus living together symbolically. Commonly found on tree trunks, old walls, bare rock, etc. Lichens are primary colonisers of rocks and important in soil formation.

Lift. To harvest root crops and potatoes, etc., by digging them from the ground, either mechanically (Potato Harvester, Sugar Beet Har-vester) or by hand, ass distinct from harvesting by cutting.

Lifting shares. A device commonly used on a Sugar Beet Harvester, consisting of a pair of triangular-shaped steel Shares, which lift the beet from the ground onto the elevator section to be depos-ited in a trailer.

Light saturation. A level of irradiance above which there is no further increase in net photosynthesis.

Light soil. Soil with a high proportion of sand, a low drawbar pull, and easier to cultivate than a Heavy Soil, and thus sometimes called boy's land.

Lignin. A complex organic substance deposited in plant tissue, particularly in woody plants, and often combined with cellulose, giving rigidity and strength to stems and tree trunks. It comprises about 25-30% of the wood of trees.

Ligule. A strap shaped organ; a minute projection from the top of the leaf sheath in grasses; the strap shaped corolla in the ray flower of Composites..

Lilacs. Lilac blooms are susceptible to cold injury. The best blooms are from stems between two and four years of age. To use blooms indoors, strip foliage off all limbs and pound stems. Place in a container of warm water. Add floral preservative, a cup of clear soda, or a mixture of one teaspoon of bleach and teaspoon of sugar. Lilac should be pruned after blooming each spring. Remove 1/4 of the plant starting the fifth year after planting to encourage new basal shoots. Remove old wood at ground level. Never cut the entire plant back by 1/4 across the top. Pruning encourages better flowering and reduces insect problems such as lilac borers which attack thicker, older wood. Pruning also allows for better air circulation, limiting the amount of powdery mildew.

Liliflorum/liliflora/liliflorus. Epithet meaning flowers like a lily; with big white trumpets.

Limb cutting. Large unrooted stem cutting used for vegetative propagation as in drumstick

Limb up. Prune off lower branches of a woody plant to (further) expose its main trunk or trunks.

Lime requirement. The amount of Lime needed to raise the pH in the top 15 cm (6 in.) of soil to about 6.5. It is expressed in tonnes per hectare (or cwt per acre) of calcium oxide or calcium carbonate. The amount varies according to the acidity and type of soil, *e.g.*, Heavy Soil require more lime than Light Soils. (Neutralising Value).

Lime. Various Calcium containing materials applied to the soil to raise pH and correct acidity (Acid Soil) and mainly derived from natural deposits of Chalk and Limestone. The main types in-clude Ground Limestone, ground chalk, Burnt Lime, hydrated lime (Calcium Hydroxide) and various by-products and waste materials mainly containing Calcium Carbonate. Applying lime to soil improves soil structure, provides calcium (and sometimes magnesium) for plant nutrition, and makes other nutrients (*e.g.*, Nitrates, Phosphates and Potash) freely available to plant, al-though if over-supplied same minor nutrients (*e.g.*, Boron, Manganese) can be made unavailable.

Limestone. (1) Various types of sedimentary rock with a high calcium carbonate content, as much as 99% in the case of chalk. Most types have an organic origin, particularly chalk, containing the hard calcareous remains of small marine organisms. Some types are relatively hard, *e.g.*, the carboniferous limestone of the Derbyshire Pennines; other are softer, *e.g.*, the magnesian lime-stone of Yorkshire. All are soluble in water containing carbon dioxide. Limestone is an important source of various types of lime. (2) A moorland breed of sheep, indigenous in the Southern Pennines, which contributed to the development of the Der-byshire Gritstone. Now numerically unimportant. (3) Old compost recipes called for the addition of limestone when creating the pile. Research has shown that the finished compost has a pH approaching Limestone needs only be added if the compost contains acidic plant material such as evergreen needles.

Limousin. A French breed of beef cattle recently imported into Britain. Red in colour with somewhat lighter tanned legs. Large bodied, rapid growing, and noted for producing a carcase with a high meat to bone ratio.

Linch fin. One of the pins used to lock an implement onto the three-point-linkagc of a tractor.

Lincoln red. A large breed of uniformly deep-red coloured cattle developed in Lincolnshire, and closely related to the shorthorn, with short, forward and downward curling horns and a short, broad head and with a long muscular body and short stocky legs. Formerly regarded as a dual-purpose breed, it is now popularly crossed with dairy breeds to produce beef calves.

Line breeding. The mating of related animals. but not those closely related such as cousins or parents, with off-spring.

Line. (1) A group of plants used for breeding. (2) A group of uniform individuals reproduced by seed from a common ancestry. It is a narrower group than a strain or variety

Liner. Nursery stock of small size planted at close spacing in a bed for rooting, ready for transplanting.

Lingonberry. Lingonberries are relatives of cranberries, grown in many Scandinavian countries. Plants are sometimes called cowberries, red whortleberry, foxberry, and northern mountain cranberry. The plant is evergreen and often found on forest floors in higher elevations. It is next to impossible to grow Lingonberries in Illinois. The temperatures are too hot during the summer and too fluctuating during the winter.

Linkage. The association of two or more Genes, usually near each other on the same chromosome, so that they tend to be inherited together. The characteristics passed on by such genes are said to be linked.

Linnaen system. A binomial system of naming plants and animals developed by the Swedish naturalist, Carl Linnaeus, in which the first name represents the Genus and the second the Species, e.g., *Hordeum sativum* (barley).

Linola. A new form of linseed known by the generic crop name Solin, which produces a high-quality edible polyunsaturated oil similar in composition to sunflower oil. It was developed and released in Australia in 1992 and first commercially grown in 1994. Linola is being produced in Australia, Canada, the U.K. and in the states of Washington and Idaho. Linola substitutes for flax in cropping rotations, is claimed to have lower production costs than canola, but brings prices comparable to canola or other edible oils. Linola is Generally Recognized as Safe (GRAS) by the Food and Drug Administration.

Linseed. Varieties of the plant *Linum usitatissimum,* selected and grown for their oil-rich seeds, which are sometimes fed after crushing, or after scalding or cooking, particularly to young calves and sick animals, hut more usually as linseed cake which puts a 'bloom' on fattening cattle.

Linters. The short fibers that remain on cottonseed after ginning. They are used mainly for batting, mattress stuffing, and as a source of cellulose.

Liquid manure. A fertilizer made by running water through or into solid manure. The urine of farm animals, Dung Liquor and drain-ings from farm buildings containing valuable Nitrogen and Potash, at one time mostly collected in tanks and distributed on fields or over solid dung heaps. Nowadays urine and faeces are commonly mixed with litter to produce Farmyard Manure, slurry is another type of liquid manure.

Liquid nitrogen fertilizers. Various nitrogen fertilizers applied in liquid form, usually less concentrated than the granulated, powder and crystal forms, e.g;., a solution of urea and ammo-nium nitrate containing 25-30% nitrogen, and often applied as a top dressing, particularly to grassland and winter cereals. Other nitrogenous fertilizers used in liquid form include gas liquor and pressurised anhydrous ammonia.

Living landscape (Sakai). Art of growing and developing plants in pots creating miniature replicas of a natural landscape: **Loam.** Soil having reasonably balanced amounts of sand, silt, and clay. Loam soils contain 7-28 per cent clay, 28-50 per cent silt, and less than 52 per cent sand.

Loam. (1) A rich soil composed of clay, sand, and organic matter. (2) A soil with a balanced soil particle mixture (approximately 25% Clay, 40% Sand and 35% Silt), having most of the advan-tages of Clay Soils and Sandy Soils and few of their disadvan-tages. Loams are easily cultivated, but can be sticky when wet, and are usually drained. (3) This is considered the best kind of soil for home gardens, with the optimum mix of sand, silt, clay and organic matter.

Lobe. (1) A part or segment of a leaf or petal.. (2) A portion of a leaf that projects outward and divides the leaf into distinct parts, but not enough to make them separate leaflets; lobes may be rounded or pointed

Locks. A type of wool oddment purchased by the British Wool Marketing Board comprising small pieces of wool detached from a fleece during shearing.

Loganberry. Loganberries, unlike the trailing forms of dewberries and boysenberries are a semi-erect type of blackberries, almost a cross between an erect plant and a trailing one. Some belief the plant is a cross between a red raspberry and a dewberry. Plants grow horizontally, compared to the erect habit of most blackberries. To grow, most plants are tied to supports or wires to limit their trailing habit. Fruits are sweeter, ripen sooner and are in looser clusters (pendulous), and ripen toward the end of summer. Loganberries are not winter hardy and need protection from cold temperatures. Plants should be untied from supports and lain horizontally after the ground starts freezing. Cover canes with straw or loose hay at least a foot or more. Prune loganberries in a similar fashion to blackberries. The most common form of loganberries are thornless.

Loganberries were developed by Judge James H. Logan in Santa Cruz, California at the turn of the century. Loganberries, like most trailing blackberries, are difficult to grow in Illinois. Choose a protected location and expect a crop every five to ten years. Be proud if you succeed more often.

Longhorn. A hardy breed of beef cattle, characterised by its long, forward and downward curving horns. Coat colour is variable including roan, red and dark plum, and many intermediate shades. There is usually a white ridge along the spine and tail, and white is also normally found on the face and thighs. A long-bodied, short-legged breed, producing a lean carcase, and milk high in buttterfat, at one time much used for cheese making.

Low-flow irrigation systems. These systems (drip, trickle, and micro sprinklers) provide water in small volumes and generally provide water to plants with less waste than furrow irrigation. Drip and trickle systems apply water through small holes in small diameter tubes placed on or below the surface of the field. Another type of system, micro sprinklers, supplies water from low-volume sprinkler heads located above the surface. Low flow systems are expensive and their use is generally limited to high-value crops such as vegetables, fruits, and vineyards.

Lucerne. A perennial leguminous plant *(Medicago saliva)* grown for both Fodder and Forage. Characterised by deep roots (with Nodules) giving it drought resistance, trifoliate leaves, distinct purple flowers, and twisted pods containing kidney-shaped seeds. It is intolerant of acid or poorly drained soils. Also called alfalfa.

Luing. An early-maturing, hardy breed of beef cattle, recently devel-oped from crosses between the Beef Shorthorn and Highland breeds. Red coloured with some roan, gold or white.

Lupins. A leguminous plant *(Lupin sp.)* sometimes grown as a green manure crop for ploughing in, particularly on poor light land in the eastern countries deficient in lime and humus. White, yellow and blue flowered varieties exist. The sweet lupin has been used for forage, being the only non-toxic variety.

Lye-peeling. Peeling of fruits using dilute acid solutions. Many fruits and vegetables are prepared for peeling and canning by exposure to caustic (NaOH). To ensure adequate and consistent peeling, the concentration of caustic in the treatment bath must be maintained at an effective strength. At the same time, if the caustic bath is too strong the fruit or vegetable can be damaged. The typical process consists of an open tank of caustic heated to 170-220°F (76-105°C) by live steam. The target caustic concentration can vary from 3 to 15%. Exposure time for the product is usually 5 minutes or less, after which product is placed on moving grates and washed by high pressure nozzles to remove entrained caustic. A significant amount of caustic is carried out by the product, so the bath must be continually replenished to maintain an adequate volume of caustic solution and a caustic concentration at the target value. This could involve addition of water, based on level control, as well as makeup caustic, based on the conductivity measurement. The best place to start is to look at how the bath behaves and how concentration is maintained manually.

Lynchet. A boundary ridge or unploughed strip. Also a terrace or bank formed on a slope due to continuous contour ploughing. Also called linch. A lynchet is a bank of earth that builds up on the downslope of a field *ploughed* for a long period of time. The disturbed soil slips down the hillside to create a positive lynchet whilst the area reduced in level becomes a negative lynchet.

M

M.C.F.A. An abbreviation for 2-inethyl-4-chloro-phenoxy-acetic acid, a Translocated Herbicide or Hormone Weedkiller, used to control many broad-leaved annual and perennial weeds. Often also include in mixtures with other translocated herbicides.

M.C.P.P. An abbreviation for 2-methyl-4-chloro-phenoxypropionic acid. A hormone herbicide used mainly to control weeds in cereals. Also called mecoprop.

Macrantha/macranthus/macranthum. Epithet meaning larger flowers.

Macro environment. The term used for the environment due to general, regional climate, traditionally measured some 4 ft above the ground and away from large obstructions.

Macro nutrients. Elements necessary in relatively large quantity for successful crop growth; Essential inorganic nutrient elements such as N, P, K, Ca, Mg and S

Macro-elements. Macro-nutrients. Elements which are required and found in relatively large quantities as natural constituents of organisms or tissues; major elements. like nitrogen, phosphorus and potassium.

Macronutrient(s). (1) A chemical element needed by crops in relatively large amounts (normally geater than 1 ppm), normally applied as a fertilizer or liming material. (2) Nutrients required by plants for normal growth. Macronutrients such as nitrogen, phosphorus, and potassium are needed in large quantities by most plants.

Maculata/maculatum/maculatus. Epithet meaning spotted or mottled in color.

Magnesium (Mg). A metallic element found in many rock types (e.g.. dolomite, magnesite) and essential in life for metabolism. A constituent of chlorophyll. Deficiency in the soil can cause Chlorosis in plants and can lead to gass staggers in livestock, and is corrected by applying magnesian limestone (Ground Limestone), Epsom Salts or Kieseritc, etc. Other Fertilizers containing magnesium include farmyard manure, kainit and basic slag. Magnesium is a key element in chlorophyll, thus is essential in the photosynthesis process. Stem strength is influenced by magnesium. *Signs of Deficiency:* Leaves become chlorotic and yield is reduced when leaves are deficient. Rose flowers may droop at the neck. Epsom salts are used a source of magnesium.

Maiden tree. A fruit tree, about 1 year old. Also a fruit tree developed from a seed as distinct from one developed from a graft.

Maiden. (1) One - year old growth, tree or shrub. (2) A rose in its first season after budding

Mailing merton series. A group of apple rootstocks resulted by crossing the Mailing series with 'Northern Spy' to incorporate resistance to woolly aphis

Mailing series. A group of apple rootstocks evolved at the East Mailing Research Station in

England, which vary very much in vigour. M27 is a popular dwarfing rootstock of the series

Main canal. In irrigation practice, it refers to a canal which extends from the source of water supply to the beginning of the distrubution system.

Maintenance ration. The amount of food which when supplied to an animal will keep it healthy and maintain a constant live weight, with no loss or gain.

Maintenance requirements. Portion of the energy of the food which *is* used in the life process of the animal for keeping the animal warm and for movement of the body.

Maize. A cereal plant (Zea Mays) now commonly grain, silage. forage and fodder. It requires good drainage, grows best on lightish soils and is frost-sensitive. Characterised by a thick.stout, tall stem, and numerous long leaves arising alternately on each side of the stem. The male flower (tassel) forms at the lop of the shoot, and several female flowers form in the axils of some of the middle-stern leaves and develop into long cylinders (cobs) of tightly packed gain. Maize is rich in Carbohydraie but low in utilisable Portein and Fibre. It is much used as a Meal for fattening livestock but is also fed flaked. Also called corn on the cob.

Malathion. An organophosphorus insecticide and acaricide available either alone or in mixtures (i.e., with D.D.T.). Used in liquid or dust from, particularly to control skin parasites of cattle, and to control various crop pests.

Malodorous. Emitting a foul smell or odor

Malting barley. The best quality barely sold for malting, represent-ing about one fifth of the crop, for which a premium price is paid. It has a high Starch content (required for obtaining Maltose) and a low Protein content (determined during testing by measuring Nitrogen content). Good quality malting barley, after being harvested well, should be dry, and should smell 'sweet' (with no suggestion of mustiness). Also the grain should be evenly size with few broken or damaged ones, evently coloured wite wrinkly surface, and free from weed seeds and disease. These qualities are important for malting, and tests with strict criteria are carried out by millers before grain is accepted from termers.

Malting. The process whereby barely or other gain in soaked in water and allowed to partly germinate (during which stored Starch is broken down to Maltose by the enzyme Diastase), and is then kiln dried, after which the roots and shoots are removed (malt culms) and the remaining grain (the malt) is used in brewing. The malt is steeped in water and the maltose extracted into solution (the wort) which is run-off, leaving the residual grain which is used as an animal Feedingstuff.

Mame bonsai (Miniature bonsai). Bonsai using very dwarf plants (5-20 cm).

Manettii. Name given to a strong vigorous type of rose which is often used as a stock in the propagation of cultivated rose varieties,

Manganese. (Mn). A metallic Trace Element required for metabolic processes as an enzyme activator, it can be deficient in Neutral or Alkaline Soils rich in Organic Matter, causing Deficiency Diseases in crops (e.g., Speckled yellows in beat, Grey Leaf in cereals), which is usually rectified by spraying with manganese-sulphate. Cattle infertility may result if it is deficient in pasture.

Mango-ginger. *Curcuma amada,* a spice with the smell of unripe mango and colour of ginger; commonly used for pickling.

Manioc. A tropical plant *(Manihot esculenta)* producing a tuber with a high Starch but low Protein content, used as a cereal substitute in animal feedingstuffs. Also called cassava or tapioca.

Manual thinning. Removal of some branches, flower buds, flowers or fruits by hand (See Chemical thinning)

Manure spreaders. Various types of machine used to distribute solid or liquid Manures on the land. Also called muck spreaders. The main types include: (a) Self-emptying trailers containing a slatted conveyor in the base, delivering the (mainly solid) manure to a revolving mechanism at the rear which shreds and flings it on the field. Also called trailer-spreaders. (b) Cylindrical, open-topped tanks with a p.t.o. driven, cen-tral, axial shaft which operates a set of attached flail chains, which throw the manure (usually containing Slurry) out to one side of the tank. (c) Gun mechanisms which distribute, under pressure, slurry conveyed via a pipe from a nearby fixed tank or from a mobile vacuum tanker. The latter are sometimes used to spread slurry directly onto the land.

Manure. A general term for anything added to the soil to increase its content of plant nutrients (Fertilizers). Normally applied to or-ganic, usually bulky, materials mostly derived from farm and animal waste products, such as Compost, Dung, Farmyard Manure, Slurry, Sewerage Sludge, etc. They are slower acting than inorganic fertilizers but have the advantage of adding Organic Matter to the soil, increasing its capacity to retain water, and improving its structure (Green Manure). Organic matter, excreted by animals, which is used as a soil amendment and fertilizer. Green manures are plant cover crops that are tilled into the soil.

Manurial residues. That proportion of a manure added to the soil but not used by the crop, or residues left in the soil by a crop, which are available to future crops.

Manurial value. Those nutrients in a feedingstuff which an animal does not take in and pass out with the faeces so as to enrich the dung. Also called the residual value or fertilizer value of a feedingstuff.

Marginal land. Poor quality land suffering from various permanent disadvantages such as steep slopes, bad drainage, poor thin soil, harsh environmental conditions, etc.

Mariculture. The form of aquaculture where fish, shellfish, or aquatic plants are cultured in a salt water environment.

Maritime climate. Climate influenced by the ocean (or other factors having the same influence) resulting in mild temperature and small daily and annual variations in temperature.

Market allocation. A quantity provision in a fruit or vegetable marketing order specifying the maximum amount of the regulated commodity that can be sold for a given use or market (such as the domestic fresh market).

Market garden. Land used to grow produce, mainly vegetable and fruit, for sale (usually to a market, e.g., Covent Garden Market) for direct human consumption, as distinct from for processing purposes. The term is often restricted to the production of vegetables on a small scale.

Market gardening. Growing an assortment of vegetables for local or roadside markets.

Marl. Calcareous clay which is used as a dressing to improve soils particularly through the addition of clay to light soils.

Marmalades. These are made usually from citrus fruits and consist of jellies of the concerned fruit containing shreds of peels suspended in them. A good jelly marmalades can be prepared by using sweet orange (Malta) and Khatta (sour orange) in the ratio of 2 : 1 by weight, or Santra (loose orange) and Khatta 2 : 1 by weight.

Marsh. An area of usually low-lying land, often flooded by fresh or saline water, with poorly drained, impermeable mineral soil usually silt or clay (as distinct from the organic soils of Fens) which is continuously waterlogged. Characterised by sedges and rushes. Many marsh area have been artificially drained and protective embankments established to prevent flooding, with subsequent improvement of their grazing value, and in some cases the introduction of arable cropping.

Mash. A mixture of animal foods, crushed and stirred to a soft pulpy state with water (wet mash) or without it (dry mash).

Mashlum, maslin. A mixture of oats and barley and sometimes wheat, sown to provide grain for feeding to livestock. Occasionally beans or peas may be added to the mixture but ripening in uneven and they are not suited to being harvested by a Combine Harvester. Also called dredge or meslen.

Mat watering (capillary watering). Irrigation of potted plants by capillarity. The pots are kept on an absorbent mat which provides a constant supply of water to the base of the plants.

Mat. A closely interwoven or tangled mass of dead grass which develops when a pasture is undergrazed, particularly on acid soils.

Mattocking. Gradual removal of the pseudostem after harvesting the banana bunches, for favouring the uninterrupted growth of the follower sucker.

Mature. (1) Chronologically speaking, it is an organ or tissue that is fully developed; for example, ripe; developmentally speaking, a plant that is able to produce flowers (reproduce); contrast with juvenile which is unable to reproduce. (2) A later phase of growth characterized by flowering, fruiting, and a reduced rate of size increase.

Maturity. (1) Stage of full growth, development or ripeness. (2) Stage at which a plant is capable of reproducing by seed

Maxi stumped budling. A budling of rubber which has been allowed to grow in the nursery for three to four years and cut back at two to three metres above the bud union, for transplanting (See Mini stumped budling)

Maximum cropping. Cropping for the highest possible production per unit area per unit time without considering cost of production or net return.

Mazzard. Wild cherry which is used extensively as a rootstock upon which commercial varieties of cherries are grafted.

Meadow fescue. A loosely tufted perennial grass *(Festucapralensis)* forming large russocks, with narrow, bright green hairless leaves, shiny on the underside, and loose, green or purple panicles. Valuable for hay and grazing, and particularlyfound on rich, well-drained, fertile clay soils, and orten abundant in water meadows. A very palatable grass which grows well mixed with white clover or timothy.

Meadow. Strictly applied to an area of grassland used to produce Hay or Silage rather than for grazing. Also used loosely for rich waterside fields which are frequently flooded by river water, either naturally or via sluices and which are grazed (water meadows).

Measure-weight test. A test which gives an indication of the relative heaviness, pulpness and quality of seeds. For this purpose, a specified volume of seed is weighed and represented as kilos per hecto litre.

Measuring weir. Device or structure which is used for measuring the flow of water. It generally consists of a broad-crested weir or a rectangular, trapezoidal, triangular, or other shaped notch, frequently located in a vertical, thin plate over which the water flows. The weir head is an index of the rate of flow.

Medfly. A shortened name for the Mediterranean fruit fly, a destructive pest of fruits and vegetables that is found throughout most of Central America. The Animal and Plant Health Inspection Service is involved in programs to keep the Medfly from spreading north into Mexico, where it could easily enter the United States on imported winter fruits and vegetables. Eradication efforts in California, Florida and Texas have prevented infestations from becoming established. Hawaii is infested with the Medfly and no eradication efforts currently are under way. Travelers returning to the continental United States from Hawaii or a foreign country are prohibited from bringing into the country fresh fruits, meats, plants, birds, and plant and animal products that may harbor pests or diseases.

Media. Substrates in which plants are grown e.g. soil, sand, peat, moss, vermiculite, humus etc.

Median Lethal Time (MLT). Refers to the time required, following administration of a specified dose of radiation, for the death of 50 per cent of the individuals in a large group of animals or organisms.

Medicinal plants. Plants yielding secondary metabolites like alkaloids, steroids, glycosides, coumarins, flavanoides, etc. which have preventive and curative properties.

Melanized soil. A soil having a very dark colour to a considerable depth in the surface layer, due to the incorporation of humus.

Melitten. Poisonous protein like substance which is present in the venom of bees.

Mercerize. Refers to the process for treating cotton fibres, yarns, or fabrics with a solution of concentrated caustic alkali to make them stronger. If kept under tension during the process they acquire a luster.

Mericlinal chimera. A chimera in which the outer layer of genetically different tissue does not extent completely around the stem, occupying only a segment of the circumference.

Meristem. (1) A tissue of actively dividing cells, from which various organs or tissues are developed. (2) Areas of actively dividing cells and plant growth; capable of developing into specialized tissues. The growing tissue area in a plant; all buds contain a meristem. Areas of actively dividing cells and plant growth; capable of developing into specialized tissues.

Merous. A suffix preceded by an Arabic number or numeric prefix such as *tri-*, indicating how many of each part a *flower* possesses; 5-merous would be a flower with 5 *sepals*, 5 *petals*, etc., while *trimerous* would have three of each part.

Mesocarp. The middle layer of cells of the pericarp or fruit wall; found between the endocarp and exocarp

Mesophytes. Plants with characteristics intermediate to xerophytes and hydrophytes and adapted where moisture and aeration are between the extremes.

Met farm income. The money and non-money income form operators realize from farming as a return for labor, investment, and management after production expenses have been paid. Net farm income is measure in two ways net farm income before inventory adjustment and net farm income after inventory adjustment. Net farm income doesn't include changes in the value of inventories such as crops and livestock at the end of the year.

Metabolite. Chemical participating in metabolism e.g.a nutrient.

Metered chop harvesters. Two kinds of forage harvester, producing a uniform short chop, having either a flywheel or cylinder type of chopping mechanism. With both types the chop length and rate of delivery by a feeder to the chopping mechanism can be controlled. These harvesters are usually trailed, but self-propelled machines are available. Also called precision chop harvester.

Methanol. A liquid alcohol (also known as methyl alcohol or wood alcohol), formed in the destructive distillation of wood or made synthetically, and used especially as an alternative fuel, a gasoline additive, a solvent, an antifreeze, or a denaturant for ethyl alcohol. As a gasoline additive it lowers the carbon monoxide emissions but increases hydrocarbon emissions.

Methyl bromide. A fumigant used for soil treatment, to control pests in postharvest storage, for killing pests on fruits, vegetables, and grain going into export trade, for plant quarantine treatment, and for fumigation of buildings. Because methyl bromide contributes to depletion of stratospheric ozone, it is subject to phase out requirements of the 1987 Montreal Protocol on Ozone Depleting Substances and of the Clean Air Act (CAA). The Montreal Protocol and Vienna Adjustments require a complete phase out in industrialized countries by the year 2010, and a future freeze in developing country use. A 1998 amendment (P.L. 105-178, Title VI) conformed the Clean Air Act phase out date with that of the Montreal Protocol. All methyl bromide regulations so far exempt quarantine and pre-shipment treatment of agricultural commodities; however, this exemption is being reevaluated after completion of additional scientific assessments. Methyl bromide is regulated as a pesticide under the Federal Insecticide, Fungicide, and Rodenticide Act (FIFRA), as a hazardous substance under the Resource Conservation and Recovery Act (RCRA), and is subject to reporting requirements under the Emergency Planning and Community Right to Know Act (EPCRA).

Micro budding. Method of budding which is commonly used in propagating citrus trees. A bud is removed from the bud-stick by giving a flat cut just underneath the bud with a sharp knife after cutting off the leaf petiole just above the bud. An inverted T-cut is made in the stock wherein the micro-bud is slipped keeping the right side up. The entire inverted T-cut including the bud is covered and exposed only after 2-3 weeks so that healing gets completed.

Micro nutrients. Mineral elements which are needed by some plants in very small quantities.

Microbe. Microscopic organism belonging to either the animal or the plant kingdom.

Microbial insecticide. Pathogen which is used for killing insects. Free from harmful residue, nonphytotoxic to crops of higher dose, harmless to beneficial insects.

Microbiology. Science and study of microscopic organisms.

Micro-budding. Budding in which the bud piece is reduced to very small size.

Microclimate. The enviroment immediately surrounding a plant; very localized climate conditions. Many different microclimates may occur at the same time in different areas of a graden. The local climatic condition around and within the plant canopy, resulting from the modifications of the general climatic conditions by local difference in relief, exposure, cover etc. Variations of the climate within a given area; a very local or small-scale climate such as valley or woodland.

Microingredients. Added vitamins, trace minerals, antibiotics, and other materials used in minute amounts.

Micro-nutrients (Minor elements or Trace elements). Plant nutrient elements necessary in extremely small amounts only (usually less than 50 parts per million) for normal plant growth. These include B, Cl. Cu, Fe, Mn, Mo and Zn. (2) Nutrients required by plants for normal growth. Micronutrients such as iron, zinc, and manganese are needed in small quantities by most plants.

Microplot thresher. Self-cleaning machine thresher operated by a gas engine or by electricity for threshing the grain from micro-plots in rapid succession without mixing the seed of two varie-ties.

Micro-plots. Small plots used for the accurate determination of the comparative performances of varieties

Micropropagation. *In vitro* propagation of plants using micro sized plant materials

Micro-species. Species of very small size and range of genetic diversity which may later be considered as genetic variants of larger species.

Midrib/midvein. The rib or central vein running down the center of a leaf.

Mildew. Various fungi, which are parasitic on plants, causing disease. The threads or hyphae of some Phycomycete fungi penetrate deeply (Downy Mildews), whilst other types of Ascomycete fungi grow more on the leaf surface with a powdery white appearance (Powdery Mildews). Also called moulds.

Mill. A machine or a building equipped with machinery for grinding (by crushing) cereal grain or other dry feedingstuffs to produce flour or meal (Crushing Mill, Hammer Mill, Plate Mill). Also to remove seed husks.

Milling wheat. The best quality wheat purchased from producers by the flour-milling industry for bread making, etc. Such wheat, normally grown under contract, must reach specified standards, and samples are subject to visual and qualitative test before acceptance by the miller, *e.g.* Grain Moisture Content, cleanli-ness and purity, damaged and diseased grain content, odour, bushel weight, protein content, Gluten quality, Aphaamylase content, etc.

Million reaction. A test for presence of tyrosine in the molecule. The Million reagent consists of mecurous nitrate in nitrous acid. Four to five drop of this reagent are added, mixed and gradually heated with 5 ml of solution. A white precipitate is formed which turns red on standing. The reaction fails to occur if the protein is not precipitable by strong acid.

Milo. A small drought-resistant grain sorghum with compact bearded heads of large yellow or whitish seeds.

Mineral matter. Mineral elements and compounds, found as a constituent of the soil and living organisms, and essential to metabolism. Mineral matter in the soil consists of fragments of rock in various stages of degradation and their decomposition products, e.g., quartz, clays, iron oxides, etc., intimately mixed with organic matter. Some of the mineral elements may be present in plants and animals as mineral salts or combined in organic compounds such as proteins. Certain parts of plants have a higher mineral content (e.g. roots, vegetative parts), and as green crops mature the mineral content of the Dry Matter diminishes. The useful mineral matter content of feedingstuffs (expressed as Ash) of animal origin is normally higher than in those of vegetable origin.

Mineral mixtures. Minerals including Trace Elements, supplied to animals to ensure an adequate supply for metabolism and to prevent deficiency diseases, infertility, lack of thrift, or poor production etc. Often added to concentrate mixtures or cakes. Also given loose as mineral mixtures in field troughs, or as hard slabs (mineral licks or salt licks) for livestock to lick at will.

Mineral. A term mainly applied to inorganic substances, usually occurring naturally in the soil or rocks as distinct from those substances which are organic, being derived from living matter. (Mineral Matter.)

Mini stumped budling. Smaller stumped budling of rubber having brown colour up to 60 cm of the scion (See Maxi stumped budling).

Miniature garden. A rock garden done on a miniature scale, laid in troughs of concrete, stone or wood.

Miniatures. Cut flowers consisting of several flowers, supported on short and branched stems.

Minimal nutritional value. Refers to foods that may not be sold in competition with the school lunch and breakfast programs. These are foods that USDA has determined contain little if any nutritional value. For example, sugar candy, soda pop without fruit juices, and chewing gum are considered to be foods of minimal nutritional value. Candy containing nuts or chocolate is considered to have some nutritional value.

Minimum access. In the Uruguay Round Agreement on Agriculture, countries are obliged to provide minimum levels of imports for products subject to tariffication. Access is assured by tariff-rate quotas.

Minimum tillage. The minimum soil manipulation necessary for crop production. Conservation tillage, reduced tillage, and no-till farming are related terms.

Minor crops. Crops that may be high in value but that are not widely grown. Many fruits, vegetables, and tree nuts come under this definition. The IR-4 program is one publicly funded program to help producers of minor crops with their unique problems.

Minor oilseeds. Oilseed crops other than soybeans and peanuts; usually a reference to the other oilseeds eligible for marketing assistance loans under the FAIR Act of 1996 (sunflower seed, canola, rapeseed, safflower, mustard seed, and flaxseed).

Mist chamber. A nursery structure where high relative humidity is maintained through intermittent spraying of water for: facilitating easy propagation of cuttings.

Mist. A gentle spray of water over the surface of plants.

Mistletoe. Mistletoe is a parasitic plant found on trees and shrubs in the south western part of the United States. The plant derives all its nourishment (food and water) from the trees and shrubs the vine grows on. Plants die out during Illinois winters. Plants bloom in summer and produce a white POISONOUS berry during the winter months... which is why the plant is often used during the holiday for decorations. Most firms will market the mistletoe with artificial berries to avoid any potential lawsuits. Historically, mistletoe was thought to cause fertility. Swedish herbalists used the plant as a poison antidote. Why is mistletoe used for kissing under? According to Nordic legends, an arrow made of mistletoe almost killed Balder, son of Freyja (Goddess of Love and Beauty). Her grateful tears of joy turned into the mistletoe berries. She, then, promised to bestow a kiss on anyone who passed beneath the mistletoe.

Mites. A sub-class of the Arachnida allied to ticks, found in many habitats (e.g. the soil decaying Organic Matter) and parasitic on both plants and animals.

Mixed bud. Bud containing primordial tissue for both leaves and flowers.

Mixed cropping. (1) Growing perennials along with the main crop. (2) Growing two or more crops simultaneously without any definite row arrangement. (3) System of sowing two or three crops together on the same land, one being the main crop and the other one or two subsidiary. Has some advantages as of following crop rotation: less of risk of total crop failures, better soil utilization and satisfying the requirements of the farmer in food and fodder. Common mixtures followed in India are upland rice, cotton and red gram, jowar with *Sateria italica, Panicum miliare and Paspalum scrobiculatum,* bajra with fibres like mesta and sannhemp, ragi with ginger, red gram, cowpea and jowar.

Mixed fertilisers. Consists of individual or straight fertilizer materials blended together to permit application in the field in one opera-tion. Mixed fertilisers supply two or three major plant nutrients. The percentages of these nutrients are expressed as fertiliser grade.

Mixed intercropping. Growing two or more crops simultaneously with no distinct row arrangement

Modified leader. System of training trees, intermediate between central leader and open centre.

Moisture penetration. Depth to which water penetrates in a soil following irrigation or rain; the downward movement of water in a soil.

Moisture release curve. Depicts the relationship between soil moisture tension and the absolute moisture content in the range from field capacity to wilting point.

Moisture. Percentage. Mode of expressing moisture in the soil; the ratio of the weight of water in the soil to the dry weight of the soil, and expressed as a percentage of dry weight.

Molar solution. One gram molecular weight of a substance dissolved in one litre of water or solvent.

Molasses. A thick treacle produced as a by-product of sugar refining, containing 20-30%

water, and rich in soluble sugars (particularly Sucrose). Used to improve the palatability of Feedings tuffs, and as a binding agent in the production of Cake. Also used as a laxative and as a wilting agent in Silage making.

Mold. A type of fungus that grows on decaying plant tissue.

Mole plough. An implement used to form mole drains, consisting of a torpedo-shaped cylinder (a mole or catridge) attached to a vertical, knife-edged coulter, which is itself fixed to a metal beam suspended between two wheels. Attached directly behind the mole is an expander (a tapered cylinder of greater diameter) which slightly enlarges the drain being cut, smoothing and compressing the drain wails as the mole is drawn through the soil.

Molisch test. A biochemical test for carbohydrate. A few drops of a-naphthol are added to a solution or suspension of the sus-pected carbohydrate and concentrated sulphuric acid is slowly poured down the side of the test tube. A violet ring at the juncture of the two solutions indicates carbohydrates.

Molybdenum. (Mo). A metallic trace element required by Legumes for Nodule formation. When present in excess (e.g., the Teart soils of the West Country) it can cause scours in Ruminants, particularly cattle.

Monetization. A P.L. 480 provision added by the Food Security Act of 1985 that allows private voluntary organizations to sell a small percentage of donated P.L. 480 commodities within the recipient country. The currency generated by these sales can then be used for such purposes as defraying the cost of food distribution within the country.

Monocrop. A cropping system in which the same crop is grown year after year in the same field.

Monoculture (Monocrop). (1) A cropping system in which the same crop is grown year after year in the same field. (2) A pattern of crop or tree production that relies on a single plant variety. (2) The growing of the same crop on a field yea rafter year. (3) Growing only one particular type of plant.

Monoecious. (1) A species that has separate male and female reproductive structures (flowers or cones) on the same plant. (2) A term applied to plants having both separate male and female flowers on the same individual plant, e.g., hazel. (3) Condition in which staminate and pistillate flowers are produced on the same plant. (4) Monoecious literally means "one house." Monoecious plants have flowers of both sexes on one plant.

Moor, Moorland. Open country, usually on high ground, having acid soil (mostly Peat) and covered with heather, coarse grasses, sedges, bracken, etc.

Moribana. Piled flowers' in which the flower arrangements are made in shallow containers (One style of Ikebana)

Morifolium. Epithet meaning leaves like mulberry, such as mums.

Morimono. Flower arrangement in combination with fruit and or vegetables.

Mosaic diseases. Various virus diseases of crops causing light and dark mottling on affected leaves, e.g., Virus Yellow. Moss. (1) A northern term for a bog. (2) A class of small lower plants (Bryophyta) with simply constructed leaves, and lacking woody tissue.

Moss. Moss can be a symptom of several problems in the turfgrass and garden: (a) Poor drainage. Aerate the soil yearly if necessary. (b) Improper soil pH. Check level and modify accordingly. The soil pH levelshould be between 6.0 and 6.5 for most turfgrass species. (c) Shade. Moss can not grow in sunny conditions. Trimming lower branches orprofessionally thinning the tree's canopy may limit moss development. (d) Too much water. Avoid over irrigation. Unfortunately, you can't prevent rain. There are no chemicals currently available to homeowners that will limit or prevent moss growth.

Mould. (1) Loose soil with a good tilth. Also soil rich in humus. (2) A general term for a variety of minute fungi which produce a powdery or downy growth. Also called mildew. (Phycomycetes)

Mouldboard. A curved steel plate on the body of a plough which turns over the furrow slice. Also called breast or shell-board. Design varies according to the type of plough. The General Purpose Mouldboard is almost flat and is twisted along its length, producing a continuous almost unbroken furrow slice. (Digger. Semi-Digger and Lea Plough). When worn, the leading edge or shin of the mouldboard may be replaced.

Mower. An implement used to cut a grass crop. The main types are : (a) Cutter bar mowers, incorporating a high speed reciprocat-ing knife mechanism. They may be fully mounted or partially supported on wheels, and driven by p.t.o. or gearing from the wheels. The crop receives a single cut, without further laceration and is deposited in a neat swath. (b) Rotary mowers. The flail type consists of flail knives mounted on a horizontal axle, which lacerate the grass, assisting drying (Haymaking), and are useful for cutting laid crops. The vertical spindle type consists of one to four vertical drums containing hinge mounted knives which rotates horizontally at high speed. (c) Cylinder mowers, usually comprising a gang of three independent horizontal cylinders, each bearing a number of cutting blades. Normally tractor-drawn and used for cutting short grass. Rarely used on farms now-a-days.

Mower-conditioner. A combination machine which both cuts the grass crop and conditions it to enhance drying. Various conditioning mechanisms are used, all of which bruise and abrade the waxy coating of the stems to increase epidermis premeability.

Mowing strip. Edging placed between a lawn and a planting bed or patio, set just below ground level to provide a flat surface for one wheel so the lawnmower blades can cut to the edge of the lawn, saving the later step of snipping the edge grass with scissors or a weed-whacker. Mowing strips are commonly made of bricks, pavers, or poured cement.

Mulch farming. (1) A system of farming in which the organic residues are left on the soil surface instead of ploughing. (2) System of a farming in which the organic residues are not ploughed into the ground, but are left on the surface.

Mulch. (1) A natural or artificial layer of plant residue or other material on the soil surface. Mulch reduces erosion, conserves soil moisture, inhibits weed growth, and can provide the soil with organic matter as it breaks down. Mulch till prepares the soil so as to leave plant residues (or other mulching materials) on or near the surface. (2) Material such as straw, leaves, sawdust, grass cuttings, loose soil, etc., applied to the soil surface to protect the soil and plant roots from drying our and from the effects of heavy rain, freezing, etc. (3) A layer of organic or inorganic material on the soil surface. Mulches help to moderate the temperature of the soil surface, reduce loss of moisture from the soil surface, suppress weed growth, and reduce run-off. (4) Any loose material placed over the soil to control weeds and conserve soil moisture. Usually a coarse organic matter, such as leaves, clippings or bark.

Multiple cropping. (1) Growing more than one annual and /or perennial crops in the interspaces of the main crop. (2) Growing two or more crops consequently in the same field in the same year.

Multiple fruit. A fruit formed from several flowers included in a single structure having a common axis. Ex. Pineapple, mulberry

Multistory cropping. Growing a friendly combination of crops, having their canopies intercepting solar energy at varying heights and roots foraging the soil at different zones.

Muriate of potash. A fertilizer containing at least 60% potassium chloride, and as much as 3% sodium chloride or common salt. Available as a straight granular fertilizer but mostly incorpo-rated in compound fertilizers.

Mutagen. (1) A physical or chemical agent that is capable of inducing mutation in organisms. (2) An agent that causes a permanent genetic change in a cell other than that which occurs during normal growth. Testing to determine mutagenicity is one component of assessing the potential chronic toxicity of pesticides and other chemicals.

Mutation. (1) A sudden heritable change in the genetic make up of the cell; can be spontaneous or artificial. (2) A change in the expected genetic or tissue makeup of a plant. A sudden change in genetic material resulting in an altered individual. Generally disadvantageous to the mutated plant's survival.

Mutual self-help housing. A program to assist groups of low-income families in building their own homes. Each family is expected to contribute at least 700 hours of labor in building homes for each other. Participating families generally have low income and are unable to pay for homes built by the contract method. The homes are generally financed by Section 502 loans.

Mutualism (Eusymbiosis). Form of symbiosis in which two organisms belonging to different species live in intimate association to the advantage of both.

Mycelium. (1) The mass of interwoven filaments (hyphae) that makes up the vegetative body of a.fungus. This is the portion of the fungus that absorbs nutrients. (2) Thread-like growth parts of many fungi, often white or gray.

Mycorrhizae. Fungi having a symbiotic, non pathogenic association with the roots of many vascular plants. They may remain superficial(ectotropic) or within the host cell (endotropic).

Mycotoxin. Toxin produced by a variety of fungi such as Aspergillus. Penicillium, Rhizopus and Fusarium. Ingestion of the toxin containing food causes the syndrome known as mycotoxicosis. Toxic substances produced by fungi or molds on agricultural crops that may cause sickness in animals or humans that eat feed or food made from contaminated crops. There are between 300 and 400 known mycotoxins, but of most concern, based on toxicity and occurrence, are aflatoxin, vomitoxin, zearalenone, funonisin, T-2 toxin, and T-2-like toxins (trichothecenes). GIPSA currently measures for aflatoxin in all exports shipments of corn, and measures for aflatoxin and vomitoxin on a voluntary basis in domestic shipments.

N

N.P.K. An abbreviation for Nitrogen, Phosphorous, and Potassium in describing the composition of a Fertilizer. (Unit of Fertilizer, Plant Food Ratio).

Nagana. Protozoan disease of animals characterized by progressive anaemia and emaciation, intermittent fever and sometimes muscular and nervous disturbances. It is caused by four types of trypanosomes viz. Trypanosoma, congolense, Trypanosoma vivax, trypanosoma, brucei and trypanosome simial.

Nail Puller. (1) A tool used in agriculture. (2) A bar used to remove a nail from a surface. The bar has jaws that grip the nail so it can be pried out.

Naked. Said of a bud which is not covered by scales; the outer pair of leaves serves the same purpose; a flower with no floral envelope (perianth).

Nana/nanus/nanum. Epithet meaning dwarf or smaller.

Narcotic. A drug that in mild doses dulls the senses, relieves pain, and induces profound sleep but in excessive doses causes stupor, coma or convulsions.

Nascent. In the act of being formed.

Nastic movements (Sleep movements). Reversible and repeatable plant movements produced in response to stimuli e.g. leaf movement in *Mimosa.*

Natural ecosystem. An eco-system that ultimately produces the climax community, which develops in the absence of any major disturbance and consists of a complex, self regulating food web of several trophic levels.

Natural rubber. Natural rubber a high molecular weight polymeric substance with visco elastic properties, obtained from the latex of plant species, mainly, *Hevea braziliensis.* Structurally it is cis 1, 4-poly isoprene (See Synthetic rubber).

Naturalism. Planting to create the effect that plants grew in that space without human intervention; gardening to keep or enhance existing natural features.

Necrosis. (1) Death of organs of a plant, either as blight or death of tissue in localized areas, usually inside fruit add stems, or die-back or death of stems or branches, or blast or a separation of a part or of organ, or antheracnose or limited lesions. Character-istic symptom in certain plant disease. (2) Death of a cell or group of cells in a tissue in animals due to various causes, such as invasion by certain microorganisms. (3) Tissue browning and drying up due to death of cells.

Necrotic. Death of the plant cells, especially when it results in the tissue becoming dark in colour. From Greek nekrOsis: localized death of living tissue

Needle. A slender, leaf usually in the shape of a needle or awl commonly found in some conifers, such as Junipers; contrast with scale-like

Negative tropism. Growth of plant parts away from the source of stimulus (Roots are generally negatively phototropic)

Nematodirus disease. A disease of lambs caused by parasitic worms *(Nematodirus spp.).* They are transmitted by eggs in the faeces which later hatch in pasture to infect the following year's lambs, causing diarrhoea and lack of thrift, and sometimes anaemia.

Net farm income. The return (both monetary and non-monetary) to farm operators for their labor, management and capital, after all production expenses have been paid (that is, gross farm income minus production expenses). It includes net income from farm production as well as net income attributed to the rental value of farm dwellings, the value of commodities consumed on the farm, depreciation, and inventory changes.

Net irrigation requirement. Depth of irrigation water, exclusive of precipitation, stored soil moisture or ground water, that is required consumptively for crop production and required for other purposes, such as leaching.

Net veined. See netted venation. With veins reticulated and resembling a fish net; the interstices close.

Nettlehead. A virus disease of hops resulting in weak deformed Bines with little ability to climb or fruit.

Neutral soil. A soil in which the surface layer is neither acid nor alkaline in reaction, *i.e.*, its pH is about 7.0 (strictly, neutrality is pH 7.0).

Neutralising value. An indication of the ability of a liming material (Lime) to neutralise soil acidity compared to the effect of Calcium Oxide as a common factor. The n.v of a liming material is the number of kgs of calcium oxide which would have the same neutralising effect as 100 kg of the material.

Ninhydrin reaction. Blue colour reaction used in the chromatogra-phic identification of amino acids. A blue compound is formed when ninhydrin is heated with alpha-amino acids or with pep-tides or proteins which contain at least one free amino and carboxyl group.

Nitrate of Potash. (KNO_3) Potassium nitrate. A fertilizer containing about 13% Nitrogen and Potash used particularly by horticultur-ists. Also called saltpetre.

Nitrate of soda. ($NaNO_3$) Sodium nitrate. A fertilizer containing about 16% nitrogen derived from natural deposits of 'Caliche' in Chile (also called Chile saltpetre). A fast acting fertilizer, sometimes used as a top dressing, particularly by horticulturists.

Nitrate. (1) A general term for any nitrogenous fertilizer. The term is specifically applied to salts containing the nitrate ion (NO3). The nitrogen in many compound fertilizers is in the form of the ammonium ion (NH4, which is quickly changed to the nitrate form by soil bacteria (Nitrogen Cycle). Nitrates ions are more readily absorbed by some plants roots (e.g., potatoes, grass, etc.) absorb nitrate and ammonium ions equally well.

(2) The nitrogen ion, NO_3-, is derived from nitric acid and is an important source of nitrogen in fertilizers. Nitrate pollution of drinking water, shallow wells being particularly vulnerable, is of concern because infants are especially sensitive. A nitrate drinking

water standard has been set under the Safe Drinking Water Act. An Environmental Protection Agency national survey of drinking water wells conducted from 1988 to 1990 indicated that 2.4% of rural domestic wells contained nitrate at or above the 10 mg/L standard. Higher rates of contamination have been found in areas of high vulnerability; for example, surveys along the upper Des Moines river indicate that 20 to 30% of wells exceed the standard.

Nitrification. (1) The process of oxidizing ammonium salts into nitrogen for plants. (2) The conversion by various soil Bacteria or organic nitrogenous compounds which are not available to plants, to nitrates which they can readily absorb. Organic matter from decaying animal and plant remains in the soil is reduced by enzyme action to amino compounds. These are broken down by *Pseudomonas* to ammonia and ammonium compounds, then to nitrates by *Nitrosomonas,* and eventually to nitrates by *Nitrobacter.*

Nitrochalk. A granular fertilizer containing ammonium nitrate blended with ground chalk, containing nowadays about 25% nitrogen. The chalk content reduces the acidifying effect on the soil of the ammonium nitrate.

Nitrogen cycle. The circulation of nitrogen atoms and compounds in nature, mainly caused by living organisms. Inorganic forms of nitrogen (mainly nitrates) are absorbed by plants and assimi-lated into complex organic compounds (e.g., Proteins). Plants either die and decay, or are eaten by animals, which produce excreta and also eventually die and decay. The nitrogenous products of excretion and decay are progressively degraded in the soil, mainly by bacterial nitrification eventually to nitrates. In addition there is secondary circulation of atmospheric nitro-gen which is 'fixed' and assimilated into proteins by certain bacteria living symbolically in root nodules of various legumi-nous plants and by various free living bacteria in the soil (Nitrogen Fixation). A certain amount of nitrogen is released from nitrogen compounds by denitrifying bacteria.

Nitrogen fixation. The combination by living organisms of free atmospheric nitrogen (N2) into nitrogenous compounds is known as biological nitrogen fixation (BNF). Several types of BNF are known, classified as symbiotic (Symbiosis)or non-symbiotic. Amongst the former, aerobic bacteria *(Rhizobiumspp)* living in the root nodules of leguminous plants, are probably the best known. These bacteria are able to 'fix' nitrogen, using energy supplied by the plants. A number of free-living bacteria (e.g., aerobic *Azotobacter* and anaerobic *Closlridium)* can also 'fix' nitrogen, using energy derived from soluble carbohydrates in Humus. In addition to bacteria several other organisms are known to 'fix' nitrogen, including certain blue green algae and yeasts. A certain amount of nitrogen in the atmosphere is combined into organic from by lighting (non-biological nitrogen fixation).

Nitrogen fixer. Some plants (notably many legumes such as clover, peas, beans, and alfalfa) have the ability to transform nitrogen from the air into a form of nitrogen that plants can use. As nitrogen is a key nutrient that often limits growth, nitrogen fixers can influence the fertility of the surrounding soil and the growth of neighboring plants. The actual "fixing" or chemical transformation is performed by bacteria that live in nodules on the roots of the host plant.

Nitrogen. (1) An odourless, colourless, gaseous element, forming about 78% of the atmosphere.

It is present in the soil in inorganic forms such as ammonium compounds (Ammonia) and nitrates, and in various organic forms such as ammo acids and protein complexes as constituents of humus, which are eventually broken down by bacterial action to nitrates (Nitrogen Cycle). As a major plant nutrient nitrogen is applied to the soil in the form of Fertilizers or Manures. Nitrogen is essential to life, mainly as a constituent of protein and nucleic acids (e.g., DNA), and in plants is vital for vegetable growth. Nitrogen deficiency results in stunted plants with yellowish leaves and poor root develop-ment. Excess produces dark green succulent foliage, but if prolonged may retard the maturing of crops, and can cause cereals to lodge (Laid Crop) due to the growth of long weak straws. (2) An element found in the air and in all plant and animal tissues. For many crops, nitrogen fertilizer is essential for economic yields. However, nitrogen can also be a pollutant when nitrogen compounds are mobilized in the environment (e.g., leach from fertilized or manured fields), are discharged from septic tanks or feedlots, volatilize to the air, or are emitted from combustion engines. As pollutants, nitrogen compounds can have adverse health effects (see nitrate and air pollution) and contribute to degradation of waters (see eutrophication). (3) Nitrogen is instrumental in increase leaf and stem growth, as well as the plant's dark green color. Nitrogen also influences the crispness and quality of leaves as well as quick early growth in the spring. *Signs of Deficiency:* Deficiency symptoms include a yellowing of foliage or a light green-yellow overall color. Plant growth is reduced and/or stunted.

Nodules. Round bacteria-filled swellings on the roots of legume plants. Swellings on the roots of leguminous plants (Legume) containing bacteria *(Rhizobium spp)* living symbolically with the root tissue and capable of converting atmospheric nitrogen into nitrogenous compounds (Nitrogen Fixation) using energy derived from carbohydrates supplied by the plant. These com-pounds are used to assimilate protein. Some are released to the soil and converted to nitrates which may be used by other plants such as the grass in a mixed clover-grass iey. The stored protein in legume crops is released to the soil on death or ploughing in, and is degraded to nitrates (Nitrogen Cycle). Legumes such as beans or clover are thus often grown prior to cereal crops to enrich the soil with nitrogen. The various strains of bacteria which form nodules are usually naturally present in the soil, but to achieve nodulation with lucerne the seeds are usually coated or 'inoculated' with the appropriate bacterial culture before sowing. The bacteria enter the plant via the root hairs.

Nonpoint source pollution. Pollutants that are not discharged or emitted from a specific "point" source, such as a pipe or smokestack. Nonpoint water pollutants are often carried from dispersed, diverse sources into water channels by rain-induced runoff. Runoff from streets, open pit and strip mines, and agricultural fields are prominent examples (see agricultural pollution). Nonpoint source air pollutants (often called fugitive emissions) include small dispersed sources, e.g., fireplace smoke, and uncontained emissions, like dust blown from fields and unpaved roads.

Non-program crops. Any agricultural commodity not covered by federal commodity programs. Program crops are wheat, corn, barley, grain sorghum, oats, upland cotton, and rice.

Non-recurrent apomixis. Apomixis in which a haploid embryo arises from the egg nucleus without fertilization.

Nonrenewable resources. Nonrenewable resources, in contrast to renewable resources, do not naturally replenish themselves within time limits that permit sustained yield (i.e., minerals and hydrocarbons, such as phosphate rock, limestone, petroleum). Nonrenewable resources may be called stock resources because of their fixed supply. Some resources, such as soil and water, can be termed either nonrenewable or renewable depending on circumstances. For example, some underground reservoirs replenish so slowly they are effectively nonrenewable, such as the Ogalala Aquifer. Soil that is eroding faster than its T value faces eventual depletion.

Non-road emissions. Pollutants emitted by non-road engines and non-road vehicles, e.g., farm and construction equipment, gasoline-powered lawn and garden equipment, and power boats and outboard motors. The Clean Air Act, §213, provides that the Environmental Protection Agency can issue regulations to limit emissions from these sources.

Normal crop acreage. The acreage on a farm normally devoted to a group of designated crops. When a set-aside program is in effect, a participating farm's total planted acreage of such designated crops plus set-aside acreage cannot exceed the normal crop acreage. The authority for set-asides was eliminated by the FAIR Act of 1996.

Normal flex acreage. A provision of the Omnibus Budget Reconciliation Act of 1990 requiring a mandatory 15% reduction in payment acreage. Under this provision, producers were ineligible to receive deficiency payments on 15% of their crop acreage base (not including any acreage removed from production under any production adjustment program). Producers, however, were allowed to plant any crop on this acreage, except fruits, vegetables, and other prohibited crops. Normal flex acreage no longer exists under the FAIR Act of 1996.

Normal moisture capacity. Amount of water which is normally held by a soil after the free or gravitational water has drained off.

Normal yield. The average historic yield established for a particular farm or area. Can also describe average yields. Normal production would be the normal crop acreage planted multiplied by the normal yield. These measures, required by previous commodity programs to calculate benefits, are not required for production flexibility contracts under the FAIR Act of 1996.

Notified kind/variety. Seed or planting materials of any kind of variety notified under Section 5 of Seeds Act, 1966.

No-till farming. A method of planting crops that involves no seed bed preparation other than opening the soil to place individual seeds in holes or small slits; usually no cultivation during crop production; chemical weed control is normally used. May also be referred to as slot tillage or zero cultivation.

Noxious Weed Seed (Primary). The seeds from any plant of those species considered to be extremely destructive or harmful to agriculture. These species are designated by law for Seed Law Enforcement.

Noxious Weed Seed (Secondary). Seeds of those noxious werd species which may be present in agricultural seeds but only in limited number *i.e.*, their presence is not completely prohibited by Seed Law Enforcement.

Noxious weeds. Undesirable plants that infest either land or water resources and cause physical and economic damage. Under the Federal Noxious Weeds Act of 1974, (P.L. 93-629, January 3, 1975), the Animal and Plant Health Inspection Service works to prevent noxious weeds from entering the country, and conducts cooperative control/eradication programs with the states. Local governments frequently impose taxes on landowners to carry out noxious weed control programs.

NPK. Nitrogen, Phosphorous and Potash; the three main food elements necessary for plant life. Nitrogen is required for foliage and growth, phosphorous for flowers and potash for roots.

Nucellar embryony (Adventitious embryony). A type of apomixis where somatic embryos arise from a cell or group of cells either in the nucellus or in the integuments, in addition to the normal embryo.

Nucellus. A tissue originally making up the major part of the young ovule in which the embryosac develops

Nurse crops. (1) Crops grown during the early years of plantation crops for providing protection against wind and frost damages. (2) Companion crops which nourish the main crop by way of nitrogen fixation.

Nurse root grafting. Temporary grafting in which the nurse root sustains the plant till the scion forms its own roots. *Nursery:* A place where young plants are raised and maintained for permanent planting.

Nursery bed. Prepared soil area of convenient size, where seedlings are raised for transplanting to the main field

Nursery business. A business that specializes in the production of planting materials.

Nursery Potting Mixture. The University of Illinois has developed a potting mixture for nursery crops such as trees, shrubs and woody groundcovers: (a) Commercial Hardwood Bark, small enough to pass through a 1/2 inch mesh screen. (b) 6 pounds of ammonium nitrate. (c) 2 1/2 pounds triple super phosphate. (d) 2 pounds sulfur. (e) 2 pounds iron sulfate. (f) 1 pound epsom salts. Combine and wet ingredients to 70 percent moisture. Compost 30 days and turn; compost another 30 days and turn; compost 30 more days (90 days total) and use. Maintain a 70 percent moisture rate throughout composting. Mixture can be combined with rice hulls at rate of 70 parts bark to 30 parts rice hulls. If using pine bark, omit nitrogen, sulfur, iron sulfate and epsom salts. Add 5 pounds dolomitic limestone. Mixture of 50 percent pine bark potting medium and 50 percent hardwood bark medium works well.

Nursery. Series of plant progenies, introductions etc. grown chiefly for individual plant testing and examination in a plant breeding programme. This should not be confused with the term *nursery* used by rice growers to designate the bed where secdings are grown for transplanting to the field.

Nurse-seed grafting. Method (asexual) of plant propagation wherein grafting operation is performed when seeds have just germi-nated. The petioles of the cotyledons are cut off transversely just at the seed, leaving the cotyledons inside. A knife point is inserted into the seed between the cut petioles, making an opening for the scion. Scions are prepared from the dormant wood of the previous season's growth.

Nut. (1) A dry, indehiscent, 1-celled, 1-seeded fruit having a hard and bony mesocarp; the outermost endocarp may be fibrous or slightly fleshy. (2) A small cube or cylindrical piece of compressed meal. (3) A type of fruit, dry, one-seeded, and usually with a hand woody shell, e.g., hazel nut. (4) A hard and single-seed fruit found in many trees and shrubs. (5) An indehiscent, dry, one seeded fruit having a hard seed coat.

Nutrient deficiency symptoms. When any essential plant nutrient is seriously lacking in the soil, plants growing on it show certain colour development in leaves and certain change in growth. These symptoms are characteristic of each nutrient. The nature of the symptom varies from crop to crop and with the degree of deficiency.

Nutrient Film Technique (NFT). A water culture technique in which the plant roots are suspended in shallow circulating nutrient solutions, delivered as a continuous film of liquid.

Nutrient pollution. Contamination by excessive inputs of nutrient: a primary cause of eutrophication of surface waters, in which excess nutrients, usually nitrogen or phosphorus, stimulate algal growth. Sources of nutrient pollution include runoff from fields and pastures, discharges from septic tanks and feedlots, and emissions from combustion.

Nutrition Assistance Programs. Federal programs in Puerto Rico and American Samoa that provide food assistance through block grant funds in lieu of food stamps, and to the Northern Marianas under a covenant governing U.S. relations with that jurisdiction.

Nutrition guidelines. Federal guidelines established for meals served in child nutrition meal service programs governing the types of foods and nutrient content required in order to be eligible for reimbursement.

Nutritive Ratio. Ratio or proportion between the digestible protein and the non-nitrogenous nutrients including tat multiplied by 2.25, in a feed or a mixture of different feeds, Nutritive Ratio (NR) is expressed as follows :

- Digestible carbohydrates+
- digestible fibre + (Digestible fat x 2.25)

Digestible crude protein When the percentage of TDN is given the nutritive ratio is computed as follows :

- TON - Digestible protein NR
- Digestible protein

Thus the nutritive ratio is the ratio between the total non-nitrogenous nutrients and the digestible protein.

Nutsedge (or nut grass). Perennial. Two types: yellow and purple. Major weed of any irrigated system (home or farm). Primarily found in southern New Mexico. Very difficult to manage.

Nyctinasty. Daily nastic movements, especially of leaves which open during the day and fold at night (See Nastic movements; Photonasty)

O

Oast. A kiln in which hops are dried. An oast-house is a building in which hops are dried and pressed into pockets, consisting of one or more kilns and a cooling room containing a press.

Oatmeal. An animal feedingstuff consisting of ground oats with the husks removed to varying extents. The husks provide a rough-age element to the diet. Fed cither dry or together with other constituents in a wet mash. Also fed as a porridge (sometimes called brose) to animals which are ill or recovering from sick-ness.

Oats. A cereal crop with a high fibre content, the grains of which are used for stock-feeding and for making bread, porridge and oatmeal. Grown mainly in the cooler and wetter areas of World, although to a decreasing extent as it yields less well than barley and is more difficult to harvest. Most cultivated oats are of the species *Avena sativa* although some *A. strigosa* is grown in the certain areas. Oats are distinguished from other Cereals by having a flowering head in the form of a (all panicle with clusters of slender spreading branches bearing Spikelets.

Ob-. (1) Used as a prefix to indicate inversion; obcordate in contrast with cordate, oblanceolate in contrast to lanceolate, etc. A cordate leaf is heart-shaped with rounded lobes at the base of the leaf. An obcordate leaf has rounded lobes at the tip. (2) A prefix meaning inversion.

Obcordate. A leaf with rounded lobes at the tip.

Objectionable weed. It refers to those weed species, whose seed is difficult to separate once mixed with crop seed or which is poisonous or injurious or has a smothering effect on the main crop. It is difficult to eradicate once established because of its high multiplication ratio. It may serve alternate host for crop disease and pests.

Oblanceolate. Widest near apex, tapering to base, longer than wide.

Oblique. Lop-sided, as one side of a leaf base being larger, wider or more rounded than the other.

Oblong. (1) Longer than broad, with the margins parallel except at the extreme basal and apical ends. (2) Much longer than wide, with sides roughly parallel for most of their length.

Obovate. (1) Egg-shaped with broader end upwards. (2) Inversely ovate, broadest above the middle. (3) Widest above the middle of the leaf or leaflet

Obscure. Not easily made out, as applied to buds, bundle scars, etc.

Obtuse. With sides forming an angle of more than 90 degrees; having a rounded and blunt apex.

Ocean ranching. A type of aquaculture, used mainly by the salmon industry, which cultures juvenile fish, releases them to mature in the open ocean, and catches them when they return as adults to spawn.

Odd pinnate. A compound pinnate leaf with a terminal or odd numbered leaflet at the tip.

Oedema. Corky portions of leaves caused by excessive humidity.

Oestrogen. A sex hormone produced by female animals which controls the development of female characteristics including the Mammary Glands and controls the oestrus or 'heat', the period during which females are receptive to mating with males.

Official seed sample. The samples which are drawn by a seed control or seed Law Enforcement Officer. These are submitted to the laboratory to determine if a seed lot being offered for sale meets the requirements of the seed act.

Officinale. Applied to medicinal plants, especially those used by herbalist.

Offset (offshoot). (1) An asexually produced sucker or basal shoot, bulbil or cormlet. (2) A young plant that appears naturally on a parent, and is easily detached and propagated. (3) Small bulb or plantlet used to propagate new plants.

Offshoot. A lateral shoot or branch which arises from one of the main stems of a plant.

Off-type. (1) Any notable deviation from the normal or standard. (2) Plant or seed significantly deviating from the characteristics of a variety.

Off-year. The year in which an alternate bearing plant does not produce fruits.

Oilseed crops. (1) Primarily soybeans, sunflower seed, canola, rapeseed, safflower, flaxseed, mustard seed, peanuts and cottonseed, used for the production of cooking oils, protein meals for livestock, and industrial uses. These specific oilseeds are eligible for nonrecourse loans. Other oilseed crops include castor beans and sesame. (2) Various crops grown for their seeds, rich in oil and extracted alter crushing for the manufacture of margarine cook-ing oil and salad oil, etc. The residues are usually processed to produce cake, rich in protein and valued for stock feeding. Only Linseed, Oilseed Rape and Sunflower are grown in Britain. Various cakes derived from tropical oilseed crops are also commonly used (e.g., Groundnut Cake, Soyahean Cake).

Oilseed rape. Certain varieties of Rape grown under contract for their seeds, rich in oil and used for the manufacturing of marga-rine, cooking and salad oils. The crop is often used as a useful break between cereals and is conveniently harvested with the same machinery A valuable cake is a byproduct alier seed crushing and oil extraction. Also known as Canola in North America.

Olericulture. (1) Branch of horticulture which deals with cultivation of vegetables. (2) The art and science of vegetable growing.

Olland. The aftermath following the harvesting of hay from a clover Ley.

Omasum. The third stomach of a ruminant in which food is ground following partial digestion in the second stomach or reticulum. Also called manyplies or psallerium.

On-farm research. Research conducted on farms with the active participation of the farmers.

Ontogeny. The process of development of an organism, organ or tissue towards maturation.

Open centre. System of training fruit trees where the main stem is allowed to gum only upto a certain height by leading it within a year of planting and inducing all subsequent vegetative growth by lateral branches, resulting in a lowhead in which bulk of crop is borne to ground that in central leader tree.

Open furrow. A type of furrow in which the furrow slices are turned in opposite directions so that they lie facing away from each other. Open furrows are used when completing the ploughing of a land or a field (the finish).

Open pollination. The free exchange of pollen between flowers and plants irrespective of flower type or cultivar

Open-pollinated seed. (1) The seed produced as a result of natural pollination as opposed to hybrid seed produced as a result of controlled-pollination. (2) Seeds produced from plants that are allowed to pollinate naturally, primarily through insects, wind and water.

Opposite. (1) Said of leaves or leaf scars that are paired on opposite sides at each node. (2) The arrangement of leaves and buds that arise from a single node on either side of a stem

Options contract. An option contract gives the buyer the right, but not the obligation, to buy (call option) or sell (put option) a futures contract at a specific price within a specified period of time, regardless of the market price of that commodity.

Orbicular. (1) Circular or disc shaped. (2) Circular or globed, as in the shape of a leaf. (3) Flat with the outline circular or nearly circular; disc shaped. (4) Rounded in two-dimensional outline as many plant leaves

Orbiculate. Circular or disc-shaped.

Orchard. An area of land used for growing fruit trees (e.g., apples, pears, plums, cherries, etc.) either on a small scale or on a more extensive commercial scale. Now-a-days commercial fruit varieties are normally grown on the more dwarfing rootstocks selected for their ability to control the size, growth, and cropping charac-teristics, since it is now economically desirable to harvest early in the life of the trees and to do so from the ground or short steps. In addition tree shape is greatly affected by the pruning system used. At the end of their useful life orchards are grubbed, the tendency being to grub trees on dwarf rootstocks at a younger age than those on vigorous rootstocks. Orchards producing desert and culinary fruit are mainly operated by specialist growers, whilst a certain number of orchards producing cider fruit are still combined with livestock grazing.

Orcharding. Commercial production of fruits, mainly from perennials.

Orf. A disease mainly of sheep, but also affecting cattle and goats, due to a primary attack by a virus which causes pustules. These are secondarily infected by a fungus *(Fusiformis necrophorus)* and develop into ulcers and scabs on the lips and mouth, and on other parts including the face, legs tail, genitals, teats, etc. Also called contagious pustular dermatitis, contagious ecthyma, necroba-cillosis, of and ulcerative stomatitis.

Organic fanning. (1) Systems of farming which avoid the use of artificial fertilizers, pesticides or herbicides, and concentrate on methods of crop rotation and the use of home-grown feeds, organic fertilizers and manures, and ley farming. (2) An approach to farming based on biological methods that avoid the use of synthetic crop or livestock production inputs; also a broadly defined philosophical approach to farming that puts value on ecological harmony, resource efficiency, and non-intensive animal husbandry practices. Farmers who wish to have their operations certified as organic so that they can label their products as organically produced currently follow standards and submit to inspection by private or state certification organizations. A National Organic Program is authorized and being developed. (3) It is a method of farming, utilising organic inputs, which maintains soil health and ecological balance

Organic fertilizers. Those Fertilizers derived from animal products (e.g. Dried Blood, Hoof and Horn Meal, Shoddy, etc.) and from plant residues (*e.g.*, malt culms, castor seed meal) all of which have a considerable nitrogen content. Also applied to more bulky organic materials such as Manures

Organic foods. Food products produced by organic farming practices and handled or processed under organic handling and manufacturing processes as defined by several private and state organic certifying agencies. Once the National Organic Program is operational, there will a national standard for what constitutes organic foods based on uniform standards for organic production, processing, and handling.

Organic gardening. (1) A method of gardening utilizing only materials derived from living or once-alive things. (2) No longer a fringe movement, this is a mainstream way to grow things without resorting to chemical fertilizers and pesticides.

Organic manure. Carbonaceous materials of plant / animal origin incorporated in soil for improving fertility and physical properties.

Organic material. Any material which originated as a living organism.

Organic matter. (1) The Organic substances in the soil, including animals and plant residues in various stages of decomposition (Humus), cells and tissues of soil organisms (*e.g.*, bacteria, fungi, insects, earthworms, etc. those organic compounds syn-thesised by such organisms, and organic materials added to the soil as fertilizers, e.g., Manures. The organic fraction of the soil is intimately mixed with the mineral matter. The organic matter content of soil varies according to soil type. On a dry weight basis it is usually in the range 2-15%, but certain organic soils consist almost entirely of organic matter (*e.g.*, peats, black Fen soils). (2) One of the two main components of the dry matter of a feedingstuff (the other being Mineral Matter) and consisting mainly of Carbohydrates, Fats and Proteins. (3) A substance derived from plant or animal material. (4) Carbonaceous materials of either plant or animal origin, which exists in all stages of decomposition in soils.

Organic. (1) A term applied to a substance containing carbon combined with hydrogen, and very often also with nitrogen, oxygen and other chemical elements, forming complex molecules. (2) Chemically, a compound or molecule containing carbon bound to hydrogen. Organic compounds make up all living matter. The term organic frequently is used to distinguish "natural" products or processes from man-made "synthetic" ones. Thus natural fertilizers include manures or rock phosphate, as opposed to fertilizers synthesized from chemical feedstocks. Likewise, organic farming and organic foods refer to the growing of food crops without the use of synthetic chemical pesticides or fertilizers; pests are controlled by cultivation techniques and the use of pesticides derived from natural sources (e.g., rotenone and pyrethrins, both from plants) and the use of natural fertilizers (e.g., manure and compost). Some consumers, alleging risks from synthetic chemicals, prefer organic food products. The FACT Act of 1990 required USDA to define organic foods for marketing purposes and implement a National Organic Program. (3) Something that was or is alive; containing carbon.

Organophosphates. Insecticides that contain phosphorus, carbon, and hydrogen. They are cholinesterase inhibitors; some are highly acutely toxic, but they usually are not persistent in the environment. Parathion is an example of an organophosphate.

Organophosphorus compounds. A group of synthetic non-persis-tent chemicals mainly used as insecticides (*e.g.* Malathion), now-a-days often in preference to the more persistent Chlorin-ated Hydrocarbons. They may have either a Systemic or a non-systemic action and mainly affect the nervous system.

Ornamental floriculture. The study and cultivation of various groups of ornamental plants which are used to decorate indoor and outdoor gardens.

Ornamental gardening. The study and cultivation of various groups of ornamental plants which are used to decorate indoor and outdoor gardens (Synonym: Ornamental floriculture)

Ornamental horticulture. The branch of horticulture specializing in the areas of floriculture, tuft grass management, nursery stock production and landscaping.

Ornamentals (Ornamental plants). (1) Plants grown for their aesthetic value rather than commercial usefulness or food value. (2) Popular abbreviation for a division of Horticultural Science dealing with plant cultivated for landscape or display value.

Orris Root. Orris root is a classified as a fixative (to keep color and fragrance longer) for making potpourri and can be purchased in small quantities at most pharmacies and drug stores. It can be poisonous and should be handled with care. Orris root is the root of a specific type of iris that is dried and ground. Drying and grinding bearded iris will not result in the same type of compound and is not recommended.

Ortet. The original mother plant or donor plant from which a group of vegetatively propagated plants is derived.

Orthodox seeds. Seeds that can be dried to a low moisture level and which show a loss of viability with the rise in moisture content (See Recalcitrant seed).

Orthophotography. Aerial photographs that more precisely show the features of the landscape, including those that might be important for agriculture such as slope or size of gullies, because they are corrected for distortion caused by tilt, curvature, and ground relief.

Orthotropic. Vertically oriented, with an upright growth habit. The antonym is plagiotropic.

OSHA. Occupational Safety and Health Administration.

Osier. Any willow grown as a coppice crop, the flexible twigs of which are used in basket making, particularly *Salex viminalis.* Also called withy.

Osmosis. The diffusion of liquid through a semi-permeable membrane.

Osmunda fibre. The root mass of *Osmunda regalis* (royal fern) used for potting, particularly for certain epiphytic orchids and bromeliads.

Other crop seed. Seeds of those plants which are grown as crops, other than the main crop.

Otto of roses. Essential oil (volatile oil) used as perfume.

Out-breeding. Crossing of genetically different plants. It is used to introduce new genes and increase heterozygosity (See Inbreeding and Heterosis).

Outcross. An off type resulting from the crossing of a parent plant with a plant other than the selected parent

Out-crossing. The mating of unrelated animals of the same breed, normally in order to introduce a desired character to the strain which is otherwise absent.

Outfall. The point where a drain discharges into a ditch, or where a watercourse discharges to a river or the sea (sometimes associ-ated with a sluice or flap valve).

Outfield, out-lying field. (1) A field situated some distance from the farm buildings usually on the farm boundary. (2) *A* Scottish term for a field under continuous arable cropping without being manured.

Outliers. Livestock kept outside during the winter.

Out-run land. The open unenclosed areas surrounding a Hill Farm, normally some distance from the farm buildings.

Out-wintering. The practice of keeping cattle in the fields during the winter as distinct from keeping them under cover.

Ovary. (1) The enlarged basal portion of a pistil which contains the ovules or the seeds. (2) The ovule-bearing part of a pistil. (3) The ovule-bearing part of a pistil; the female part of a flower containing immature seeds (ovules).

Ovate. (1) Egg-shaped with the broadest part towards the base. (2) Widest below the middle of the leaf three or more leaftlets radiating fan-like from a common base (for example, Horsechestnut); may also refer to lobes of some leaves (for example, Sugar Maple).

Oven dry soil. Soil is considered to be oven dry when it has reached equilibrium with the vapour pressure of an oven at 105°C. The oven dry condition corresponds to a relative humidity of approximately 0 per cent or a pF near 7.

Over budding. Budding done at higher levels for converting an existing bud wood nursery of a clone to another clone, without replanting.

Over irrigation. Application of more water than is necessary for the needs of vegetation, resulting in loss of water through seepage and leaching with the resultant loss of humus, nitrogen and other mineral elements of the soil.

Overstocked. Condition of overcrowding in tree crops, leading to retarded growth.

Ovoid. Said of a bud which is egg-shaped, with the broadest portion near the base.

Ovule. (1) The egg-containing unit of an ovary, which after fertilization becomes the seed. (2) The structure which develops into a seed after the fertilization of the contained egg

Oxidase. A type of Enzyme which acts as a catalyst in the oxidation of substances, removing hydrogen which then combines with oxygen to form water.

Oxidation. The combination of oxygen with a substance, or the removal of hydrogen from it. In cellular respiration, oxygen reacts with carbohydrates such as Glucose yielding carbon dioxide and water as by-products and liberating energy. This process is comparable to the oxidation of substances by burning, in which energy in liberated as heat. The term is also used to describe chemical reactions in which electrons are lost from atoms such as when ferrous iron (Fe2+) is oxidised to ferric iron (Fe3+).

Oxygen (O). An odourless, colourless, gaseous element, forming about 20%, of the atmosphere and a constituent of water. Essential to most living organisms for respiration and necessary for combustion.

Ozone (O_3). A highly reactive molecule composed of three oxygen atoms. Environmentally, ozone is important in two completely separate contexts—one, as a naturally occurring screen of harmful radiation in the outer atmosphere (i.e., stratospheric ozone), and two, as a component of polluting smog formed from emissions resulting from human activities (i.e., urban smog). In the stratosphere 7 to 10 miles above the Earth, naturally occurring ozone acts to shield the Earth from harmful radiation. In the 1970s and 1980s, it was discovered that emissions of certain chemicals catalyze destruction of stratospheric ozone, allowing more radiation to reach the Earth's surface. The U.S. is a signatory to the 1987 Montreal Protocol on Ozone Depleting Substances, which bans or limits uses of chemicals whose emissions deplete stratospheric ozone. Among the chemicals being phased out as ozone depleters are chlorofluorocarbons used in refrigeration and air conditioning and methyl bromide, a pesticide. In the lower atmosphere (troposphere), ozone is a major air pollutant that contributes to smog, adversely affects human health, and is toxic to some plants, damaging forests and crops. Tropospheric ozone forms from reactions between nitrogen oxides and volatile organic compounds in the presence of sunlight. The precursor pollutants are emitted by combustion sources such as motor vehicles and utilities, use of solvents, and petrochemical facilities. Tropospheric ozone is regulated under a National Ambient Air Quality Standard.

P

Packhouse. A building in which fruit, vegetables, eggs, etc., are packed into boxes or containers for despatch from a farm to a market, retail outlet, store, etc. Casual Workers are usually employed seasonally for packhouse work.

Paddle. A long-nandled, spade-like tool used to remove mud from plough shares and mouldboards, etc. In some areas called spud.

Paddock grazing. A system of managing grassland in which grazing areas are divided into paddocks by permanent or semi-perma-nent fences. Each paddock is grazed in turn and then allowed to rest.

Paddy cleaner. Machine for removing foreign seeds, immature grains, foreign materials, such as stones, nails, dirt and dust etc. It is used prior to shelling.

Paddy weeder. Equipment for interculture, used in paddy cultiva-tion. It is used for uprooting weeds and burying them in puddled soil between rows of standing paddy crop. It improves the aeration of soil.

Paid diversion. A program, repealed by the FAIR Act of 1996, under which farmers were paid to voluntarily take acreage out of production. The diverted land was devoted to approved conservation practices. Unlike acreage reduction and set-aside programs, participation in a paid diversion program was not normally a condition of eligibility for other support program benefits.

Pale, paling. A wooden stake driven upright into the ground for fencing. Also a wooden board nailed vertically with others to horizontal pieces to form pale fencing or paling.

Pallet(s). (1) Low portable bench used to grow and transport bedding and potted plants. (2) Wooden or iron frame structures used to support a stack of bags, to prevent them from touching the floor and absorbing moisture.

Palmate. (1) Lobed, divided in a palm-like manner. (2) Radiating, fan-like from a common point; as of leaflets of a palmately compound leaf or veins of a palmately-veined leaf.

Pan. A hard layer in the soil that is (a) compacted, such as a plough pan caused by continual ploughing to the same depth, (b) indurated, such as an iron pan in which the soil particles have become cemented together by precipitated iron and silicates, or *(c)* has a very high clay content. Pans often impede drainage and root growth, and are broken down using a subsoiler. Also called hardpan or sole.

Panicle. (1) A branched cluster of stalked flowers. (2) An indeterminate branching racemose inflorescence with stalked flowers.

Papain. Water soluble proteolytic enzyme (protease) present in the latex of unripe papaya fruit. It is used as a meattenderizer and has industrial and pharmaceutical applications.

Paprika (Hungarian paprika, Sweet pepper). A non-pungent variety of chilli belonging to *Capsicum annuum.*

Paraquat. A contact herbicide which disrupts photosynthesis causing chlorosis. Used to control a wide range of annual broad-leaved weeds and grasses. It is very poisonous to animals.

Parasite. (1) An organism that lives and derives its food from another plant or animal. (2) An organism that lives on or in another living organism (called a host). The parasite obtains its food supply from the host. (3) An organism which lives in or on another (the Host) from which it derives food, for all or part of its life. Often, though not always, harmful to the host. Animal parasites include internal worms *(e.g.,* Roundworms, Tapeworm, etc.) and external Flies, Lice, Ticks, etc. Some are parasitic on a secondary host during part of their life cycle *(e.g.,* tapeworm, Liver Fluke). Plant parasites include fungi and bacteria (causing such diseases as Mildew, Rust, Fire Blight, etc.), various plants such as dodder and mistletoe, and numerous insects and their larvae, *e.g.,* Aphids, Fruit fly, Eelworms, etc. A semi-parasite is one which feeds partly on its host but also manufacturers some of its own food *(e.g.,* mistletoe).

Parasitoid wasp. Many species of wasp prey on other insects. They deposit their eggs into the larvae or pupae of the host insect, and the wasp larvae then consume the host. These wasps are correctly termed parasitoid rather than parasitic, since they kill their hosts. Parasites or parasitic animals live off a host without killing it, in a relationship that could be viewed as a type of symbiosis. Read more about specific *parasitoid insect species.*

Parathion. An extremely poisonous Organophosphorus Compound (diethylpara-nitrophenyl-thiophosphate) used to control a vari-ety of insect pests including Mites, Aphids, Eleworms, etc.

Parent stock. The individual plants or animals from which others have been bred, often in large numbers.

Parity price. A measurement of the purchasing power of a unit of a particular commodity. Originally, parity was the price per bushel, bale, pound, or hundredweight that would be necessary for a unit of a commodity today to buy the same quantity of other goods (from a standard list) that the commodity could have purchased in the 1910-14 base period. Under permanent law, prices of some commodities would be supported at 50 to 90% of parity through direct government purchases or nonrecourse loans. In 1948, the parity price formula was revised to make parity prices dependent on the relationship of farm and nonfarm prices during the most recent 10-year period for nonbasic commodities. Basic commodities, including wheat, corn, rice, peanuts, and cotton use the higher of the historical or the new formula.

Parthenogenesis. (1) Seed development without fertilization. (2) The growth of an ovum into a new individual without fertilization. It occurs naturally in some plant *(e.g.,* dandelion) and animals *(e.g.,* Aphids), and the offspring are normally all diploid and genetically identical to the parent.

Partners in quality. An AMS documented quality assurance system for fresh produce packing houses. Those that incorporate the program's rigorous quality standards and requirements (monitored by periodic unannounced AMS audits) into their ongoing daily operations may not have to undergo the traditional, more costly, and less flexible end-of-the-line inspection that AMS conducts before awarding a quality grade to the company's products, according to the agency.

Parts per million (ppm). Unit of measurement used for expressing the quantity of material in a solution or mixture. In metric units, one part per million is equal to one milligram of material in one litre of the solution or mixture.

Pasture. An area of grassland used only for grazing as distinct from one cut to produce Hay or Silage (a Meadow).

Pastureland. Land used primarily for the production of domesticated forage plants for livestock (in contrast to rangeland, where vegetation is naturally-occurring and is dominated by grasses and perhaps shrubs). Rotation pasture or cropland under winter cover crops is not included in this definition. The 1992 national resources inventory recorded 126 million acres of pastureland, 9% of all nonfederal rural lands.

Pathogen. (1) An organism capable of causing a disease. (such as fungus, bacterium, or virus) (2) An organism or substance which causes disease. (3) Infectious agent that causes disease.

Pathogen; pathogenic. Pathogens are infectious or toxin forming microorganisms causing disease. A food borne pathogen is a microorganism that causes illness through the ingestion of food.

Pears. Various high-quality varieties of pear *(Pyrus communis)*, particularly Conference and Cornice, are grown commercially in Orchards in the East and South-East of England. Perry pears are grown in the West and South-West of England. The area grown (except Perry pears) has declined slightly in recent years and, according to the f980 Annual Review White Paper, was estimated at 4500 hectares (c. 11 100 acres) in 1979.

Peas. A leguminous crop *(Pisum sp.)* with a long trailing stem, compound leaves and leaf stalks terminating in climbing ten-drils.

Peat moss. An organic soil additive from Sphagnum and related mosses. The partially decomposed remains of various mosses. This is a good, water retentive addition to the soil, but tends to add the acidity of the soil pH.

Peat soil. Soil acidic, pH 3.9 and below, 10 to 40 per cent organic matter with partly decomposed raw plant material, suitable for paddy when water recedes.

Peat. (1) It consists of the remains of aquatic, bog. marsh or swamp vegetation which has been preserved under water in a partially decomposed state. (2) Partially degraded vegetable matter found in marshy areas. Peat is commonly used as asoil amendment.

Pectin. Soluble gelatinous polysaccharides in ripe fruits. Used as a setting agent in jams and jellies.

Ped. A natural unit of soil structure as distinct from clods or soil fragments formed artificially by tillage.

Pedate. A palmately divided or compound leaf whose two lateral lobes are again cleft or divided.

Pedicel. (1) Flower stalk. (2) The stalk of a flower or fruit when in a cluster or when solitary. (3) The stalk which directly supports a flower or fruit.

Pedigree registered animal. One whose forebears have been re-corded with a Breed Society.

Pedigree selection. The selection of animals in stock-breeding on the basis of desirable features or qualities in their ancestors.

Pedigree. The recorded line of ancestry of cultivated plant varieties or animals, much used in determining programmes.

Peduncle. (1) The main supportive stalk of a flower or fruit cluster. (2) The stalk of a flower cluster or a single flower when the flower is solitary, or the remaining member of a reduced inflorescence. (3) The stalk of a flower cluster or a single flower when the flower is solitary, or the remaining member of a reduced inflorescence. (4) Inflorescence stalk.

Peeler. A plant which has high nutrient demand and in the absence of fertilizers would impoverish the soil.

Peeper. Very young banana sucker, bearing scale leaves only, as it appears above the soil surface.

Pellet application. Fertilizers are applied in the form of pellets one to two inches deep between the rows of paddy.

Pellet feed. Agglomerated feeds formed by extruding individual ingredients or mixtures by compacting and forcing feed through die opening by any mechanical process.

Pelleted seed. Layer of particulate matter around the seed. The pelleting is usually made from finally, pulverized limestone or mineral phosphate. Use of synthetic or arabic gum helps in holding the layer of these materials on the surface of the seed.

Pendula/pendulous/pendulum/pendulus. Weeping or drooping with branches hanging down from a central stem; a weeping or spreading cultivar.

Pendulous. (1) Hanging or drooping, can refer to plant form or parts, like branchlets or inflorescences. (2) More or less hanging or declined.

Percolation. The movement of water downward and radially through subsurface soil layers, usually continuing downward to groundwater. The rate at which soils permit percolation is a measure of the vulnerability of groundwaters to contamination by surface waters as well as a determinant in the siting of septic fields.

Perennation. (1) Surviving from year to year. (2) Persistence of fruit long after its usual season of maturity.

Perennial plant. A plant that lives for more than 2 years, often living for many years. Almost all woody plants and many herbaceous plants are perennials.

Perennial. (1) A plant that dies down to the ground during winter and survives to grow again each spring; plants that persist for more than two years. (2) A plant that lives for more than two years. (3) Of three or more seasons duration. (4) Plants with stems which persist for more than 2 years.

Perennials. (1) Live through several growing seasons, and can survive a period of dormancy between growing seasons. These plants regenerate from root systems or protected buds, in addition to seeds. (2) Plants which do not perish after flowering but continue to live for at least three or more years or seasons.

Perfect flower. (1) A flower having both functional stamens and pistils; a plant with both functioning male and female parts. (2) A perfect flower contains both the male and female parts (stamen and pistil). The flower may or may not contain sepals and petals.

Perfect stage. The stage in the life cycle of a fungus in which spores are produced after sexual fusion. Also known as sexual stage.

Pergola. (1) An extension of an arch using climbers in a garden. (2) An arbor with an open roof of rafters supported by posts or columns; an overhead covering for an outdoor structure. An open framework over a walkway or path, which is usually designed to be covered in climbing plants.

Perianth. (1) A collective term for the floral envelopes, usually the combined calyx (sepals) and corolla (petals). (2) The two floral envelopes of a flower; a collective term embracing both corolla and calyx as a unit. (3) The two floral envelopes of a flower; a collective term embracing both corolla and calyx as a unit; often used when it is not possible to distinguish one series from the other (as in most monocots) and the parts then called tepals.

Pericarp. The fruit wall, consisting of three distinct layers *viz.*, the exocarp, the mesocarp and the endocarp

Periclinal chimera. Chimera in which genotypically or cytoplasmically different tissues are arranged in concentric layers

Perlite. (1) A light weight white-colored soil additive from volcanic materials. (2) Small pieces of expanded volcanic rock, used in potting media. Permanent sod. A soil management system in which a sod is periodically mowed but no tillage is carried out.

Permaculture. The maintaining of a permanent *horticulture* or agriculture by relying on renewable resources and compatibility with the local *ecosystem.*

Permanent vegetative cover. Trees, or perennial grasses, legumes, or shrubs with an expected life span of at least 5 years. Permanent cover is required on cropland entered into the Conservation Reserve Program.

Permanent wilting percentage. The percentage water content of a soil in that condition when a crop has removed so much water from it, that which remains is insufficient to meet the require-ments of the crop, as a result of which it wilts.

Permearneter. Device for measuring the permeability of soils or other material. It usually consists of two reservoirs or tanks, connected by a conduit containing the material under investigation, water being passed from one reservoir under varying conditions of head etc. through the connecting conduit.

Permitted acreage. The acreage on which a farm program participant is permitted to grow a program crop after satisfying acreage reduction requirements. For example, when a 10% acreage reduction program is in effect for wheat, a farmer with a 100-acre wheat base may grow wheat on 90 acres, the permitted acres. Limits on production are eliminated by the FAIR Act of 1996 through the year 2002.

Persicifolia. Having leaves shaped like the peach.

Persistent chemicals. Those which remain active for long periods after they are applied as herbicides, insecticides, etc.

Persistent pesticides. Pesticides that do not readily break down in the environment. Becoming long-lived components of the ecosystem, these chemicals may have enduring effects at low concentrations or may bioaccumulate, posing hazards to higher predators.

Persistent. Not deciduous, as applied to leaves; not disappearing, as applied to pith, pubescence, epidermis, etc. Adhering to a position instead of falling, whether dead or alive.

Pest(s). (1) Animals, insects, fungi, weed, etc., which are troublesome to crops or livestock, or cause harmful diseases, and result in considerable losses in agricultural or horticultural production. (2) Any insect or animal which is detrimental to the health and well being of plants or other animals.

Pest Resistance Management (PRM) plans. To protect the continued use of biopesticides, the Environmental Protection Agency is requiring companies developing transgenic crops (see genetic engineering) to submit and implement pest resistance management (PRM) plans as a requirement of product registration. If they are exposed to a toxin excessively, most insect populations can develop resistance, making pest control products less effective. With new biopesticide technologies comes the concern that pests will rapidly develop resistance to natural insecticides. A resistance management plan is intended to sustain the useful life of transgenic technology.

Pest scouting. Inspecting a field for pests, including insects, weeds, and pathogens. Pest scouting is a basic component of integrated pest management programs. It is used to determine whether pest populations are at levels that warrant control intervention and also may help to determine the most appropriate method of control.

Pesticide. (1) A substance used to kill, control, repel, or mitigate any pest. Insecticides, fungicides, rodenticides, herbicides, and germicides are all pesticides. Environmental Protection Agency regulates pesticides under authority of the Federal Insecticide, Fungicide, and Rodenticide Act (FIFRA). In addition, under FIFRA, a substance used as a plant regulator, defoliant, or desiccant is defined as a pesticide and regulated accordingly. All pesticides must be registered and carry a label approved by EPA. (2) Used to destroy pests. Fungicides (destroys fungus), herbicides (destroys plants), insecticides (destroys insects), and nematicides (destroys nematodes) are all pesticides. (3) A chemical used to kill an organism considered a pest. Various categories of poisonous chemicals used to kill Pests, including fumigants, fungicides, Herbicides, Insecticides and rodenticides. Most are synthetic organic chemicals, pro-duced in concentrate form but diluted for application with various substances such as water, talc, clays, kerosene, etc., to ensure even distribution. They are available in a variety of forms including dusts, granules, solutions, suspensions and emulsions. Some are persistent in the environment (Chlorinated Hydrocar-bons) and are dangerous to wild-life populations. A consider-able amount of information about pesticides is now available to users, and the introduction and use of pesticidal chemicals is now controlled under various voluntary schemes.

Petiole. (1) A stalk that attaches the leaf to the stem; a leafstalk. (2) Leaf stalk. (3) The attachment structure between a leaf blade and the stem.

Petiolule. (1) In compound leaves, the attachment structure between a leaflet and the stem-like structure of the compound leaf (rachis). (2) Leaflet stalk. (3) Stalk of a leaflet. Negative logarithm of hydrogen ion concentration

Petroculture. The growing of crops for fuel. Following experiments in Brazil in the 1930s, alcohol from sugarcane and Manioc is now commonly added to petrol. Maize, Beet, and Sunflower are also under investigation in various parts of the world as fuel crops.

pH Value. A measure of the hydrogen ion concentration of a solution, and therefore an indication of its acidity or alkalinity. Expressed on a scale from 0 to 14 pH 7 is neutral, less than 7 is acid, more than 7 is alkaline.

pH. (1) A measure of acidity or alkalinity. A pH of 7.0 is neutral. Smaller numbers indicate acidity, larger numbers indicate alkalinity. Rhododendrons, like other members of the heath family, favor soils which are mildly acidic with a pH around 5.5. (2) An expression of the intensity of the basic or acidic condition of a liquid or of soil; the logarithmic scale ranges from 0 to 14, where 0 is the most acid, 7 is neutral, and above 7 is alkaline. Natural waters usually have a pH between 6.5 and 8.5. Plants have differing tolerances for acidity and alkalinity. (3) The scale used to denote the acidity or alkalinity (lime content) of a soil or growing compost. 6.5 pH is considered neutral and supports most plant life.

Phagocytes. White blood corpuscles which engulf Bacteria as part of the body defence mechanism of animals.

Phased planting. Spreading the planting of a crop over an optimum period in order to extend the availability of a produce for a protracted period or to minimise competition among the component crops in intercropping (See Staggered planting)

Phenology. (1) Science which relates climate to periodic events in animals and plant life. The phenological data for crops include for example, dates of sowing, germination and emergence, dates of flowering, ripening and harvest. (2) Science (hat deals with the time of appearance of characteristics periodic events in the life cycle of organisms in nature, especially as those events which are influenced by environmental factors.

Phenotype. (1) The external physical appearance of an organism (without reference to its genetic nature) resulting from an interaction between the genotype of the individual and the environment. (2) The visual appearance of a plant regardless of genetics.

Phloem. (1) Inner bark tissue responsible for translocation of foodstuffs (e.g. sugars and starches) produced by green tissue to the roots and other parts of the plant. (2) Nutrient-conducting vessels found throughout the plant. Phloem vessels transport nutrients produced in the foliage down through the stems, branches, or trunk to the roots. (3) Phloem vessels transport food (carbohydrates) and waste products throughout the plant. Phloem cells, like xylem cells, harden off and die after several months, forming bark tissue on woody plants.

Phosphates. Various salts of phosphoric acid applied to the soil as phosphatic fertilizers to supply phosphorus to crops.

Phosphorus. (1) An essential nutrient for plants and animals that is commonly applied to crops as a phosphate fertilizer. Phosphorus can contribute to the eutrophication of lakes and other water bodies. Sources of excess phosphorus include sewage and agricultural runoff. (2) A chemical element occurring naturally only in compounds, mainly as calcium phosphate in the mineral, apatite. Essential to life as a constituent of certain proteins and for many metabolic (Metabolism) reactions. In animals it is important in the formation and maintenance of bones and teeth, and for efficient Carbonhydrate utilisation, and is mainly derived from Bone Flour, milk, cereals, and oil-seed Cake. In plants phospho-rus is necessary for root development. It also enchances crop ripening and counteracts the weakening effect of excessive applications of nitrogen. It is applied to the soil in the form of various phosphatic fertilizers.

Photoblastic seeds. Seeds in which the germination is influenced by light. Germination is stimulated by light in positively photoblastic seeds, and inhibited in negatively photoblastic seeds

Photo oxidation. Oxidation of certain cell constituents when light intensity exceeds a limit, resulting in the bleaching of chlorophyll and inactivation of enzymes

Photomorphogenesis. The non-photosynthetic influence of light on germination, growth, development and reproduction

Photonasty. Nastic movements as influenced by light e.g. opening and closing of flowers at different parts of the daily cycle (See Nastic movements)

Photoperiodism. (1) The developmental responses of plants to the relative lengths of light and dark periods. (2) The effect of the length of day and night on plant flowering. Some plants are long-day plants requiring 14-16 hours of sunlight per day to flower, e.g., Wheat, lettuce, Beet, etc. Short-day plants require only 8-9 hours of sunlight to flower, e.g., kidney beans, rice, soyabean, chrysanthemum, etc. Some plants are unaffected by day length (day-neutral plants) e.g., tomato.

Photorespiration. Respiratory utilisation of photosynthetic products during day time or light period. Observed in C3 plants, unlike in C4 plants which have little or no photo-respiration.

Photosynthesis. (1) The process in green plants by which carbohydrates are synthesised from water and carbon dioxide using the energy of sunlight absorbed by chlorophyll. Also called carbon assimi-lation. (2) The process plants use to change air and water into food, using the sun's energy. (4) Synthesis of carbohydrate from carbon dioxide and water in the presence of chlorophyll, using light energy. (5) The internal process by which a plant turns sunlight into growing energy. The formation of carbohydrates in plants from water and carbon dioxide, by the action of sunlight on the Chlorophyll in the leaves. (6) The process by which plants use the sun's light to produce food (carbohydrates).

Photosynthetically Active Radiation (PAR). That part of light spectrum (between 400 and 700 nm) which is primarily absorbed by plants and used in photosynthesis.

Phycomycetes. A class of fungi, mainly single-celled and found in damp situations, *(e.g.,* powdery Mildews). Those possessing thread-like hyphae lack cross walls. Many attack crops *(e.g., Phytophthora infestans,* the cause of Potato Blight).

Phylogenetic system. System of classification of plants according to their evolutionary pedigree.

Phymatotrichum root rot. Occurs in pistachios and grapes (sometimes in alfalfa or cotton). Can kill plants within two weeks. Plugs water-conducting tissue of the plant. Only found in three states: Arizona, New Mexico, and Texas.

Physical analysis. The determination of the composition of a sub-stance by physical means. In Soil Analysis, for example, samples, are dispersed in water or simple solutions and the particles allowed to settle. After oven drying the soil is shaken through sieves of varying mesh size to separate Sand, Silt and Clay, and the proportion of each component then determined. This tech-nique is used to accurately determine the texture of a soil sample.

Physiological drought. (1) Condition when a plant is unable to take in water, because of the low ground-temperature, or because it holds substances in solution which hinder the absorption of water by the plant. (2) Non availability of water to the plant due to unfavourable physiological conditions.

Physiology. (1) The study of chemical function of any organism. (2) The study of the internal functioning of living organisms.

Phytology. (1) Study of plants. (2) Phytology (botany) is the *scientific study* of plant *life.* As a branch of *biology,* it is also sometimes referred to as plant science(s) or plant biology. Botany covers a wide range of scientific disciplines that study the *growth, reproduction, metabolism, development, diseases, ecology,* and *evolution* of *plants.* As with other life forms in biology, plant life can be studied at a variety of levels, from the *molecular, genetic* and *biochemical* level through *organelles, cells, tissues, organs,* individuals, plant *populations,* and *communities* of plants. At each of these levels a botanist might be concerned with the classification (*taxonomy*), structure (*anatomy*), or function (*physiology*) of plant life. Botanists studied all organisms that were not generally regarded as *animal.* Some of these "plant-like" organisms include: *fungi* (studied in *mycology*); *bacteria* and *viruses* (studied in *microbiology*); and *algae* (studied in *phycology*). Most algae, fungi, and microbes are no longer considered to be in the plant kingdom. However, attention is still given to them by botanists; and bacteria, fungi, and algae are usually covered, somewhat superficially, in introductory botany courses.

Phytopathology. (1) Science of plant disease and their control. (2) The study of plant diseases.

Phytophthora root rot (chile wilt). Occurs in chile. Causes complete plant loss late in the season. Severe under conditions of excess moisture. Develops rapidly and can kill whole field if conditions are favorable.

Phytoremediation. The systematic use of plants to treat environmental contamination. It is being investigated as a potential low-cost technology to help meet environmental regulations. For example, it has been discovered that young seedlings of Indian mustard (*Brassica juncea*) grown in aerated water are very effective at removing toxic metals from water.

Picker combine. A machine used to harvest Maize, the picking mechanism consisting of snapping rollers which remove only the cobs from the stems. 2-, 3-, or 4-row models are used.

Piecework. A basis of employment often used on farms by which seasonal workers such as fruit pickers are paid according to the amount of work done (e.g., per basketful picked), as distinct from according to an hourly rate.

Pigments (Photosynthesis). Pigments such as carotenoids and phycobili-proteins which obtain light energy and transfer it to the reaction centres.

Pinching (Pinching back). (1) Removal of shoot apex to overcome apical dominance and promote lateral shoot development. (2) Nipping off the very tip of a branch or stem. Pinching promotes branching, and a bushier, fuller plant

Pink root. Occurs in onions. Fungal disease that attacks roots, reducing bulb size and quality.

Pinna. (1) A leaflet of a compound leaf; when applied to ferns, the primary division attached to the main rachis; feather-like. (2) Primary leaflet of a compound leaf.

Pinnata/pinnatum/pinnatus. Having pinnate or feather-patterned veins or leaflets. Pinnatio means many small, lateral branches growing off a rib or axis.

Pinnate. (1) A compound leaf having leaflets (pinnae) arranged in two ranks, one on each side of a leaf. (2) A compound leaf with leaflets or segments along each side of a common axis or rachis; feather-like. (3) Leaftlets or lobes along each side of a common axis. (4) Resembling a feather; having parts or branches arranged on each side of a common axis: *a polyp with a pinnate form; pinnate leaves.*

Pioneer crop. A crop or mixture of crops (e.g., Italian ryegrass, rape and turnips) grown on ploughed-up grassland which has be-come worn out and the soil impoverished, prior to reseeding the grassland with a good seeds mixture. Pioneer crops are fertilized and grazed under heavy stocking (to accumulate dung and urine) and ploughed in, having the effect of improving the fertility of the soil for reseeding.

Pistil. (1) The female part of a flower composed of ovary, style and stigma. (2) The ovule producing part of a flower which is composed of stigma, style and ovary. (3) The pistil is the term for the collective female parts — the stigma, style and ovary. The stigma is the sticky receptacle or "landing pad" for the pollen. The style supports the stigma and is tube-like, leading to the ovary where the eggs/seeds are. (4) The seed-bearing organ of a flower, consisting of the ovary, stigma, and style.

Pistillate. An imperfect flower with a pistil, or seed organ, but having no functional stamens (male pollen producing organs).

Pitch pole. A type of Harrow with double-ended tines mounted on central shafts, so that they are able to be turned through 180°. This enables clogged or obstructed ends to be turned and the 'clean' ends to be brought into play.

Pitchfork. a farming implement with a long handle and two or three widely spaced, slightly curved prongs, used for lifting and moving hay.

Planker. An implement used to crush clods on land where a roller cannot be used, consisting of a number of fixed overlapping planks, shod with iron bars along the working edges, which is pulled over the lands. Also called float, rubber or scrubber.

Plant breeding. Applied branch of botany dealing with the crop improvement and production of new improved crop varieties far better than original in all aspects.

Plant food ratio. The ratio of nitrogen to phosphate and potash in a fertilizer. Thus a compound fertilizer containing 20% N, 10% P, and 10% K has a ratio of 2:1:1. Such ratios are a guide to the type of compound. A number of fertilizers may have same ratio and therefore be usable for the same purpose, but the quantity needing to be applied will vary in relation to the actual percent-age nutrient content.

Plant growth hormone. Organic compound, produced in very small amounts in one part of a plant and transported to another part where it promotes, inhibits or modifies physiological processes.

Plant growth substances (Plant growth regulators): Organic compounds (natural or synthetic) other than nutrients which in very small amounts promote, inhibit or modify physiological processes in plants.

Plant introduction. A method of plant breeding in which a variety or a species is brought to an area where it is not grown before

Plant murals. A hanging patch of green plants (herbaceous perennials) on the wall like a hanging picture, suitable for congested apartments in cities.

Plant positions in container gardens. A plant can fulfill different positions in a design depending on the size of the container.

Plant quarantine. A technique for insuring disease (and pest) free plants by isolating them during a period while performing tests for latent diseases. Often used when importing new cultivars.

Plant regulator. A chemical that affects the physiological behavior of plants, for example through accelerating or retarding the rate of growth or maturation of produce. Typically the definition of plant regulator excludes nutrients. Plant regulators must be registered as pesticides under the Federal Insecticide, Fungicide, and Rodenticide Act.

Plant tissue culture. *In vitro* cultivation of plant parts, whether a single cell, tissue or organ, under aseptic conditions in defined or semi defined media.

Plantation crops. A group of extensively cultivated perennial crops, demanding plant to plant attention.

Plantation. (1) A self contained settlement or estate on which plantation crops are grown. (2) In *forestry*, plantations of *trees* are typically grown as an even-aged *monoculture* for *timber* production, as opposed to a natural *forest*, where the trees are usually of diverse species and diverse ages. A plantation is not a natural *ecosystem*. Plantations are also sometimes known as "man-made forests" or "tree farms", though this latter term more typically refers to specialist tree nurseries which produce the seedling trees used to create plantations. More generally a plantation is a forest land where trees are grown for commercial use, most often in a naturally regenerated forest but could also be a planted. In the *United States*, the term "Tree Farm" is a *trade mark* of the American Tree Farm system, a third party verification system for certifying sustainable forestry. The American Tree Farm system dates back to 1941 as a progam to improve forestry practices on farms. The term tree farm is also sometimes used to describe the sale of live trees for landscaping. A plantation is usually made up of fast-growing trees planted either to replace already-*logged* forests or to substitute for their absence. Plantations differ from natural forests in several ways (a) Plantations are usually monocultures. That is, the same species of tree is planted in rows across a given area, whereas a conventional forest would contain far more diverse tree species. (b) Plantations may include introduced trees not native to the area, including (in a few cases) unconventional types such as *hybrid* trees and genetically modified trees. Since the primary interest in plantations is to produce wood or pulp, the types of trees found in plantations are those that are best-suited to industrial applications. For example, pines, spruces and eucalyptus are widely used because of their fast growth rate, and are good for paper and timber production. (c) Plantations are always young forests. Typically, trees grown in plantations are harvested after 10 to 60 years, rarely up to 120 years. This means that the forests produced by plantations do not contain the type of growth, soil or wildlife typical of old-growth natural forest ecosystems. Most conspicuous is the absence of decaying dead wood, a very important part of natural forest ecosystems.

Plantations are grown by state forestry authorities (for example, the Forestry Commission in Britain) and/or the paper and wood industries and other private landowners (such as Weyerhaeuser and International Paper in the United States). Christmas trees are often grown on plantations as well. In southeast Asia, rubber, oil palm, and more recently teak plantations have replaced the natural forest. By convention, plantations of fruit-bearing trees are termed orchards, even if grown on scales that occupy a landscape to the horizon. Plantations of grapevines are termed vineyards.

Planter. (1) Machine used for precision drilling, hill dropping or check - row planting. (2) A farmer who owns large estate or farm. (3) In its earliest sense, *planter* meant a person who helped "plant" or found a colony, often called a *plantation*. Thus, the founders of Plymouth Plantation called themselves "adventurers and planters" in a document drawn up on the eve of their departure for New England in 1620. But by then the modern meaning of *planter* had already begun to develop in Virginia. Ten years of precarious existence seem to have taught the English adventurers there one lesson: The way to get rich was tobacco. It was the one export crop that earned big money, and those who were planters of this crop began to become a wealthy elite. In 1619 the Virginia House of Burgesses used *planters* in this sense: "Provided first that the Cape Marchant do accept of the Tobacco of all and everie the Planters here in Virginia." As they sold more tobacco and bought more land, these planters needed more laborers, so they imported African slaves in ever-increasing numbers. *Planter* thus became the name for an owner of a large estate worked by slave labor. At first it referred to tobacco growers in Virginia and Maryland, but by the end of the seventeenth century it was applied to owners of PLANTATIONS (1645) in general, regardless of the crop, and anywhere in the South—in fact anywhere in the tropical and sub-tropical English-speaking world. There was another more modest early meaning for *planter*: any individual farmer, regardless of size or type of holdings. An article in the *South Carolina Gazette* of 1732 refers to "the poorer sort of Planters." But they were overshadowed, in terminology as well as trade, by wealthy owners. In the nineteenth century, as human labor began to be replaced by mechanical, *planter* was the name given to a device for planting seeds. Although labor-intensive plantations of the old sort are long gone, people still evoke the fortunate lifestyle of the plantation owner when they make planter's punch, a rum-based cocktail that is a twentieth-century invention.

Planting board. Device for planting trees in exact position. The planting board is usually 1.5 m long, 10 cm wide and 2.5 cm thick with a central notch and one hole on either end

Plant-pesticide. As proposed by the Environmental Protection Agency (November 23, 1994), plant-pesticides are all substances responsible for pest resistance in plants, as well as the genes needed for production of these substances. EPA has further proposed that plant-pesticide traits introduced into plants using recombinant DNA techniques should be registered under legal requirements of FIFRA and FFDCA. Exempt from tolerance requirements would be those defense substances and genes evolved naturally or transferred to the plant by traditional plant breeding methods.

Plashing. (1) The cutting or bending and intertwining the branches of small trees, as in hedges. (2) The dashing or sprinkling of coloring matter on the walls of buildings, to imitate granite, etc.

Plastochran. The time interval between two successive and similar occurrences e.g. the rhythmic initiation of leaves by the apical meristem.

Plate mill. A machine which coarse grinds grain, consisting of two circular plates of equal diameter positioned one above the other, with an adjustable gap between them to control the fineness of grinding. The lower plate is fixed whilst the upper one rotates against it. Plate diameter varies between 15 and 45 cm (6-18 in.)

Pleiotropy. (1) The condition in which several characteristics are affected by a single gene. (2) The control by a single gene of several distinct and seemingly unrelated phenotypic effects. (3) Pleiotropy is the multiple effects of a single gene. In other words, it is when a single gene controls several phenotypic traits. These traits may be seemingly unrelated. A classic example is the human disease PKU (phenylketonuria). A mutation in a single gene that codes for an enzyme needed for the metabolism of the amino acid phenylalanine causes mental retardation, and reduced hair and skin pigmentation. The term pleiotropy comes from the Greek pleio, meaning "many", and tropo, meaning "changes". The inverse of pleiotropy is epistasis, whereby the interaction of multiple genes affects a single phenotypic trait.

Ploidy. A general term indicating the number of sets of chromosomes e.g. diploid = two sets; triploid = three sets; tetraploid = four sets; pentaploid = five sets; hexaploid = six sets.

Plough sole placement. Method of fertilizer application in which fertilizer is placed in a continuous band on the bottom of the furrow in the process of ploughing. Each band is covered as the next furrow is turned, usually during the growing season.

Plough. One of the oldest types of tillage implement which breaks up the land, turning over the soil into ridges and furrows, burying surface vegetation and manure and exposing the new soil surface to the air in preparation for sowing seed. The earliest types were wooden and of simple design. The principal parts of modern ploughs comprise a steel frame to which are fixed a number of bodies, each consisting of a Coulter which cuts a vertical slice, and a Share which makes a horizontal cut under the furrow slice which is then turned by the Moulboard. The lateral accuracy and stability of each body is maintained by a Landside. Three main categories of tractors-drawn ploughs are in common use on farms, viz, *(a)* fully mounted types, attached to the three-point linkage and raised or lowered hydraulically, (b) semi-mounted types, in which a rear wheel partially supports the weight, and (c) trailed types, pulled by a drawbar.

Plough-in. To bury a crop, stubble, weeds, etc., by turning them under the soil with a plough.

Ploughing. (1) Tilage operations carried out with the help of tractor drawn or bullock drawn implements known as plough, before the crops are sown. (2) Tillage operations carried out for prepar-ing the seedbed of crops. Ploughing essentially means opening the upper crust of the soil, breaking the clods and making the soil suitable for growing seeds.

Ploughland. (1) Arable land. (2) The amount of land that could be tilled in a year by a plough drawn by a team of eight oxen, together with a proportionate amount of pasture, used historically as a basis for determining taxes.

Plumule. (1) The major young bud of the embryo within a seed from which develops the aerial portions of the plant. (2) Part of the embryonic axis above the cotyledons.

Pneumatic separator. Seed processing machine in which seed separation is made by air on the basis of differences in terminal velocity. It has a fan at the intake end and a pressure greater than the normal atmospheric pressure is builtup within the machine. The air blast is responsible for seed separation.

Pneumatophores. Vertical extensions of the buried roots of certain trees (e.g. mangroves) in marsh or swamp habitats. They help for exchange of gases between the atmosphere and the submerged roots

Poaching. The trampling of land when wet, mainly by cattle, so that the soil becomes churned and muddy. Poaching is a particular problem of heavy land, and leads to the deterioration of soil structure.

Pocket book plants. Calceolaria, sometimes called the "Pocket Book Plant", produces balloon like flowers. Most plants are grown from seed. Plants should be kept in bright light, but out of direct sun. Temperatures should be no warmer than 65 degrees F during the day and 50 degrees F at night. Avoid allowing the soil to dry as plants wilt easily. Most plants are treated as annuals and are best discarded after blooming.

Pocket. (1) An elongate sack into which hops are ressed, about 2 m (6 1/2 ft) tall and 60 cm (2 ft) in diameter. Also the quantity of hops contained in such a sack, about 76-89 kg (1½ 11/4 1/4cwt). (2) A measure of wool equal to 76 kg (168 Ibs).

Pod. (1) A dry dehiscent fruit; capsule, legume, follide. (2) Dry dehiscent fruit which splits along two sutures e.g. the fruit of a legume. (3) Fruit from a member of the legume (bean) family.

Podsol. A type of soil in which minerals have been leached from the surface layer and deposited in the subsoil, often forming a hard pan impenetrable by roots. The upper leached layer is usually extremely poor in mineral salts and consists of bleached sand. It is often covered with very acid, decomposing plant material. Such soils are common on quick draining sandy heaths, and under conifer forest, particularly in the cool, wet areas.

Poinsettia flowers. The colored leaves of a poinsettia are not the flowers, nor the petals of the plant; the colored leaves are more correctly called bracts. The actual poinsettia flower is the small yellow "ball" in the middle of the colored bracts. Flowers are petal-less and often fall off indoors due to low humidity and light levels. Removing the small flowers will not make the plant last longer, nor will it make the bracts more colorful. Flowers may be removed to limit them falling on the floor or furniture throughout the holiday. Flowers, like the rest of the plant, are not considered poisonous. However, the poinsettia is not a food crop plant and should not be eaten.

Poinsettia identification. Poinsettias can be identified by the color of their petiole, or leaf stem. Red poinsettias will have a bright red petiole; pink plants will have a pink petiole and white plants will have a green petiole. Multi-colored plants such as 'Jingle Bells' and 'Marble' will usually have a pink petiole.

Poinsettia. *Cut Fowers:* Poinsettias can be cut and used in flower arrangements, provided the stems are sealed. Cut the "blooms" with at least four inches of stem. Immediately seal the cut end by dipping in boiling water or holding over a flame for fifteen seconds. Make sure hands are protected with a kitchen mit. Sealing prevents the sap from oozing from the cut and thus, preventing the cut stem from wilting. "Blooms" should last a week or more. Make sure the cut end is in water or a wet florist block. Discard flowers when wilted and leaves start falling. This method works for plants accidentally damaged during transport or carelessly broken. Poinsettias are thought to be poisonous and cause death. THIS IS NOT TRUE. No case of poisoning by the plant is recorded. In fact, members of the Society of American Florists commonly eat poinsettias for the press each year to dispel the rumor. However, this doesn't mean you should eat the plant. No plant should be eaten that isn't grown to be eaten. The milky sap might cause an allergic skin reaction, though it does not affect everyone. Allergic conditions to poinsettias are similar to allergic conditions to chocolate, milk, wheat and poison ivy — some people are affected but not everyone.

Reblooming: Poinsettias should be kept in a bright light humid environment. Keep the soil moist but not soggy. After the colored bracts (the blooms are small yellow flowers in the center of the bracts) fall, cut the plant back by half. Place outdoors during the summer, on the east or north side of the house. Fertilize and water regularly to stimulate growth. Pinch regularly to maintain a short, stalky plant. In September, bring the plant indoors, place in a southern window and continue to water and fertilize. Repot in large pot and loose houseplant soil if needed. Check for insects. Beginning October 1, poinsettias should be kept in absolute darkness from 5 p.m. until 8 a.m., or there abouts. During the day, give the plant bright light and continue watering and fertilizing practices. Plants can be covered with a cardboard box painted black on the inside, or placed in a closet or unlighted room. A street lamp or night lamp can disrupt the night schedule. Plants should start turning color by November. Continue with regular practices. Watch humidity levels — keep high. Avoid misting plants — leaf spots will appear on leaves and bracts. Poinsettias can be identified by the color of their petiole, or leaf stem. Red poinsettias will have a bright red petiole; pink plants will have a pink petiole and white plants will have a green petiole. Multi-colored plants such as 'Jingle Bells' and 'Marble' will usually have a pink petiole.

Point source pollution. Pollutants that are discharged or emitted from discrete "point" sources, such as pipes and smokestacks. Both the Clean Water Act and the Clean Air Act focus control requirements on point sources and both require permits for major sources of discharges from point sources. While much agricultural pollution is nonpoint source, some agricultural activities are affected: for example, feedlots of over 1000 animal units are considered point sources requiring permits under the Clean Water Act. However, irrigation return flows, although considered point sources, are expressly exempted from the permit requirement.

Point. (1) A measure of price change equal to 1/100 of one cent in most futures contracts traded in decimal units. In grains, it is one cent; in T-bonds, it is one percent of par. (2) A share or tine.

Poisonous plants. (1) It is difficult to create a list of poisonous plants. While some plants are outright toxic to anyone who eats or comes in contact with the plant (Jimson Weed, nightshade), others may cause a reaction in some but not in others. The following is a listing of plants normally thought to be poisonous. People should not eat plants not grown as food crops — vegetables, fruits, cereal grains. Children should be taught never to eat non-food crop plants. Poisonous plants vary between animals. Don't assume that since a bird, squirrel or pig ate a plant with no problems, humans can do the same. If problems are encountered, or for a more complete list of poisonous plants, contact a doctor or the emergency room of your local hospital. Never hesitate contacting a physician if there is the smallest doubt. Unless otherwise stated, consider all parts of the plant poisonous.

- Agave (leaves).
- Amaryllis (bulbs).
- Apple (seeds).
- Azalea.
- Bird of Paradise (seeds).
- Black Locust (seeds).
- Bloodroot.
- Boston Ivy (berries).
- Boxwood (leaves/twigs).
- Buckeye (seeds).
- Buckthorn (berries).
- Buttercup (leaves).
- Caladium.
- Calla (rhizome).
- Castor Bean (seed).
- Christmas Rose.
- Cone Flower.
- Crown of thorns.
- Daffodil.
- Daphne.
- Delphinium.
- Dumbcane (Dieffenbachia).
- Eggplant (plant)
- Elderberry (unripe berries).
- Elephant Ear.
- English Ivy (berries).
- Flowering Tobacco.
- Foxglove.
- Holly (berries).
- Horsechestnut (nuts).
- Hyacinth.
- Iris.
- Jack-in-the-Pulpit.
- Jerusalem Cherry.
- Jimson Weed.
- Jonquil.

- Lantana.
- Larkspur.
- Lily-of-the-Valley.
- Lupine.
- Mayapple.
- Mistletoe (berries).
- Morning Glory seeds.
- Mustard (root).
- Narcissus.
- Nicotiana.
- Nightshade.
- Oleander.
- Philodendron.
- Poison Hemlock.
- Poison Ivy.
- Potato (green).
- Privet (berries).
- Ranunculuc.
- Rhododendron.
- Rhubarb (leaf blade).
- Rosary Pea (seed).
- Snow-on-the-Mountain.
- Sweet Potato.
- Skunk Cabbage.
- Sweet Pea (seeds).
- Tansy.
- Tomato (leaves).
- Tulip.
- Virginia Creeper (berries).
- Water Hemlock.
- Wisteria (seeds/pods).
- Yew (berries)

Poisonous. Poinsettias are thought to be poisonous and cause death. THIS IS NOT TRUE. No case of poisoning by the plant is recorded. In fact, members of the Society of American Florists commonly eat poinsettias for the press each year to dispel the rumor. However, this doesn't mean you should eat the plant. No plant should be eaten that isn't grown to be eaten. The milky sap might cause an allergic skin reaction, though it does not affect everyone. Allergic conditions to poinsettias are similar to allergic conditions to chocolate, milk, wheat and poison ivy — some people are affected but not everyone.

Poke. A loosely woven sack for green hops, able to contain 8-10 Bushels (0.29-0.36 cubic m). Also called greenbag, greensack.

Polarity. The phenomenon by which stem cuttings form shoot at the distal end and roots at the proximal end and root cuttings form roots at the distal end shoots at the proximal end.

Pole. An old measure of length equal to 5 1/2 yds (4.6 m), orof area equal to 30 1/4 sq.yds (25.3 sq.m.). Also called rod or perch.

Pollard. A tree having had its crown cut off, and with new branches growing from the top of the stem.

Pollarding. (1) A training system used on some large-growing trees to keep them to a modest size or to give the landscape a formal look. These trees are severely headed back annually or every few years. (2) The cutting back of trees or shrubs close to the main trunk or ground level resulting in the production of a number of vigorous new shoots (See Coppicing).

Pollards. A term once given to a category of offal from wheat milling, consisting of fine bran and other products. All wheat offals, apart from bran, are now sold as weatings, and are used as a Feedingstuff.

Pollen grain. Dust like material in the anther which gives rise to male sperm cells.

Pollen. The male cells or microspores produced by the stamens.

Pollination. (1) Movement or transfer of pollen from the anther to the stigma. (2) Process of transfering pollen from a stamen to a pistil. (3) The transfer of pollen from the stamen (male part of the flower) to the pistil (female part of the flower), which results in the formation of a seed. Hybrids are created when the pollen from one kind of plant is used to pollinate and entirely different variety, resulting in a new plant altogether.

Pollinator. (1) The agent of pollen transfer e.g. insects or wind. (2) Plant or cultivar needed for fertilization and fruit set in another plant of the same species. (3) A tree planted in an Orchard to provide Pollen for the fertilization of surrounding trees. Pollinators are usually of a different variety to the rest of the trees in the orchard. (4) Cultivar, often a male clone, proven to pollinate other cultivars.

Pollution. Alteration of the environment, as through the introduction of hazardous or detrimental substances, heat, or noise whose nature, location, or quantity produces adverse health or environmental effects. Under Section 502 of the Clean Water Act, for example, pollution means the man-made or man-induced alteration of the physical, biological, chemical, and radiological integrity of water.

Polyculture. The cultivation of two or more species of crops in such a way that they interact agronomically (biologically).

Polyembryonic. (1) Containing more than one embryo. (2) The presence of more than one embryo in an ovule.

Polyethylene (polythene). A plastic material. It is extensively used for covering greenhouses and making containers for potting plants

Polygamo-dioedous. With hermaphrodite and unisexual flowers on different individuals of the same species.

Polygamous. (1) Bearing staniinate, pistillate and hermaphrodite flowers at the same time. (2) Bearing unisexual and bisexual flowers on the same plant.

Pome. (1) A type of fleshy fruit represented by the apple, pear or related genera. A fruit with a papery or bony core at the center and with sepals or scars from which the sepals have fallen at the blossom end. (2) Inferior, indehiscent, many-seeded fruit, the greater part of which is developed from receptacle of the flower and not from the ovary e.g. apple.

Pork bellies. One of the major cuts of the hog carcass that, when cured, becomes bacon. Futures contracts for pork bellies are traded in the futures market.

Posterior. At or toward the back; opposite the front; nearest the axis; away from the subtending bract.

Postharvest. Refers to activities in the food and fiber sector that occur after agricultural products are sold from, or leave, the farm or ranch. In total, about 75% of the retail cost of the market basket of foods is added in postharvest activities.

Pot ale. A syrupy residue (Syrup Feeds) from the first distilling process of whisky which, is now being used by stock farmers as a supplement to complement low-quality roughage, such as straw, or low-quality hay. The syrup is a good source of energy and protein and also contains substantial amounts of phosphorous, potassium and copper. It has an M.E. content of dry matter higher than that of barley. Methods of feeding the syrup vary from inclusion in complete diets or pouring directly over roughage along feed troughs, to self feeding from open containers. Ball feeders are generally unsatisfactory due to the syrup's viscosity, except when the dry matter content is low.

Potash fertilizers. Potash fertilizer, organic or inorganic material containing one or more of the nutrients—mainly nitrogen, phosphorus, and potassium, and other essential elements required for plant growth. Added to the soil or other medium, fertilizers provide plant nutrients that are naturally lacking or that have been removed by harvesting or grazing, or by physical processes such as leaching or erosion. Organic fertilizers include animal and green *manure*, fish and bone meal, and *compost* (see also *humus*). Microorganisms in the soil decompose organic material, making its elements available for use by plants. Inorganic or artificial fertilizers (also called chemical or mineral fertilizers) are formulated in appropriate concentrations and combinations for various crops and growing conditions. The most popular inorganic fertilizers include: anhydrous ammonia, a gas that is 82% nitrogen; urea, a solid compound containing 46% nitrogen; superphosphate; and diammonium phosphate, containing 18% nitrogen and 46% phosphate. Fertilizers may be spread over the soil surface or plowed under, drilled into deep or shallow layers of the soil, applied in bands under the rows where the seeds are to be sown, drilled into the bands at the time of planting, or side-dressed between planted rows. Nitrogen fertilizer washing from farms into surface waters promotes overgrowth of aquatic vegetation, which degrades water quality and can cause eutrophication. Use of inorganic nitrogen suppresses nitrogen-fixing soil bacteria, making agriculture increasingly dependant on artificial fertilizer.

Potash nitrate. A fertilizer containing about 15% Nitrogen and 10% Potash, derived from natural deposits in Chile, and mainly used in market gardening as a top dressing.

Potash. A term applied to potassium carbonate (K2CO3), potassium hydroxide (KOH), potassium oxide (K2O), and to potassium salts in general.

Potassium. (K). A metallic element found naturally in various salts and present in the soil in Clay Minerals. It is essential to life, influencing both Carbohydrate and Protein metabolism, and the movement and utilisation of water in living organisms. In plants it ensures healthy growth, improving both resistance to drought and fungal disease, and winter hardness. It assists in the transportation of the products of photosynthesis and counteracts the effects of excessive applications of nitrogen. In animals, it contributes to the maintenance of the osmotic balance of the body fluids, particularly intracellular fluid, and influences muscular and nerve activity. Continuous cropping, lack of manuring, excessive dressings of nitrogen, and certain crops with a high potassium requirement (e.g. mangolds, lucerne, potatoes, sugar-beet, etc.) may deplete the soil of potassium. It may also be deficient in sandy, peaty or chalk soils. It is applied in the form of various potash fertilizers.

Potato blight. A fungal disease *(Phytopthor a infestans)* of potatoes which devastated the British crop in 1845 and caused the Irish famine in 1846. Characterised initially by dark green, turning to dark brown, leaf patches, resulting in death of the haulm and loss of yield. Tubers, which become separately infected, show brown staining and usually rot.

Potato planter. A tractor-drawn machine, similar to a drill, which plants Seed Potatoes in widely spaced rows by feeding them down tubes and covering them with a ridge of soil. Various types exist including *(a)* handfed planters, in which an operator feeds the tubers into a rotating seed wheel which drops them down the tubes, *(b)* semi-automatic planters, in which the operator chan-nels the potatoes into cups which more rapidly feed the tubes, and (c) automatic planters, which require no operator to feed the cups.

Pot-bound (Root-bound). Having closely packed mass of roots on a plant that has grown too large for its container.

Potential crop yield. Yield of a crop grown under the most ideal environmental and management conditions.

Potting soil. A growth substrate suitable for container gardening. A soil mixture designed for use in container gardens and potted plants. Potting mixes should be loose, light, and sterile.

Potting. The act of planting a cutting or seedling in a pot containing potting medium (See Depotting and: Repotting).

Poultry science. Study of principles and practices involved in the production and marketing of poultry and poultry products. It includes breeding incubation, housing, brooding, curing dis-eases, feeding, marketing, management etc.

Poundage quota. A quantitative limit on the amount of a commodity that can be marketed (also called a marketing quota) under the provisions of a permanent law. Once a common feature of price support programs, this supply control mechanism now only restricts the production and/or sale of tobacco and peanuts.

Powdery mildew. (1) Occurs mostly in chile and grapes. Foliar disease that causes leaf curling and premature leaf drop. This exposes the fruit to the sun, resulting in sun scald. (2) Those white or grayish spots on leaves are probably this common fungus, which usually weakens but doesn't kill. Powdery mildew likes humidity. Plant resistant plants or use fungicides.

Power harrow. A type of Harrow consisting of a number of p.t.o. driven transverse bars bearing Tines which reciprocate in a rotary manner. Used mainly to produce a fine tilth in a single opertion, particularly for potatoes.

Precipitation. (1) Rain, drizzle, dew, snow, hail, fog or any other form of water from the sky. (2) amount of precipitation expressed in depth of water which would cover a horizontal plane if there is no ran off, infiltration or evapo-transpiration

Precision farming. Farmers use global positioning (GPS) technology involving satellites and sensors on the ground and intensive information management tools to understand variations in resource conditions within fields. They use this information to more precisely apply fertilizers and other inputs and to more accurately predict crop yields.

Precocious. (1) Appearing or developing early. (2) Developing, occurring, or appearing before the expected time.

Pre-emergence weedicide. A chemical applied to the land to destroy weed seeds before they germinate.

Preharvest. Refers to activities on the farm or ranch that occur before crop or livestock products are sold. "Preharvest food safety activities," for example, is a term often used to describe USDA's efforts, through research and cooperative work, to foster changes in on-farm production that can reduce public health risks in live animals before they are sent to slaughter.

Pre-packaging. Packing a product in a consumer package by the wholesaler rather than by the retailer.

Preservatives. (1) A preservative is a natural or synthetic *chemical* that is added to products such as foods, pharmaceuticals, paints, biological samples, etc. to retard spoilage, whether from microbial growth, or undesirable chemical changes. Preservatives may be added to wood to prevent the growth of fungus, as well as repel insects and termites. Typically copper, borate, and petroleum compounds are used. For more information on wood preservatives see *lumber*. Preservative *food additives* are often used alone, or in conjunction with other methods of *food preservation*. A distinction is sometimes made between anti-microbial *preservatives* which function by inhibiting the growth of *insects, bacteria* and *fungi*, and *antioxidants*, which inhibit the *oxidation* of food constituents. Common anti-microbial preservatives include *sodium nitrate, sodium nitrite, sulfites, (sulphur dioxide,* sodium bisulfate, potassium bisulfate, etc.) and *disodium EDTA*. *Antioxidants* include *BHA* and *BHT*. Other preservatives include *formaldehyde* (usually in solution), *glutaraldehyde, diatomaceous earth* (kills insects), *ethanol* and *methylchloroisothiazolinone*. The benefits and safety of many artificial food additives (including preservatives) are the subject of debate. Some methods of food preservation involve the use of *salt, sugar* or *vinegar*, which are sometimes considered to be foods rather than additives. (2) Substances used in food processing, capable of inhibiting, retarding or arresting the growth of micro-organisms. (3) A substance added to food products or to organic solutions to prevent decomposition due to chemical change or bacterial action.

Prevented planting acreage. Land on which a farmer intended to plant a program crop or insurable crop, but was unable to because of drought, flood, or other natural disaster. Used in the calculation of disaster payments and crop insurance indemnity payments.

Prey crop. The crop grown for eliminating undesirable plants through physical or allelopathic effects

Price index. Current price expressed as a proportion to the same price in an earlier time period, commonly called the base period. Monthly price indexes computed by the National Agricultural Statistics Service are the index of prices received by farmers and the index of prices paid by farmers for commodities and services, interest, taxes, and farm wage rates. The ratio of these two indexes is referred to as the parity ratio.

Prickles. (1) Slender, sharp outgrowths of the stem tissues beneath the epidermis. (2) A small sharp point, spine, or thorn.

Prilled fertilizer. A fertilizer in the form of small spherical granules, about 2-3 mm in diameter.

Primary branch. A branch arising from the trunk of a tree.

Primary flower, fruit. The terminal flower or fruit in the central axis of the flower or fruit cluster. Major nutrients.

Prime farmland. Land that is best suited to and available for the production of food, feed, forage, fiber, and oilseed crops. It can be cropland, pastureland, rangeland, forestland, or other land. It has the soil quality, growing season, and moisture needed to produce high yields of crops each year economically, if managed according to acceptable farm practices. Prime farmland produces the highest yields with minimal expenditure of energy and economic resources and does so with the least damage to the environment. Of the 334 million acres of prime farmland, 216 million are in cropland use, according to the 1992 national resources inventory. See unique farmland.

Prior appropriations. The system of water allocation used primarily in the arid western United States, where water is scarce. Under this system, earlier claims have priority over later claims, and claims are associated with specific volumes of water. Rights to water can be lost if they are not used.

Prior converted wetland. Under the swampbuster program, these are wetlands that were converted to cropland before swampbuster was enacted on December 23, 1985, and meet wetland criteria for saturated soils or water-loving plants. Under swampbuster, there are no restrictions on either drainage maintenance or additional drainage on prior converted wetlands, which are estimated to total more than 50 million acres.

Privacy (information). The right to control the disclosure of personal information; often referred to as confidentiality, but more accurately might be called data protection. Many landowners, including farmers, have grown concerned that information about the condition of their land could adversely affect market and other values.

Processing technology. A branch of Horticulture, dealing with the post-harvest handling, storage and processing of horticultural produces.

Procumbent. (1) Lying flat on the ground but the stem not rooting at nodes or tip. (2) Trailing over the ground, but not rooting.

Production expenses. Measures the aggregate business cost of farming. The two main components are current farm operating expenses and overhead costs. Farm production expenses are accounted for differently in calculating farm income measures. Only production expenses paid in cash are deducted from gross farm income to derive net cash farm income. Gross farm income and net farm income include both cash and noncash production expenses. Non-cash expenses include such overhead costs as charges for depreciation and other capital consumption associated with farm buildings and other structures, motor vehicles, farm machinery and equipment, and expenses associated with farm operator dwellings.

Productivity. (1) A measure of technical efficiency, typically expressed as the added output for an additional unit of input or the average output per unit of input, i.e., labor, land, capital productivity. (2) The yield or biomass production capacity of a plant per unit area per unit time.

Program crops. Crops eligible for deficiency payments under commodity program provisions that ended in 1995. These were wheat, feed grains, cotton, and rice. These same crops are now called contract commodities because they are eligible for production flexibility contract payments. Other commodities receiving price support include extra long staple cotton, soybeans and other oilseeds, tobacco, peanuts, sugar, and milk.

Program yield. The farm commodity yield of record, determined, in general, by averaging the yield of a particular commodity for the past 5 years, dropping the high and low years. This yield was used in calculating deficiency payments under prior laws, for example, and for calculating contract payments for production flexibility contracts under the FAIR Act of 1996. Program yields remained frozen at their 1985 levels from that date forward.

Projected yield. The number of bushels (or pounds or hundredweight) per acre that, based on current weather estimates and other factors, USDA analysts estimate farmers will harvest.

Proliferous. (1) Bearing adventitious buds on the leaves or in the flowers which can root and form separate plants. (2) Freely producing buds or offshoots, especially from unusual places, as shoots from flowers or fruits from fruits.

Promotion program. Any program by an approved commodity promotion board or marketing order, including paid advertising, to present a favorable image of an agricultural commodity to the public to stimulate sales and improve the competitive position of the commodity in the domestic marketplace. Funds are collected through a mandatory check-off on marketings to finance the promotion activities.

Propagate. Any structure, sexual or asexual that becomes separated from the parent plant and serves as a means of propagation.

Propagation. (1) Method of increasing the number of plants by dividing roots, layering or start from cuttings. (2) The reproduction of a species. The term is used in horticulture to mean the artificial multiplication of plants by vegetative means, including the taking of cuttings, Layering, Budding, and grafting etc. (3) Various methods of starting new plants ranging from starting seeds to identical clones created by cuttings or layering..

Propagulam. Any of various usually vegetative portions of a plant, such as a bud or other offshoot, that aid in dispersal of the species and from which a new individual may develop.

Propogation. (1) Means of reproducing plants, such as by seeds, cutting, budding or grafting. (2) One of the most rewarding aspects of herbgardening is raising new plants. I myself have never been a gardener of any sort, but the idea of rearing a new plant from scratch, and ultimately using that plant for all sorts of things is extremely satisfying. The following methods may be used: (a) Growing from seed ie Chives, fennel, feverfew, lovage, rue, salad burnet, pot marygold, winter savory and chillies. A lot of herbs selfseed easily in situ. (b) Layering is a slow but foolproof method of inducing the stem of a woody perennial such as Sage, Lavender, Rosemary or honeysuckle to make roots of its own while it is still nourished by the parent plant. (c) Cuttings are small portions of stem or root, trimmed off and induced to grow roots of their own.

Proportion. Use larger and/or more plants in larger containers, and less and/or smaller plants in small containers. Rule of thumb is the height of the tallest plant should not exceed 1X-2X the height of the container excluding pedestals and "air-fairy" sprigs. Best to use odd numbers:one, three, five or seven plants or plugs of each cultivar.

Proprietary clones. Clones distributed under agreement with certain estates or institutions which can be multiplied but are not re-saleable.

Prorate. A quantity provision in a fruit or vegetable marketing order that is intended to even out weekly shipments (or shipments for some other specified periods of time); it aims to prevent too much of the regulated commodity from entering the commercial market at the same time and thereby depressing prices.

Protandry. Condition where the male organ matures earlier than female organ in bisexual flowers.

Protectants. Materials applied to the seed to protect it from soil fungi. Certain organic mercury compounds like ethyl mercury *p*-toluene sulphamide, methyl mercury dicyan diamide, hydroxy mercuric chlorophenol etc. are highly effective protectants. Another group of fungicides, which are nonmercuric organic compounds like chloranil, thiram, ferbam, captan, zinc trichlorophenate etc. are less hazardous for operator and for the seed, but are somewhat less effective.

Protected cultivation. Cultivation of crops in artificial structures providing ideal micro climate and acceptable growing conditions.

Protein equivalent. An indication of the value of a feedingstuff expressed as the percentage of True Protein plus half the percentage of non-protein nitrogen, on the assumption that the latter is capable of conversion to protein by animal.

Protein. A class of complex organic compounds, containing Carbon, Oxygen, Nitrogen (about 16%), Hydrogen, and some also containing Phosphorus and Sulphur. They consist of thousands of Amino Acids linked in polypeptide chains and sometimes folded in a complicated manner. About 21 different amino acids are found in proteins, the sequence of their arrangement in the chains being specific for each protein and giving it its characteristic properties. Proteins may be simnple, containing only amino acids, or conjugated, being combined with other compounds *e.g.*, Carbohydrates (muco or glycoproteins), Fats (lipoproteins), nucleoproteiris (phosphorus containing proteins combined with DNA and forming the Chromosomes of cell nuclei) etc. With water they form the basic constituents of Protoplasm, and play an important role in the structure of organisms. Plant proteins are often stored in seeds and other organs and include globulins, albumins, gluteins and prolamins. They are digested by animals into their component amino acids which are then used to synthesise animal proteins. Examples include Albumen in eggs, osein in bones, Collagen in connective tissue., Casein in milk globin in Haemoglobin, creatine in muscle, etc. Protein may be provided to livestock in feedingstuffs or animal origin *(e.g.,* Fish Meal, Meat And Bone Meal, etc.), or of vegetable origin (e.g., oilseed Cake, peas, beans, etc.), although the latter tend to be deficient in one or two 'essential' amino acids. Enzymes are also proteins.

Prune. (1) Any kind of plum that can be dried without spoiling. (2) The partially dried fruit of any of several varieties of the common plum, *Prunus domestica.* (3) Prune, popular name for a dried *plum*. Fruits of the many varieties of *Prunus domestica*, which are firm-fleshed and dry easily without removal of the stone, are gathered after falling from the tree, dipped in lye solution to prevent fermentation, dried in the sun or in kilns, and then "glossed" with a steam, glycerin, or fruit-juice bath to produce a sterile, glossy skin. Most of the commercial product comes from the Pacific coast states. A type of prune was used by Native Americans as a staple item of diet.

Pruning paint. Pruning paint is often used on trees and shrubs to cover wounds. However, research has shown pruning paint to be of LITTLE BENEFIT to the tree or shrub. Insects will not be prevented from infesting a tree, and will attack the wound whether it is sealed or not. Disease organisms will have already infected the cut surface. In fact, pruning paint may actually seal disease organisms inside the wound, preventing weather from affecting the organism. Pruning paint could be classified as a "placebo" — making the user feel good but doing little for the plant. Wounds will heal properly if pruned correctly.

Pruning. (1) Selective removal of plant parts such as buds, shoots and roots to maintain a desirable form and to improve plant health and fruit yield. (2) Reducing the growth on a shrub/tree to promote new flowers, stems or foliage. Also for cosmetic reasons. The cutting and trimming of plants to remove dead or injured wood, or to control and direct the new growth of a plant.

Pseudo-bulb. Specialised storage structure of many orchid species consisting of an enlarged fleshy section of stem with one or more nodes.

Pseudo-carp. Fruitlike structure that incorporates tissue that is not derived from the ovary wall. The additional tissues may be derived from floral parts such as the *receptacle* and *calyx*. For example, the coloured, fleshy part of a strawberry develops from the receptacle and the true fruits are small *achenes* – the 'pips' embedded in its outer surface. Rose hips are a type of pseudocarp that consists of a hollow, fleshy receptacle containing a number of achenes within. Different types of pseudocarp include pineapples, figs, apples, and pears. A coenocarpium is a fleshy, multiple pseudocarp derived from an *inflorescence* rather than a single flower. The pineapple has a thickened central axis surrounded by fleshy tissues derived from the receptacles and floral parts of many flowers. A fig is a type of pseudocarp called a syconium, formed from a hollow receptacle with small flowers attached to the inner wall. After fertilization the ovaries of the female flowers develop into one-seeded achenes. Apples and pears are *pomes*, another type of pseudocarp.

Pseudo-stem. (1) A type of shoot arising from an underground rhizome, consisting of overlapping petioles. (2) The shoot of a banana plant.

Psychrometer. (1) Instrument to measure the humidity of the atmosphere. (2) An instrument that uses the difference in readings between two thermometers, one having a wet bulb and the other having a dry bulb, to measure the moisture content or relative humidity of air.

Public lands. As defined in the Federal Land Policy and Management Act, public lands are any land and interest in land outside of Alaska owned by the United States and administered by the Secretary of the Interior through the Bureau of Land Management. In common usage, public lands may refer to all federal land no matter what agency has responsibility for its management or may refer even to state- and local municipality-owned lands.

Puncture vine (goat heads). Annual plant notorious for its seed pods that can puncture automobile tires and lower the quality of hay and forage crops. Move by attaching to anything that comes by. Difficult to manage.

Pungency (Piquancy). Sharp, stinging or biting quality of a spice or condiment, while tasting.

Pupa. A resting stage in insect Metamorphosis between the Larva and young adult, during which movement and feeding ceases and a great change in form occurs. Also called chrysalis.

Pure Live Seed (PLS). Percentage of pure germinating seed deter-mined by multiplying pure seed percentage by its own germina-tion percentage and dividing the product hundred.

Pyrene. (1) A hard, stony endocarp enclosing one or two seeds. (2) The stone of certain fruits, such as the cherry.

Pyrethrins. Insecticidal compounds present in the flowers of the pyrethrum plant. The general term includes four compounds that occur together in the flowers, namely pyrethin I, pyrethin II, cinerin I, and cinerin II. The pyrethrins are effective and quick-acting against many insects and have very little toxicity warm blooded animals.

Pyrethrum. (1) A genus of composite plants, the powdered flowerheads of which are used as an insecticide. Now less commonly used due to the introduction of synthetically produced insecticides. (2) An insecticide prepared from the dried flower - heads of *Chrysanthemum cinerarifolium* and *C roseum*

Pyrophyllite. Mineral extensively used as a carrier or dilutent for insecticide and fungicide dusts. It is so named because its plate or leaftike structure (phyllon) will separated when heated (pyro). The mineral is ground to such finess that 85 per cent will pass through a 325 mesh screen. The particles have a flat or platelike structure, so they adhere quickly to plant surfaces dusted with them. The mineral looks, feels and acts like talc. But it is chemi-cally inert or neutral to most materials used as insecticides and fungicides. It does not absorb moisture from the air.

Q

Quality seed. Good quality seeds are genetically true to species; or variety; have capacity for high germination, are free from dis-eases and insects, are free from mixture with other crop seeds, weed seeds and inert and extraneous material. The germination, and purity of the seed can be determined by conducting a seed test on a small representative sample drawn from the seed lot in question.

Quality. (1) Various factors including flavour, texture, colour, appearance, odour and food value that influence the produce. (2) Characteristic that helps in differentiating a product regarding its acceptability to the consumer.

Quarantine. (1) The isolation of plants to determine whether diseases or insect pests are present, in order to prevent infestation to healthy plants. (2) Regulation forbidding sale or shipment of plants, to prevent disease or insect invasion in a specific area. (3) A restraint on importation of certain animals or plants from areas where pests or contagious diseases are endemic. or isolation of animals or plants suspected of carrying pests or diseases. to prevent the spread of those pests and diseases. States as well as the federal government may impose quarantines or may operate quarantine facilities. (4) Quarantine, isolation of persons, animals, places, and effects that carry or are suspected of harboring communicable disease. The term originally referred to the 40 days of offshore wait during which incoming vessels could not discharge passengers or cargo in the era when *plague* and other great epidemics swept across Europe. The practice has been changed by developments in medical science. Usually the word of the ship's officer that the passengers are free of disease and presentation by the passengers of certificates of inoculation against certain diseases are now sufficient to permit passage of travelers from one country to another. Some nations still maintain extended periods of quarantine for cattle and household pets coming from another country to guard against such diseases as foot-and-mouth disease and rabies. Plant life may also be held for assurance that fungus and other plant diseases are not being introduced. Local quarantine regulations are also in effect to guard against the spread of communicable disease. Public health laws require that physicians report certain infections to the authorities. The patients (and those who have come in contact with them) may be isolated and their effects disinfected, condemned, or destroyed, if it is in the public interest, since quarantine laws supersede even property rights. Although antibiotics, vaccinations, and other treatments have greatly reduced the use of quarantine in public health, persons with newly recognized or hard to treat communicable diseases may still be isolated by health officials. For example, quarantine was used effectively to control the spread of SARS (severe acute respiratory syndrome), a sometimes deadly pneumonialike illness, in 2003.

Quartz. (1) A very hard mineral composed of silica, SiO_2, found worldwide in many different types of rocks, including sandstone and granite. Varieties of quartz include agate, chalcedony, chert, flint, opal, and rock crystal. (2) Quartz can be used as a landscape mulch. Quartz comes in various colors and will not alter the soil's pH. (3) quartz, one of the commonest of all rock-forming minerals and one of the most important constituents of the earth's crust. Chemically, it is silicon dioxide, SiO_2. It occurs in crystals of the hexagonal system, commonly having the form of a six-sided prism terminating in a six-sided pyramid; the crystals are often distorted and twins are common. Quartz may be transparent, translucent, or opaque; it may be colorless or colored. Varieties are classified as crystalline and cryptocrystalline. Crystalline varieties include ordinary colorless crystallized quartz, or rock crystal; rose quartz; yellow quartz, sometimes used as imitation *topaz;* smoky quartz, or cairngorm stone; milk-white milky quartz; aventurine quartz, which contains scales of hematite or mica; and *amethyst*. Varieties of cryptocrystalline quartz, the crystal structure of which can be seen only under the microscope, if at all, are *chalcedony, flint,* hornstone, and chert. Colored varieties of chalcedony known by special names are *carnelian,* sard, chrysoprase, *agate, onyx,* sardonyx, and *jasper*. Clastic quartzes are *sand* and *sandstone.*

Queen of spices. Cardamom *(Elettaria cardamomum)*. Cardamom is a sweet, exotic-tasting spice. The small, round seeds, which grow inside husk like pods, are best purchased whole. They can be ground with a spice grinder or a mortar and pestle.

Quiescence (Eco-dormancy). Period of non-visible growth caused by unfavourable environmental conditions like cold temperature or drought (Dormancy is subdivided into rest and quiescence)

Quonset type greenhouse. (1) Greenhouse having semi-round roof. Ideally, 30m long and 10 m wide with 2.5 m sides and 4.0 m centre and covered with fibre glass or poly sheets. A structure, primarily of glass, in which temperature and humidity can be controlled for the cultivation or protection of plants. (2 enclosed glass house used for growing plants in regulated temperatures, humidity, and ventilation. A greenhouse can range from a small room carrying a few plants over the winter, to an immense heated glass building called a hothouse or conservatory, covering acres of ground and used for forcing fruits or flowers out of season. Greenhouses have long been used for holding plants over cold seasons and for growing tropical plants and less hardy fruits, but only in this century has the greenhouse been used for forcing vegetables. Now millions of dollars' worth of plant products are raised yearly in greenhouses.

R

R. Raceme. An inflorescence consisting of a long main axis, bearing stalked lateral flowers, the lowest opening first.

Rabbing. Process of buring of the residues of the crops in the field after harvesting of the crop. Rabbing is followed with an idea to keep the land free from weed seeds and pest etc.

Rabi winter season. An important crop season of North India; refers also to the group of crops including cereals, especially wheat, fodders, rapeseed etc. Sown in October-November and har-vested in March-April.

Rabies. A notifiable disease caused by a virus which infects the centralnervous system. All mammals are susceptible. It is trans-mitted in the saliva mainly by biting. Affected cattle, sheep, pigs and goats show abnormal changes in behaviour and develop any of the following symptoms; anxiety, aggressiveness, frequent bellowing, excessive saliva with choking, difficulty in eating, grinding of teeth and constipation or diarrhoea with staining and tail swishing. Progressive paralysis develops with death occur-ring on the 3rd-7th day. Also called hydrophobia.

Raceme. (1) A simple indeterminate inflorescence with pedicel led flowers. An elongated, unbranched cluster of stalked flowers. (2) An inflorescence having stalked flowers arranged singly along an elongated unbranched axis, as in the lily of the valley.

Rachis. (1) The axis along which the leaflets of a pinnate leaf are arranged. (2) The stem or petiole-like central axis of a pinnately compound leaf, connecting the petiolules to the petiole

Raddle. A flexible piece or strip of wood used for weaving between uprights in the making of fences or hurdles.

Radical. (1) Proceeding from or pertaining to the roots. (2) Of leaves arising from the base of a stem, or a rhizome. (3) Arranged around and spreading from a common center. (5) Of or pertaining to the root. (5) Arising from the root or its crown: *radical leaves.*

Radicant. Rooting, with reference to roots developing from stems or leaves.

Radicchio. Radicchio is a salad green, similar to Belgian Endive or whitloff chicory, with the exception that leaves have a reddish, burgundy color with white midribs. The head should be tight and firm, and usually pointed when purchased. Some will be rounded depending on variety. Ideal heads will be slightly smaller than head lettuce. Leaves can remain fresh for up to two weeks in the vegetable crisper. Plants are grown in a manner similar to "forced" chicory or Belgian Endive (Whitloff Chicory).

Radicle. (1) The part of a plant embryo that develops into a root. (2) The rudimentary root of the embryo, which forms the primary root of the young seedling. (3) Lower portion of hypocotyl. (4) The embryonic root of a seed.

Rafter. To plough land so that each furrow is overturned onto an unploughed strip.

Ragwort. Composite plants of the genus *Senecio*, particularly the coarse yellow-flowered *S. jacobaea* which is a common weed of pastures, and is poisonous to livestock. It is scheduled as an Injurious Weed and must be controlled either by cutting or treatment with weedkillers.

Rainforest. Tropical forest developed under fairly high rainfall. A rainforest is a *forested biome* with high annual *rainfall* due to the *Intertropical convergence zone.* Some cite a minimum normal annual rainfall of 2500 mm (about 100 inches or 250 centimetres), with normal rainfall at least 60 mm during each of the twelve months of the year. Others set the minimum annual rainfall barrier as low as 1700 mm (about 67 inches). The soil can be poor because high rainfall tends to leach out soluble *nutrients.* This type of biome is found in both *temperate* and *tropical climates.* As well as prodigious rainfall, many rainforests are characterised by a high number of *resident species* and tremendous *biodiversity.* Rain forests are home to two-thirds of all the living animal and plant species on the planet. It has been estimated that many hundreds of millions of new species of plants, insects, and *microorganisms* are still undiscovered and have no names yet. Tropical rain forests are called the "jewel of the earth", the "Earth's lungs", and the "world's largest pharmacy" because the large amount of natural medicines discovered there. Despite the growth of flora in a rainforest, the actual quality of the soil is quite poor. *Oxisols,* infertile, deeply weathered and severely leached, have developed on the ancient *Gondwanan shields.* Rapid bacterial decay prevents the accumulation of *humus.* The concentration of *iron* and *aluminium* oxides by the laterization process gives the oxisols a bright red color and sometimes produces minable deposits (e.g. *bauxite*). On younger substrates, especially of volcanic origin, tropical soils may be quite fertile. The undergrowth in a rainforest is restricted in many areas by the lack of sunlight at ground level. This makes it possible for people and other animals to walk through the forest. If the leaf canopy is destroyed or thinned for any reason, the ground beneath is soon colonised by a dense tangled growth of vines, shrubs and small trees called *jungle.*

Rake. (1) Long-handled toothed gardening tool: a tool with a long handle and a head with long teeth, used for gathering leaves or cut grass, or for smoothing or loosening the surface of the soil in a garden (2) A hand implement with a long handle attached to a crossbar bearing several prongs, used for scraping, gathering material (e.g. cut grass) together, smoothing a seed bed etc. (3) A tractor-drawn wheeled implement with long, curved, spring tines, used for gathering hay, scraping weeds into heaps, etc. (4) A very thin horse. (5) A northern term for a track, particularly one on a pasture or hillside, or in a gully. (6) A pasture. Also called raik. (7) To keep a flock of sheep moving from pasture to pasture.

Rangeland. Land on which the natural potential (climax) plant cover is principally native grasses, grasslike plants, and shrubs. It includes natural grasslands, savannahs, certain shrubs and grasslike lands, most deserts, tundra, alpine communities, coastal marshlands, and wet meadows. It also includes lands that are re-vegetated naturally or artificially and are managed like native vegetation. The United States has 399 million acres of non-federal rangeland, about 30% of all non-federal rural lands, according to the 1992 National Resources Inventory. The BLM manages approximately 167 million acres of federal rangelands, and the Forest Service manages approximately 95 million acres of federal rangelands.

Rape. A plant *(Brassica napus)* related to the turnip and grown for Forage, particularly for sheep, during the autumn and early winter. There are giant and dwarf varieties which produce rapid, succulent, leafy growth, something weeds. It is sometimes included in grass Seeds Mixtures as a Cover Crop and grazed early. It is also often grown as a Catch Crop after Peas or early Potatoes, or in cereal stubble. Rape is very susceptible to Club Root and Mildew. Also called forage rape, cole or coleseed. Certain varieties possess seeds rich in oil.

Ratoon. (1) New tillers, suckers or plants arising from the root, crown or rhizome of plants, which will be utilised for a subsequent crop. (2) To send up new shoots after being cut down or cropped. (3) Emergence of new crop tillers from the root in soil stocks specially in sugarcane and sorghum is known as ratoon and the practice of keeping the same crop for the next season is termed as ratooning.

Ratooning. An intensive cropping in which more than one harvest is obtained, as the ratoons are retained after the harvest of the first crop.

Raw agricultural product. The Federal Food, Drug, and Cosmetic Act defines this term as "any food in its raw or natural state, including all fruits that are washed, colored, or otherwise treated in the unpeeled natural form prior to marketing." The nonregulatory definition generally means any agricultural commodity that has undergone little or no processing.

Ray fungus. A bacterium *(Actinomyces bovis)* which forms radiating threads in grasses and cereals and may cause Actinomycosis if eaten by livestock.

Readily available moisture. The volume of water held in the soil between Field Capacity and Permanent Wilting Percentage which is capable of being removed by plants. As a soil dries between these two reference levels the growth rate and yield of a crop are reduced.

Real value of seed. Value of seed sample in comparison to another defined as 'the percentage of pure germinating seeds'. It is calculated by the formula. Purity % x Germination

Recalcitrant seeds. Seeds that lose viability when dried below a certain critical moisture level, which is comparatively high for ordinary seeds. Recalcitrant seeds are short lived (See Orthodox seeds).

Reclaim. To recover land from an undesirable state and put it in a desirable state; to undertake the reclamation of land; for ex-ample, to drain land or irrigate arid land to make it useful for agricultural purposes.

Reclaimed rubber. Obtained by treatment of vulcanised rubber to make it reusable as a raw material in the same manner as raw rubber.

Reclamation. The process of rehabilitating disturbed lands, or converting unproductive lands to productive uses. The term is also used for the process of recycling or reusing water. In the context of the Reclamation Act and reclamation law, it means putting arid lands to use through irrigation.

Recommended Dietary Allowances (RDAs). Levels of essential nutrients that, on the basis of scientific knowledge, are judged by the Food and Nutrition Board (FNB of the Institute of Medicine at the National Academy of Sciences), to be adequate to meet the known nutrient needs of practically all healthy persons. They are designed to provide a safety factor appropriate for each nutrient and exceed the actual requirements of most persons whose individual nutrient needs vary.

Recurrent flowering. The successive production of several crops of flowers during one season.

Recycle / reuse. Minimizing waste generation by recovering and reprocessing usable products that might otherwise become waste (i.e., recycling of aluminum cans, paper, and bottles, etc.).

Red clover. A deep-rooting, short-lived, species of clover, with an erect hairy stem, untoothed leaflets, short, broad, sharp-pointed stipules, and pale rose flowers.

Red fescue. A species of grass *(Festuea rubra)* normally under-sown in thinly sown spring cereal crops. Best suited to relatively poor upland and marginal soils.

Reddzlagg. A trade name for heat-treated calcined calcium alumin-ium phosphate—one of the main forms of phosphatic fertilizer now in use. It is a finely ground powder containing 32% phosphoric acid (P2O5) and a range of minor elements. It can be used on soils at all pH's.

Reed. (1) A tall grass *(Phragmites communis)* which grows exten-sively in thick reed-beds in swampy conditions. The dried steins are used in East Anglia for thatching. (2) The uterus of a Ewe

Reentry interval. The period of time immediately following the application of a pesticide during which unprotected workers should not enter a field.

Referendum. In agriculture, referendum generally refers to a vote by farmers on whether to approve or disapprove a farm program, such as mandatory production or marketing controls, assessments for generic commodity promotion, or marketing orders.

Rcforest. Establishment of forests on land from which trees have been removed by felling, fire etc.

Reforestation. The planting of trees on land where a forest has previously stood, but has been destroyed, e.g. by a forest fire. (Afforestation). Reforestation is the process of restoring *tree* cover to areas where *woodlands* or *forest* once existed but was removed by *logging* for wood products; if this area never returns to its original state of a forest this destructive process is called *deforestation*. In order to maintain a sustainable forest industry reforestation is necessary. Reforestation occurs in many ways. In many temperate zones such as the eastern *United States*, reforestation occurs quite naturally as the native hardwood forests are so resilient that, given any opportunity, they quickly re-establish themselves. However urban sprawl, and agriculture required permanently deforested lands which is leading to forest reduction in the area. From poor *logging* practices and/ or if the soil quality is too poor and erosion prone for natural regeneration, artificial reforestation is required. In various *arid*, *tropical*, or sensitive areas, forests cannot re-establish themselves without assistance due to a variety of *environmental* factors. One of these factors is that, once forest cover is destroyed in arid zones, the land quickly dries out and becomes inhospitable to new tree growth. Other critical factors include *overgrazing* by *livestock*, especially animals such as *goats*, and over-harvesting of forest resources by native populations or outside businesses. Together these may lead to *desertification* and the loss of *topsoil*; without soil, forests cannot grow until the very long process of soil creation has been completed - if *erosion* allows this. In some tropical areas, the removal of forest cover may result in a *duricrust* or duripan that effectively seal off the soil to water penetration and root growth. In many areas, reforestation is impossible above all because the land is in use by people. In these areas, reforestation requires the planting of tree seedlings, *treeplanting*. In other areas, mechanical breaking up of duripans or duricrusts is necessary, careful and continued watering may be essential, and special protection, such as fencing, may be required. One debatable issue in artificial reforestation is whether or not the succeding forest will have the same *biodiversity* as the original forest. If the forest is replaced with only one species of tree and all other vegetation is prevented from growing back, a *monoculture* forest similar to agricultural crops would be the result. However, most reforestation involves the planting of different seedlots of seedlings taken from the area. More frequently multiple species are planted as well. Another important factor is the natural regeneration of a wide variety of plant and animal species that can occur on a *clearcut*. In some areas the suppression of forest fires for hundreds of years has resulted in large single aged and single specied forest stands. The logging of small clearcuts and or prescribed burning, actually increases the biodiversity in these areas by creating a greater variety of treestand ages and species. Reforestation need not be only used for recovery of accidentally destroyed forests. In some countries, such as *Finland*, the forests are *managed* by the wood products and *pulp and paper industry*. In such an arrangement, like other crops, trees are replanted wherever they are cut. In such circumstances, the cutting of trees can be carefully done to allow easier reforestation. In *Canada*, the wood product and *pulp and paper industry* systematically replaces many of the trees it cuts, employing large numbers of summer workers for *treeplanting* work. The sustainable management of forest resources is called *forestry*.

Regionalization. Countries generally have prohibited imports of a foreign agricultural product if it has been associated with an unwanted pest or disease in the exporting country. Until recently, importing countries would not permit any of that product from the exporting country, even if it came from a region that did not have the disease or pest. Regionalization is a fundamental principle in the Sanitary and Phytosanitary provisions of recent trade accords. It provides for the acceptance of such imports if the exporting country can demonstrate that they are from a disease-free or a pest-free area.

Regionally appropriate. Plants that are winter- and summer-hardy in your area (check the USDA plant hardiness zone rating, a number between 1 and 11) and are not considered *invasive* by state or federal authorities.

Registered seed. Progeny of breeder's or foundation seed handled under procedures acceptable to the certifying agency to main-tain satisfactory genetic purity and identity.

Registration. The Environmental Protection Agency approval of a new pesticide for specific uses before it can be sold or distributed in the United States. Under the Federal Insecticide, Fungicide, and Rodenticide Act, the Environmental Protection Agency is responsible for registering (pre-market licensing) pesticides based on scientific evidence that the approved use will result in no unreasonable adverse effects on human health or the environment when applied according to approved label directions. The label on a registered pesticide specifies the legal obligations of the user in terms of uses, timing, and rates of application.

Rejuvenate. Renewal of growth and vigour of plants by selective pruning, fertilisation, soil management, plant protection measures, treatment with plant growth substances etc.

Relative Humidity (R.H.). (1) The amount of water vapour present in the air, expressed as a percentage of the maximum water vapour that the air would hold at the same temperature and pressure. (2) The measurement of the amount of moisture in the atmosphere.

Relay intercropping. Growing two or more crops simultaneously during part of the life cycle of each.

Remote sensing. The act of detecting objects when the sensor is not in direct contact; commonly refers to using aerial photographs to observe conditions on the Earth's surface. In agriculture, this technology can be used to determine what plants are being grown and their condition.

Rendzina. A shallow type of soil developed over calcareous rocks, with a brown to black horizon of Mull Humus, often as the only Soil Horizon, over the parent rock. A neutral to alkaline soil with a variable content of free calcium carbonate.

Renewable resources. Natural resources, sometimes called flow resources, that replenish themselves within time limits that permit sustained use, in contrast to nonrenewable resources. Congress recognized the renewable nature of resources managed by the Forest Service and Bureau of Land Management (including timber, livestock forage, recreation, water, and wildlife and fish) by requiring both agencies to follow the principles of multiple use and sustained yield.

Re-potting. Transfer of a plant from one pot to another filled with fresh potting mixture. Actively growing houseplants need repotting from time to time. The frequency of repotting depends on the growth rate of the particular plant. Fast-growing plants may need repotting annually, while slow-growing plants may require repotting every two to three years. Some plants (e.g., *amaryllis*) thrive when potbound. When new growth starts in the spring, turn the potted plant upside down and remove the plant, if possible. A plant can be removed easily from its pot if the lip of the upside down container is knocked against any solid objεct. Hold your hand over the soil, straddling the plant between the fore- and middle-fingers, holding back the soil, and gently knock the rootball out of its present container. If roots are in a solid mass around the rootball, the plant needs to be repotted. A plant also needs to be repotted if roots are growing out of the drainage hole, or if the plant requires frequent watering or wilts shortly after watering. Repot only as needed during spring and summer while the plant is actively growing. Do not repot ailing or dormant plants or those beginning to flower. Remove the plant from its pot and gently disturb the root system so that roots are not in a tight rootball. If the roots are too tight to loosen, score the rootball with a knife to loosen them. Select a pot that is 1 or 2 inches larger in diameter than the current pot. Place a small piece of broken clay pot over the drainage hole to prevent soil from washing out. Do not add a layer of clay pieces or rocks because this actually slows water movement through the pot. Place enough dry potting mix in the bottom of the pot so that the top of the rootball is within 1 inch of the container top. Do not add soil above the original level on the rootball. Fill around the rootball with mix. Do not pack the soil to firm or settle it; *gently* press the soil with your fingers. Water well so excess water drains out of the pot. Repeat many times, especially if the soil mix contains peat. Never pack wet media. For plants in large containers that are impossible to repot, remove the top 2 or 3 inches of soil and replace this with fresh mix every other year or so. This is called topdressing.

Re-seed. To re-establish a Ley by sowing seeds of grassland Species, usually in the form of a seed mixture. Reseeding may be carried out after ploughing up a previous ley which has deteriorated, or as part of a rotation

Reserve pool. A quantity provision in a fruit or vegetable marketing order that requires that some marketable supplies be withheld from the fresh market to be used instead in a so-called secondary market such as for frozen or processed forms, for a nonfood use, or stored for sales in a future marketing year.

Residual supplier. A country that supplies the world market only after importers have met their initial needs from preferred suppliers. A residual supplier is not initially competitive because of higher prices or lower quality. The United States used to be considered a residual supplier of grains because its commodity support programs kept U.S. grain prices higher than those of competing exporters.

Residue. That which remains after a part is taken or separated. Agricultural residues refers specifically to the dry portion of agricultural crops as distinct from the fruit, grain or other ingradient for which the crop is grown e.g. stalks, straws, cobs, seed hulls, nutshells, fruit bits. Vegetable wastes, such as green pea vines and bean vines, are sometimes called vegetable residues.

Resins. Water insoluble complex oxidation products of various essential oils which normally oozes out through the plant bark and hardens on exposure to the air.

Resistance, pesticide. The evolutionary capacity of pests exposed to a pesticide to develop resistance to that pesticide. Some genetically engineered crops now being marketed are accompanied by pest resistance management plans to prevent or substantially retard the development of resistance.

Respiration. Respiration is the opposite of photosynthesis. Sugar and oxygen are broken down into carbon dioxide and water. If respiration is higher than photosynthesis, a plant doesn't grow or mature. Respiration usually occurs during the night when light is not present, though respiration can occur during the day in some plants, particularly some grasses. Respiration increases as temperatures increase. It is because respiration uses oxygen that plants and flowers are not encouraged in hospital intensive care wards. Respiration is also a direct cause why tomatoes fail to ripen during hot, summer days. Sugars are not allowed to accumulate as respiration uses them up just as fast as they are produced.

Rest. Condition in which growth will not occur due to physiological reasons, despite the presence of favourable environmental conditions (See Dormancy).

Restorative crops. Crops which are grazed whilst growing so that livestock return organic matter and nutrients to the soil in the form of faeces and urine, e.g., a ley, folder roots, Kale, etc.

Restricted-use pesticides. A pesticide may be classified for restricted use if it requires special handling because of its toxicity. These may be applied only by trained, certified applicators or those under their direct supervision.

Reticulate. (1) Like a net, the interstices closed. Usually referring to the veins of a leaf. (2) Net-like appearance; for example, referring to a leaf having veins that branch many times in a dense pattern, often more prominently noticed when the veins are impressed.

Reticulum. The second stomah of a ruminant which receives harder food material passed on from the first stomach or rumen and in which accidentally eaten small stones normally accumulate. Also called the honeycomb stomach after its internal mucous membrane lining which has a honeycomb-like structure.

Retting. (1) Process of fermentation for separating fibre from coconut husks by submerging them in water in tanks, ponds or streams. (2) The exposure of harvested flax to moisture so that it partially rots and the pectin, which binds the fibre bundles to the woody stem, is weakened, allowing the fibres to he more easily sepa-rated from the rest of the stem.

Return flow. Surface and subsurface water that leaves the field following application of irrigation water. While irrigation return flows are a "point source," they are expressly exempted from permit requirements under the Clean Water Act.

Revenue Assurance (RA). A form of revenue insurance that protects a grower of an insurable crop whenever low prices, low yields, or a combination of both causes revenue to fall below a guaranteed level selected by the producer. It differs from other revenue insurance programs in that it allows a farmer to use the posted county price, rather than a national price, in determining a target level of revenue.

Revenue Insurance. A crop insurance program that provides coverage to producers against lost revenues (or incomes) caused by low prices, low yields, or a combination of low prices and low yields. An indemnity is paid to a producer when any combination of yield and price results in revenue that is less than a pre-specified revenue guarantee. The FAIR Act of 1996 mandates a USDA pilot program for crop years 1997-2000 under which producers of feed grains, wheat, soybeans, and other crops in specified areas may elect to receive insurance against loss of revenue. The two largest pilot programs to date are the Crop Revenue Coverage (CRC) program and the Income Protection (IP) program. The two programs are similar, except that the CRC allows farmers to insure their revenue at either planting-time prices or harvest prices, whichever is higher. The full cost of the premium associated with CRC and IP is paid by the participating farmer.

Reversible plough. A popular type of plaugh having two sets of bodies, used for one-way ploughing. One set of bodies, turning furrows to the right, are used for ploughing in one direction; the other left-handed set for ploughing in the other direction. Auto-matic, hydraulic or mechanical mechanisms are used for turning over the sets of bodies. The reversible plough has the advantage that headland use is much reduced; no ridges, depressions or finishes are left in the field and the surface is left level. They are mainly used for deep ploughing before preparing seed beds for root and vegetable crops. Also called one-way plough.

Reversion. (1) The process by which soluble monocalcium phosphate, a constituent of superphosphate, is rapidly converted in the soil to the less soluble di-calcium salt, producing a finely divided precipitate which is easily available to plants. (2) A viral disease affecting blackcurrants causing the development of narrower and darker leaves than normal and reducing the amount of fruit produced.

Rhizobium. A bacterium which can infect legumes in a symbiotic relationship and fix atmospheric nitrogen

Rhizome. (1) A horizontal creeping underground stem growing just below the soil surface, from which axillary buds produce new stems and roots giving rise to new separate plants, acting as a foodstore and means of vegetative reproduction, e.g. Couch Grass, raspberry. (2) An underground portion of a stem, producing shoots on top and roots beneath; different from a root in that it has buds, nodes, and scaly leaves; rootstock. (3) An underground stem, capable of storing food, and from which roots and stems of new plants can emerge. An underground stem distinguishable from a root by the presence of nodes, buds or scale-like leaves. (4) Rhizomes are thick, fleshy, horizontal underground stems. Some rhizomes may lie directly on the soil surface (bearded iris). Foliage, roots and flowers grow from buds on the rhizome. Rhizomes generally creep or spread, unlike tubers. Rhizomes are usually divided so pieces contain at least one fan or group of foliage and significant amount of roots. The following is a list of rhizome plants according to type of storage organ.

- Iris
- Oxalis
- Zantedeschia (Calla Lily). (5) Underground horizontal stem, serving as storage organ and capable of producing new shoots and roots at the nodes. (6) Roots that travel underground laterally and send up new plants. An example is the perennial sweet woodruff (*Galium odorata*), which travels laterally an inch or so underground, then breaks the surface to grow a new stem.

Rhododendron. Genus of shrubs in the Ericaceae family (from Greek *rhodo* meaning rose and *dendron* meaning tree). In common usage, "rhododendron" usually refers to members of the subgenera *Rhododendron* and *Hymenanthes*, as distinct from the azaleas.

Rhombic. (1) A four-margined leaf that is diamond-shaped, having three prominent tips, two on the side and one at the top. (2) With four nearly equal sides, but unequal angles, diamond shaped.

Rhombic. With four nearly equal sides, but unequal angles, diamond shaped.

Rhubarb (frost damage). Rhubarb hit by a frost or freeze can still be eaten PROVIDED the plant/stalks are still firm and upright. Leaf injury may be noticeable with some brown or black discoloration on the edges. Severe cold injury may cause the oxalic acid crystals in the leaves to migrate to the stalks when poisoning problems can occur. If in doubt concerning the safety of eating the stalks, DON'T. Cut those stalks off and compost. Allow new stalks to develop before eating. Rhubarb is a perennial vegetable whose leaf stalks (technically "petioles") are cut and used in pies, jams, jellies, sauces and juice. The blade or green leaves of the plant contain oxalic acid crystals which can cause serious problems when eaten. Crystals can cause the tongue and throat to swell, preventing breathing. Therefore, the leaves should be removed before using. Rhubarb can be eaten after being hit by a frost or freeze PROVIDED the plant/stalks haven't wilted. If the stems appear soft and mushy, do not eat. Rhubarb plants will occasionally send up flowers and seed stalks in the middle of the plants. The flowers and seed stalks can occur due to winter's chilling conditions. Stalks can also occur due to natural maturity of the plants. Flower and seed stalks reduce the vigor of the plant. Energy is funneled into the stalks instead of new growth. Therefore, flower and seed stalks should be cut out as soon as they start forming. However, the plant may still continue to produce the flower stalks. Keep cutting. Plants do NOT become poisonous after flowering starts. Leaf stalks can still be cut and used.

Riddle. A large coarse sieve. Also called screen. Riddles of specified mesh size are used to sort Ware Potatoes from a potato crop.

Ridge. (1)The soil thrown up by a plough between two furrows. (2) A raised strip ploughed to mark out lands in a field before it is systematically ploughed, consisting of a balanced number of furrow slices leaning from opposite directions towards the centre.

Ridger. A type of plough bearing one or more double mould boards, each throwing two furrows simultaneously, one to each side.

Rill erosion. An erosion process in which numerous small channels, typically a few inches deep, are formed. It occurs mainly on recently cultivated soils or on recent cuts and fills.

Rimose (Rimous). Having numerous clefts of tissues as the bark of trees

Ringbone. A callus of bony tissue which sometimes develops on the pastern bone of a horse's hoof.

Ringworm. A contagious skin disease characterised by ring-shaped patches caused by certain fungi growing on the skin surface or in hairs growing in the affected skin. In livestock young store cattle are most commonly affected.

Riparian buffer. A strip of vegetation along the bank of a body of water which slows the rate of flow of runoff from adjoining uplands, causing sediment and other materials to fall out onto the land before the runoff enters and pollutes the body of water.

Riparian rights. The system of water allocation used in the humid eastern portion of the United States. Water may be used only by riparian landowners and it is recognized that all users will experience shortages periodically. In contrast with the prior appropriations system used in the arid West, water is not acquired by use, and access to it cannot be lost by lack of use.

Riparian. (1) Land adjacent to a river bank. Also the owner of such land. (2) Pertaining to or situated on or along the bank of a stream or other body of water. Often referred to in the context of cattle grazing and protection of streams for fish and wildlife habitat, and water quality purposes. Riparian rights refers to the entitlement of a land owner to certain uses of water on or bordering the property, including the right to prevent diversion or misuse of upstream waters (generally a matter of state law). (3) Pertaining to the area along the banks of a river, stream, or lake.

Ripening. Composite changes occurring in fruits after the end of the maturation period and by the beginning of senescence.

Risk-benefit analysis. Comparison of the short- and long-term risks to the overall societal benefits of an activity, chemical use, or technology. When risks and benefits are expressed in monetary terms, this is effectively cost-benefit analysis. Both the Federal Insecticide, Fungicide, and Rodenticide Act and the Toxic Substances Control Act direct the Environmental Protection Agency to base regulatory decisions on a risk-benefit or cost-benefit basis.

River rock. River Rock is an ideal landscape mulch. Make sure the rock is clean and not containing pieces of limestone.

Rodenticide. A pesticide used to destroy, control, or deter mice and rats or other rodent pests from damaging food, crops, etc.

Roguing. (1) Removal or cutting of individual plant from the seed plot which deviate in significant manner from the plants of the variety being multiplied. This is a step in the maintenance of purity in an established variety. (2) The process of removing undesirable or off-type plants from a standing crop.

Roller crusher. A machine used in haymaking which passes freshly cut grass between steel or rubber faced rollers rotating at high speed, bruising the grass and allowing sap to be removed subsequently by the action of sun and air. The rollers may be cleated or grooved to provide more effective bruising.

Roller. (1) The traditional machine for macerating tea. (2) A tractor-drawn implement consisting of heavy rotating cylinders. Used to crush clods, consolidate soil, flatten and smooth the surface, break crusts, or prepare fields for sowing seeds.

Rolling. Method of firming and smoothing ground surface, following seeding of a lawn. *Roof gardening (Terrace gardening)* Gardening on the flat roofs or terraces of buildings.

Root anchor. A large root that holds a plant in a physical soil mass feeder; a small surface root which easily absorbs nutrients and water.

Root ball. (1) The clump consisting of the main roots of a plant and the soil (or other growing medium) clinging to them. The root ball should be kept intact when transplanting. (2) The network of roots, with attached soil, of any given plant.

Root cap. A protective mass of cells located at the root tip which help protect root cells in that area as the root pushes through the soil.

Root crop. Crop plants in which the reserve foods are stored primarily in enlarged roots e.g. sweet potato

Root crops. Certain biennial plants grown for their edible swollen roots which act as food reserves, commonly in the form of sugar. The Mangel, Turnip, Swede and Fodder Beet are mainly grown as stock feed. The total area grown has decreased during this century due to high labour demands, but they are increasing again in popularity with mechanisation. Sugar Beet and Car¬ots and also the Potato (usually classified as a root crop) are mainly grown as cash crops. The 'tops' or foliage of root crops (except carrots and mangels) are valuable fodder, and sheep or pigs are often folded (Fold) on them. Root crops are often grown follow-ing or between cereal crops in rotation. Also called roots.

Root medium. A suitable substrate in which plant roots can grow. It consists of one or more mineral and/or organic components mixed together.

Root pressure. Pressure generated in the root and stem of plants, partially accounting for the rise of water in plants

Root stock (under stock). A seedling or other plant part on which the scion is grafted or budded.

Root zone. The area of ground under which a given tree's (or other plant's) roots spread. Often the root zone covers the area encircled by the *drip line*; that is, the roots often spread underground to the same distance that the tree's branches extend above.

Root-hairs. Unicellular outgrowths from the epidermis of young roots, immediately behind their root tips, useful for absorbing water and nutrients from the soil.

Rooting hormone. (1) A growth hormone in powder or liquid form, which promotes the formation of roots at the base of a cutting. (2) Chemical used to encourage faster rooting or encourage rooting of more cuttings e.g. IBA, NAA, IAA etc.

Rootstock. (1) The plant that supplies a root system for a grafted plant. Gardeners graft every thing from cacti to fruit trees. The grafting process involves bringing together the upper part of one planti and the rooted part of another. (2) The portion of a grafted plant that provides the root; grafted plants typically consist of a scion, which develops into the shoot or crown and a rootstock that provides the root system; rootstocks may include a signicant length of stem, called standards, commonly used for weeping trees or shrubs.

Rosette. (1) A radiating cluster of basal leaves. (2) Bunchy or rosette appearance in many plants due to: profuse leaf development with retarded internode growth. (3) A crown of leaves radiating from a stem, and at or close to the surface of the ground.

Rotation. A cropping system in which two or more crops are grown in a field in a fixed sequence. If a ley is included it is known as 'alternate husbandry' or 'ley farming'. The benefits of rotation include reduced accumulation of disease and pests which ac-company monoculture, weed control, the maintenance and improvement of fertility, spreading the risk of specific crop failure, and the even distribution of labour requirements over the year. In recent years farming has moved away from rigid traditional cropping programmes to more simplified systems due to various developments. These include the production of pesticides and artificial fertilizers, improved and increased mechanisation and guaranteed crop prices.

Rotational grazing. A method of managing grassland in which successive areas are intensively grazed for a period, in which defoliation is rapid and complete, followed by a rest period during which regrowth occurs. This method (also called the on-off principle) is used in a number of grazing systems as a means of matching livestock requirements with grass growth.

Rough grazing. A term generally applied to grassland of poor quality with a low stock-carrying capacity, particularly that covering rough land, Heath, Moorland and hill areas, where cultivation and the use of fertilizers is impractical, and which is character-ised by coarse grasses, sedges, heather, bracken or mosses, etc. Such land is usually grazed by hardy hill breeds of sheep and some cattle.

Rough stalked meadow grass. A close-growing perennial grass *(Poa trivialis)* with short creeping stolons, a rough, flat stem, and an erect panicled flower head, reddish, purplish or green in colour. It is common in lowland meadows and pastures, particu-larly on rich moist soils. It is included in Seeds Mixtures for permanent pastures, particularly those on wet heavy soils, and grows vigorously producing a low closely knit sward which remains green through the winder. It is very palatable to live-stock and is also useful for hay.

Row crops. (1) The rows or planting beds are far enough apart to permit the operation of machinery between them for cultural operations. (2) Those crops planted in widely-spaced rows facilitating cultivations between the rows, e.g. rootcrops, potatoes, cab-bages, etc.

Row intercropping. Growing two or more crops simultaneously where one or more crops are planted in rows.

Row-crop tractors. Tractors designed to work in growing Row Crops. They usually have high, narrow wheels giving good ground clearance, an adjustable track width, and a narrow turning circle. They are used with special tools for cultivating between the rows. Most models are four-wheeled although some have only three wheels.

Rubarth's Disease. Acute contagious disease affecting mainly young dogs, characterised by fever, vomitting, diarrhoea and some-times convulsions. It is caused by a filterable virus.

Rubbed seed. Sugar beet seeds used for sowing which have been separated from the clusters in which they are naturally fused together. Also called graded seed.

Ruddle. A colouring material (often red ochre) applied to the chest of a ram so that those ewes which have been covered (mated with) are marked. Also called keel, raddle or reddle.

Runner. (1) A horizontal, specialised above-ground stem that gives rise to new plants from the nodes touching the ground. (2) A slender trailing shoot that usually roots at the tip and some nodes; a long offshoot of a plant. A slender stem growing out from the base of some plants, which terminates with a new offset plant. The new plant may be severed from the parent after it has developed sufficient roots.

Russet. (1) Brownish rough area on the skin of fruits due to abnormal production of cork, caused by disease. (2) Apple varieties having natural russet coloured skin

Rusts. An order of small parasitic Basidiomycete fungi (Uredinales), the spores of which from spots, often rust coloured, on the leaves and stems of infected plants, *e.g.* Black Rust of barley.

Rye. The hardiest of the Cereal crops *(Scale cereale)*. It has a similar structure to Wheat, with single spikelets attached to each notch of the Rachis, but also has small narrow Glumes and conspicu-ous large outer pales tapering to long whiskery awns.

S

S. Saddle grafting. Grafting in which the root stock is shaped in the form of a wedge and a cleft is made in the scion for effecting successful graft union.

Sack. (1) A measure of capacity, particularly for grain, equal to 4 Bushels. Also called coomb. Now in disuse in the grain trade since most grain is now transported in bulk by lorry. (2) A large bag of coarse material.

Saddle grafting. Method (asexual) of plant propagation. The top of the stock is cut first transversely and then two upwards cuts are given on either side to form a wedge and cleft is made in the scion so that it fits tightly over the stock, after fitting the scion, the union is sealed.

Sainfoin. A perennial leguminous fodder plant *(Onobrychis viciifo-lia)*, once extensively grown for both hay and silage, with Aftermath grazing, mainly in chalky areas, usually on its own, but sometimes included in seed mixtures. It has a similar nutritive value as lucerne which it is sometimes substituted for on potash-deficient soils, being somewhat drought resistant. It has a strong, woody, long root, a branching stem with pinnate leaves, pink flowers in a conical raceme, and single-seeded pods.

Salometer. Instrument (special hydrometer) for measuring the per-centage of saturation of a brine solution with respect to salt. Used in picking and meat packing industries. Also called a salimeter or salinometer.

Salt Injury. Salt injury occurs due to an increase in the soil's sodium content. De-icing salt is the major culprit. Salt readily absorbs moisture, tying it up in the soil and extracting it from plant roots. Sodium and chloride ions also travel throughout the plant, accumulating in various plant parts such as leaves and stem at a toxic level. Symptoms of salt injury include stunted, yellowed foliage with brown edges, premature leaf color and death of the leaf margin; the later gives the plant a scorched appearance. Affected evergreen foliage turn yellow or brown in the early spring. Damage usually occurs only on the side of the plant facing a road, driveway or sidewalk. Avoid injury by leaching the soil thoroughly with water. Alternatives to salt such as sand, cinders or light gravel can be used. Kitty litter can also be used for traction. Urea fertilizers can be used to melt ice when necessary. Excessive applications of urea can cause nitrogen burning or unusual turfgrass growth in the spring.

Sample grade. In commodities, usually the lowest quality of a commodity, too low to be acceptable for delivery in satisfaction of futures contracts.

Sand culture. A kind of hydroponics where plants are kept in sand medium and supplied with a nutrient solution.

Sandy soils. Soil containing a high proportion of sand particles and very little clay, usually less than 5%. Such soils have a low water holding capacity, drain freely, and due to leaching are usually poor in plant nutrients. They are also well aerated and organic matter quickly decomposes. They therefore require adequate applications of fertilizers and manures and are consequently often called 'hungry soils'. Liming is also needed to correct acidity. They tend to dry out quickly allowing them to be worked throughout the year. They are often used for market gardens using irrigation. Crops suited to sandy soils include barley, carrots, peas, potatoes, sugar beet, etc. Sometimes called 'light soils' because they are relatively easy to cultivate compared with heavy or clay soils.

Scab. (1) A fungal disease of various kinds of fruit and vegetables, causing the growth of scaly crusts, e.g., Common Scab of potatoes *(Streptomyces scabies)* and apple scab *(Venturia in-inequalis).* (2) A fungal scale disease that leaves spots on foliage and fruit.

Scabrous. (1) Rough or gritty to the touch; rough-pubescent. (2) Rough to the touch as with many hairy leaves or stems.

Scaffold branches (Scaffold limbs)*:* The main support branches of a tree arising nearest to the ground.

Scaling. A method by which Forest Service personnel, or an independent third-party to measures the volume of timber actually removed as a result of a timber sale. Scaled sales are a timber sales that use scaling for billing purchasers at the stumpage price for the timber cut.

Scalper. Seed processing machine designed to use the bulk of trash from seed so as to facilitate elevating and processing operations. It is fabricated with a vibrating or rotating screen or sieve.

Scalping. (1) Close mowing of lawn to remove the major portion of foliage. (2) Removing an excessive quantity of functioning, green leaves at any one mowing; exposes crowns, stolons, dead leaves and even bare soil resulting in a shabby appearance.

Scape. (1) A leafless peduncle arising from the basal rosette of a few or not basal leaves; sometimes a few scale-like leaves or bracts may be borne on it; a scape may be one or many-flowered. (2) The leafless peduncle of a flower which arises at or near the ground, as in the tulip.

Scarification. (1) Chemical or physical treatments given to some seeds in order to weaken the seed coat for facilitating germination. (2) Deliberate wounding of a seed to allow water to enter. (3) Process of mechanically scaring or roughing the hard seed coat to make it more permeable to water.

Scarifier. Acultivator, the action of which breaks up and stirs the soil without turning it over. Also called scuffler.

Scarp slope. A short steep slope to one side of a line of hills *(e.g.,* North Downs), as distinct from the long, gentle dip slope to the other side.

Scientific name. Internationally accepted Latin name of an organism consisting of genus, species, varieties and forms and their subdivisions which avoids confusion while using common names (See Binomial system of nomenclature)

Scion. (1) Shoot or bud grafted or budded on a root stock. (2) The upper part of the union of a graft. (2) A short length of stem, taken from one plant which is then grafted onto the rootstock of another plant; the top "slip" or desirable propagated tissue in a grafted plant (3) The shoot portion of a rootstock-scion graft.

Scooping. Removing the basal plate of a bulbous plant with the help of a scalpel or blade and exposing the cut surface of the bulb scale for propagation.

Scorch. Injury to plant parts due to burning by wind or water loss; sometimes used as a synonym of scald, such as burning by the sun.

Scrambler. A plant that climbs without twining or grasping in some way.

Scrapie. A fatal, neurological disease which affects most breeds of sheep. Sheep display uncharacteristic behaviour (i.e., suddenly leaving the flock), become increasingly nervous as the disease progresses, tremble and go into convulsions. When rubbed they make chewing and nibbling movements. At a later stage rubbing against walls or posts increases and large areas of skin are exposed which become ulcerated by repeated nibbling. Death occurs after a few weeks or months of the first symptoms.

Scratch. (1) A term for two light furrows ploughed to mark the setting out or opening of a plot, which form the base of the first two furrows (the crown furrows). Also called scribesod. (2) To rub the rotted stems of flax, hemp, or other fibre crop with a comblike instrument so that the stems are broken and the fibres laid bare for further processing.

Scur. A loose, horny knob that sometimes develops in homed cattle at the size where a horn would normally grow. Also called snag.

Scurfy. Describes a surface covered with bran-like or flaky particles.

Scythe. A hand implement with a long wooden handle with two hand grips (nibs), bearing at its lower end, fixed at right-angles, a large gently curved blade. Used for mowing by sweeping it through the vegetation. Sometimes spelt sithe.

Seams. The angles or shaped sections of the slice turned over in ploughing where they touch each other.

Season. (1) One of the four climatic divisions of the year, *e.g.*, spring. (2) The oestrus or 'heat' period of a female animal. (3) To prepare a seed bed.

Secondary compounds. The compounds produced in the plant body as by-products of primary metabolic pathway e.g. hydrocyanic acid (HCN) in cassava.

Secondary dormancy. Dormancy observed for a second time in a seed when it is exposed to unfavourable environmental conditions after becoming imbibed.

Secondary metabolites. See Secondary plant metabolites.

Secondary plant metabolites. Plant metabolites, produced as the by-products of the primary metabolic pathway, used for protection and better adaptation to the environment e.g. alkaloids, terpenes, glycosides and tannins.

Secondary plant nutrients. Calcium, magnesium and sulphur are called secondary plant nutrients as they are not included in straight commercial fertilisers.

Seconds. Those grains of a cereal crop intermediate in size between the largest ones (head corn) and the smallest ones (tail corn) threshed out during harvesting with a combine harvester.

Sedges. A large family of grass-like plants *(Cyperaceae)*, particularly the genus *Carex*. They are distinguished from grasses by having a solid, usually triangular, stem, and one scale beneath each flower (grasses have two).Flowering heads are greenish, brownish or purplish. Sedges are common in wet or badly drained areas and are regarded as weeds.

Sedges. Plants of the genus *Carex*, which includes over 1,500 species; sedges look similar to grasses.

Sediments. Particulate material, including soil, sand, and minerals, transported and deposited by water or wind (see erosion). Waterborne sediments cloud the water diminishing sunlight available to aquatic plants; sediments deposited in reservoirs, rivers, and harbors destroy fish and wildlife habitat and may fill shipping channels. Farming, mining, and building and construction activities that expose soil to wind and rain are major sources of sediments.

Seed bank. A facility used for the preservation and dissemination of seed, particularly varieties that are not in commercial use and that may be threatened with extinction. The USDA administers the U.S. National Plant Germplasm System.

Seed barrow. A hand-pushed or tractor-mounted implement used for sowing seeds, consisting of a long narrow box with a perforated base, mounted laterally on a wheel barraw-like frame with a single wheel. The seeds were brushed through the holes. Also called shandy barrow.

Seed bed. (1) A well prepared land of convenient dimensions for sowing or planting. (2) An area of land or a field cultivated to a level, fine tilth, in which seeds are sown.

Seed blowers. Mechanical device by which light weight is removed by blowing.

Seed borers. Many insects bore into seeds, either to feed on them or to lay their eggs in them. Some attack fresh growing seeds: other dried seeds such as stored grain. Chalcid wasps bore holes in seeds with their long ovipositors and deposit eggs inside larvae already present within the seed. Codling Moths such as *Las-peyresiapomonella* by there eggs on leaves and flowers of apple tree and the newly hatched eaterpillars bore into the developing seeds; weevils such as the Grain weevil and the Bean weevil use their snouts for boring into dried seeds.

Seed certification. (1) Field inspections and seed tests to ensure genetic and physical purity and good germination of seed lots, and issuance of a certificate to that effect. (2) Maintaining and making available to the public, the seeds of high quality and superior propagating materials so as to ensure genetic identity. This is achieved through continu-ous inspections of fields and seeds and by regulations for checking on the production, harvesting and cleaning of each lot of seed.

Seed cleaning. The removal from seed, usually by machine, of impurities such as weed seeds, broken leaves, insects, etc.

Seed cum fertilizer drill. Seed drill fitted with fertilizer dropping attachment, distributes the fertilizer uniformly on the ground.

Seed disinfection. Ridding the seed surfaces of organisms which are potentially disease producing.

Seed dressing. The chemical treatment of seeds, particularly cereals, with fungicides and sometimes insecticides, to protect them against soil and seed-borne diseases and pests. A dye may be added which enables treated seed to be distinguished.

Seed drill. A tractor-mounted or trailed machine which sows seeds in rows. It consists of a seed hopper incorporating seed metering units which are generally driven by gearing or chains from the land wheels. These units, which are adjustable to suit various seed size and sowing rates, transfer the seed from the hopper down coulter tubes to a groove cut in the soil by coulters. Drills for cereals and grasses normally sow the seed at random, while precision drills used mainly for sugar beet and vegetable crops, place individual seeds at specific pre-set intervals in the rows.

Seed hardening. Subjecting seeds before sowing to alternate cycles of wetting and drying to induce tolerance to drought

Seed house. A commercial company providing seed varieties for crop production.

Seed inoculation. The dressing of seeds of leguminous plants with a culture of nitrogen-fixing bacteria, usually when the soil is deficient in the bacterial strain appropriate to the legume concerned and would otherwise restrict root nodule formation.

Seed processing. Process involved in the preparation of harvested seed for marketing.

Seed quality. The quality of crop seeds has two principal components. Firstly, the value of different varieties for crop production, assessed by yield trials and tests for disease reaction and produce quality. Secondly, the value of different 'seed lots' for sowing assessed by laboratory analysis of samples for purity, germination, weed content and seed-borne diseases.

Seed rate. The quantity of seed required for sowing in a unit area of land.

Seed royalties. Fees paid under licence for the use of seed of protected (patented) plant varieties, and usually included in or added to the sale price of seed. Royalties are paid to breeders and are used to finance continued plant breeding work.

Seed sample certificate. Form of international seed Analysis Cer-tificate (Blue Certificate) use when sampling from the seed lot is not under the responsibility of a member station.

Seed stalks. Rhubarb plants will occasionally send up flowers and seed stalks in the middle of the plants. The flowers and seed stalks can occur due to winter's chilling conditions. Stalks can also occur due to natural maturity of the plants. Flower and seed stalks reduce the vigor of the plant. Energy is funneled into the stalks instead of new growth. Therefore, flower and seed stalks should be cut out as soon as they start forming. However, the plant may still continue to produce the flower stalks. Keep cutting. Plants do NOT become poisonous after flowering starts. Leaf stalks can still be cut and used.

Seed technology. (1) Science which studies the production, harvesting, processing testing, packaging, storing and marketing of seeds. (2) The science dealing with the production, harvesting, processing, testing, packaging and marketing of seeds.

Seed testing. It provides information to meet legal standards, deter-mines seed quality, and establishes the rate of sowing fora given stand of seedling.

Seed. (1) A reproductive structure of flowering plants, conifers and various other plants. It develops from a fertilized ovule and comprises an embryo and a food reserve contained in a protec-tive coat or testa. (2) To sow seeds of a crop in a field. (3) A fertilized ripened ovule that contains an embryo. (4) A mature ovule consisting of an embryonic plant together with stored food and surrounded by a protective coat

Seed-borne diseases. Those plant diseases which may be transmitted from one generation to the next, either within or on the surface of seeds, e.g., many fungal diseases.

Seedless hay. Hay derived from the remains of a grass seed crop after threshing out the seed heads with a combine harvester.

Seedling rootstock. A stock grown from a pip or seed of a wild crab apple. Also called seedling crab.

Seedling. (1) A young plant, grown from seed as distinct from one grown from a cutting or by a graft. (2) Young plant grown from seed.

Seeds harrow. A light type of Zig-Zag Harrow, with short, straight tines, usually comprising several sections attached to a frame. Used to prepare seeds beds and over seeds after sowing, and often hitched behind the drill. Sometime also used for light weeding.

Seeds mixtures. Mixtures mainly of grass and clover seeds, some-times including other herbage legumes, sown to produce both short-term and long-term Less. Mixtures are precisely com-pounded with varieties suitable to the conditions and needs for which the ley is required. They may be prepared to order by a merchant or standard mixtures can be purchased from seeds firms. Also called small seeds.

Seedstock. Pedigreed or well-bred livestock which is maintained for breeding purposes. A specially selected strain of plants or seeds which are to be used as parents for future generations.

Segmented seed. The 'seeds' of certain rootcrops (e.g., Mangel, Sugar Beet) naturally borne in clusters of fruit each containing a single true seed, which are often separated for sowing. If the cluster itself is sown, several seedlings may develop which makes singling a slow operation.

Selection harvest. Harvesting individual trees or small groups or trees at periodic intervals (usually 8 to 15 years) based primarily on their vigor and age. Selection harvesting perpetuates and uneven-aged stand.

Selection. The basis of animal and plant breeding in which individuals are selected for breeding on the basis of specific desired charac-teristic or qualities.

Selective cutting. System of cutting in which single trees, usually the largest, or small groups of such trees, are removed and reproduction secured under the remaining stand and in the openings.

Selective herbicide. A herbicide capable of killing or stunting the growth of weeds growing in a crop but having little or no harmful effect on the crop itself.

Self pruning. Natural death and fall of branches, especially the lower branches, from the live plants due to causes such as light and food deficiencies, decay, insect attack, snow and ice.

Self. To cross a plant with its own flowers or one of identical type; said of flowers of one color as opposed to a bicolor.

Self-feed silage. A management system whereby stock are allowed to graze silage *in situ,* the amount taken being controlled by an electric fence or movable barrier set a short distance from the silage.

Self-mulching soil. A soil, the surface of which is so well aggregated that it does not crust over under rainy conditions and is able to act as a surface mulch when the soil dries.

Self-pollination. (1) A plant capable of pollinating its own flowers. (2) Pollination with pollen from the same flower, from other flowers of the same plant, or from other flowers of plants with the same genetic constitution (from the same clone)

Semi-digger. A type of mouldboard with a gently concave curvature and twisted along its length, sometimes with a renewable lead-ing edge or shin, and producing a broken furrow slice. It is in general use for various operations including moderately deep ploughing (*e.g.*, for rootcrops), ploughing-in farmyard manure and surface trash, rapid seed bed preparation in the spring, and winter ploughing on well drained soils. It ploughs to a depth of 30-35 cm (12-14 in.) and is therefore intermediate between the digger and general purpose mould-board.

Semi-evergreen. (1) A plant whose foliage is retained all year in warm climates but not in colder ones; some of the foliage is retained in most climates where hardy. (2) Having some leaves deciduous and some leaves persistent throughout the winter, with the degree of persistent leaves determined by the harshness of the winter season.

Senescence. (1) The aging process of an organ or organism. (2) Irreversible changes associated with the ripening of a fruit leading to its death.

Sepal. (1) A segment of the calyx, or outermost floral part, which is normally green in colour. (2) Modified leaves forming the outer floral envelope. (3) One of the green leaf-like parts of a flower which form the calyx and surround the petals. (4) The divided part of a calyx or outer whorl of a flower.

Septic system. An on-site system designed to treat and dispose of domestic sewage from a residence or business not connected to a sewer line. A typical septic system consists of a tank that receives waste and holds it while bacteria decompose solids, and a system of tile lines or a pit for disposal of the liquid effluent. The sludge that remains in the tank after decomposition of the solids must be pumped out periodically. By using the assimilative capacity of the land, a properly operating septic system has minimal impact, but an improperly functioning system can be a source of nitrogen pollution and of groundwater contamination.

Service sample. Seed sample submitted to the Central Seed Labora-tory, or a State Seed Laboratory and the results of which are to be used as information for seedling, seeling or labelling pur-poses.

Sessile. (1) Without an obvious attachment structure, such as a leaf attached to a stem without an obvious petiole a leaf with a single leaf blade (not compound). (2) Without stalk

Set-up. To mark out the headland in a field with a market furrow, and the various lands with ridges, before commencing ploughing.

Sewage sludge. A partially dried residue from sewage treatment works sometimes used as a fertilizer. Compared with Farmyard manure it is poor in potash and less beneficial on soil texture.

Sex dimorphic palm. In coconut, palms which produce pure male inflorescence during certain months and during other months they produce both male and female flowers

Sex reversal. Phenomenon in plants like papaya where the sex of plants are changed, permanently or alternatively

Shade trees. Fast growing trees which provide shade to many crops in the tropics like coffee, tea and cocoa and to some orchards for their growth and development

Shaping. A horticultural operation in perennial herbs and foliage plants, involving pruning, clipping or trimming of young branches for keeping them compact and elegant looking

Share. A pointed steel or cast iron blade attached to the front of a mouldboard on the Body of a plough which makes a horizontal cut under the furrow slice which is then turned by the mould-board. Also called plughshare or point.

Shrub land. Ecosystem in which relatively small (up to 6 m), woody, usually multistemmed, densely branched vegetation dominates.

Shrubby. Applied to woody plants which at maturity are less than 20 ft. high, have more than one stem, and no definite crown shape (a purely arbitrary definition); See also "Arborescent".

Side graft. A type of graft in which the scion is attached to the side of a branch rather than into the cut end of a stock.

Side grafting. A method of grafting where an angular cut is made on one side of the stock, and the scion wood having slanting cut on both sides is inserted into this cut.

Sideling. Land which slopes sideways in relation to the line of work, e.g., when ploughing along the contour.

Sieve. (1) A frame containing a mesh or perforated base for sifting material. Normally finer than a Riddle. (2) A basket for fruit with a bushel capacity.

Silage. A feedingstuff consisting of forage crops (e.g., Grassland Species, kale, beet tops, pea haulm, etc.) cut or harvested in the green state and preserved in a silo in a succulert condition for later use. The principle of silage making is the fermentation, by bacteria, of carbohydrates in the plant material to organic acids, and of Proteins to amino acids, which act as preservatives. Well made silage is yellowish-brown in colour and contains mainly lactic aid (as much as 20% of its fresh weight) produced by *Lactobacilli,* with some acelic acid and with a pH or 4.0 or less. Crops contaminated with faeces and soil tend to produce a dull olive green silage containing undesirable butyric acid, produced by *Chlostridia Lactobacilli* grows best when the silage material has a high sugar content. Crops are ideally cut after dry sunny weather which stimulates photosynthesis, and

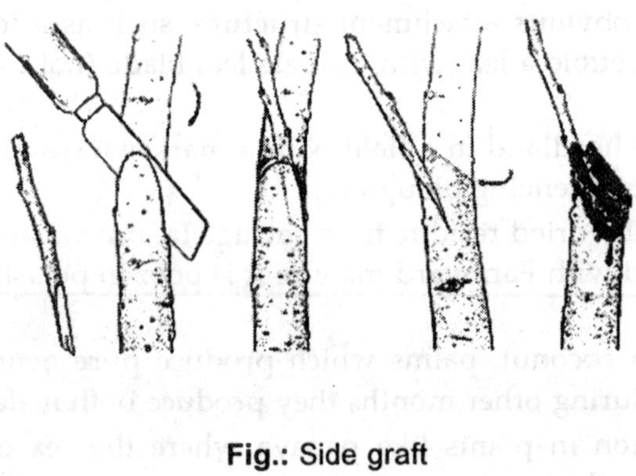

Fig.: Side graft

therefore sugar production, and wilts the plant, increasing sugar concentration. Cut crops may also be crimped or lacerated to increase wilting. When sugar content is low various additives are sometimes included, e.g., molasses, which provides supplementary sugar and formic acid, which increases acidity and suppresses *Chlos-tridia.*

Silo. A container in which silage is made and stored. A number of types are in use including the various forms of clamp on the surface, in a pit, or in a large sealed polythene 'bag' evacuated of air. Silage may also be prepared in wooden, concrete or steel towers (sometimes under airtight conditions) in which carbon dioxide production restricts respiration and allows good fer-mentation. The term silo is sometimes applied to a moist grain storage tower.

Slender foxtail. An annual grass commonly found on heavy and poorly drained land. An arable weed, particularly in early sown winter cereal crops. A tufted, smooth-sheathed, green or pur-plish grass, with a slender, cylindrical, tapering panicle. Also called black grass.

Slips. (1) Herbaceous or softwood cuttings. (2) Shoots borne on the stem below the fruit of pineapple: and which are used for vegetative propagation.

Slot seeding. A method of re-seeding pasture without destroying all existing turf. A special drill was produced which opens up slots about 25 mm (1 in.) wide and deep, sprays a band of herbicide to kill off adjacent competition and sows the grass seeds into the base of the slot. Other machines have now been developed with multiple sets of rotary cultivators and seed spouts which open and seed larger slots. The rotary action copes better with thick mats of old grass and roots, and the floating action of the individual units copes with uneven ground.

Sod-forming grass. A grass that spreads by creating new plants from existing plants by running roots across or under the ground (*stolons* or *rhizomes*, respectively), creating a dense turf. Also called *spreading grass.* Contrast with *clump-forming grass* (also called *bunchgrass*), in which single plants get larger over time but do not create new plants.

Sodium adsorption ratio. A ratio of sodium, magnesium and calcium that is used to express the relative activity of sodium ions in exchange reactions with soil. When the SAR exceeds a certain level (i.e. 10 for most woody plant) the exchangeable soil sodium would be toxic to the plant.

Sodium. (Na). A metallic element, an essential constituent of the tody fluids in animals. It is mainly present as various salts, mainly sodium chloride (NaCl) or common salt, its concentration in the blood being finely controlled with excesses excreated. Reduced performance results from sodium deficiency and livestock diets are usually supplemented with salt in mineral mixtures. Sodium chloride is also used as a salt fertilizer.

Soft fruit. (1) Loosely berried Fruit with relatively soft flesh, divided into bush fruit (e.g. black, red and white currants, goose-berries, etc.) and cane fruit (e.g. raspberries, blackberries and hybrid berries), and also strawberries. The latter are the most widely grown soft fruit, the principal maincrop areas being in Kent and East Anglia. Blackcurrants are also widely grown in Britain, mainly for the preparation of juice for the manufacture of soft drinks and flavours for confectionery. The largest concentration of raspberry plantations in the world is located around Perth in Scotland. (2) See Small fruits.

Soft pinch. The removal of only a portion of a shoot terminal at a point where the tissue has not become woody

Soil air. The gaseous content of the soil (i.e., that part of the volume of the soil other than solids or liquid) occupying part of the capillary spaces between soil particles and the soil pores. Soil air differs from atmospheric air, having a higher concentration of carbon dioxide due to respiration by plant roots and micro-organisms, and is normally saturated with water vapour. The carbon dioxide dissolves in the soil moisture forming carbonic acid, which assists in dissolving mineral matter to release plant nutrients.

Soil analysis. The determination of the composition of soil by various laboratory chemical and physical methods. Mechanical analysis is used to accurately determine soil texture, an important factor in soil and land use classification. The analysis of soil in terms of its chemical composition is important in revealing its state of fertility and in recommending the amount and type of fertilizer treatment required.

Soil classification. The systematic arrangement of soils into various categories according to their characteristics. This may be (a) according to texture, e.g., sandy loams, clays, etc., *(b)* on a geological basis, e.g., chalk soils, sandstone soils, etc., or (c) on a pedological basis. The basic pedological unit of classification now used for soil mapping is the soil series. The five most important soil groups are brown earths, calcareous soils (e.g., rendzinas, brown calcareous soils), Gley soils, Peat soils and podsols.

Soil conditioner. An organic material like humus or compost that helps soil absorb water, build a bacterial community, and take up mineral nutrients.

Soil conservation district. A legal subdivision of state government, with a locally elected governing body, responsible for developing and carrying out a program of soil and water conservation within a geographic boundary, usually coinciding with county lines. The nearly 3,000 districts in the United States have varying names. soil conservation districts, soil and water conservation districts, natural resources districts, resource districts, resource conservation districts.

Soil horizon. A layer of soil approximately parallel to the surface with fairly distinctive properties such as texture, colour, struc-ture, mineral or chemical composition, etc. A vertical section of soil (a soil profile) will reveal the various horizons present from the surface to the parent material (if soil development has been *in situ).*

Soil loss tolerance (T value). For a specific soil, the maximum average annual soil loss expressed as tons per acre per year that will permit current production levels to be maintained economically and indefinitely. T values range from 2 to 5 tons per acre per year. According to the 1992 national resources inventory, about 63 million acres of highly erodible cropland are still eroding at more than their T value, including 21 million acres that are still eroding at three times T.

Soil moisture deficit. The amount of water which a soil requires to be added to return it to field capacity. The size of this deficit, together with the condition of the crop, provide an indication of whether irrigation is required.

Soil scoop. Equipment used in excavating ditches, clearing drains and moving soil over short distances.

Soil series. (1) A group of soils having horizons (or layers) similar in characteristics and arrangement in the soil profile, except for the texture of the surface portion. They are given proper names from place names within the areas where they occur. Thus, Norfolk, Miami, and Houston are names of some well-known soil series. (2) The basic unit of soil classification and mapping consist-ing of soils which are identical in all the main profile character-istics, except surface texture, and predominantly confined to one parent material. The soil series can be considered as compa-rable to the genus in the classification of living organisms. It can be subdivided into lower categories (comparable to Species) on surface texture, depth, slope, etc.

Soil solarization. Fumigating and warming soil by covering it with black plastic. This is an alternative pest control technique being investigated as an alternative to the use of methyl bromide (a chemical fumigant used in Florida tomato production and for other speciality crops, which will be phased out of use because of its ozone depleting effects).

Soil sterilant. (1) A chemical applied to the soil to kill pests, diseases or weeds. (2) A chemical that temporarily or permanently prevents the growth of all plants and animals, depending on the chemical. Soil sterilants must be registered as pesticides.

Soil survey. A program of the Natural Resources Conservation Service to inventory soil resources as a basis for determining land capabilities and conservation treatments that are needed, provide soil information to the public (primarily through maps), and provide technical support to those who use soils information. About 90% of the private lands have been mapped. In FY1996, maps were prepared or updated on more than 17 million acres.

Soil texture. (1) The' feel' of a soil, e.g., soapy, gritty, sticky, etc. Soils containing small clay particles, with small interparticle spaces are called close-textured, as distinct from open-textured soils containing larger particles (e.g., Sand) with larger air spaces. (2) A soil sample may be describe as belonging to a particle class of soil texture (e.g., sandy loam, silty clay, sandy clay loam, etc.) following Physical Analysis to determine the relative propor-tions of clay, sand and silt. The various textural classes have strict compositional limits, e.g. sandy clay loam containing 20-35% clay, less than 28% silt, and 45% or more sand.

Soil. (1) The unconsolidated material covering the surface of the earth in which plants grow, and in which many animals (e.g. insects, worms, beetles, bacteria, etc.) live and derived food. It consists mainly of particles of Sand, Silt and Clay, closely associated with Organic Matter, the relative proportions of each determin-ing the soil type. Soil mineral matter is derived from the weathering and erosion of rock. Water percolating through the soil depletes the surface layers of both slouble and fine insoluble substances, and has an effect on the hydrogen and hydroxzyl ion content of the soil, thus determining its acidity. From the agriculture viewpoint the top horizons, which crop roots pene-trate to obtain water and nutrients, are the most important. (2) In the United States, about 70,000 kinds of soil are recognized in the nationwide system of classification. Each has a unique set of characteristics and a potential for use. These characteristics are important in designing a conservation plan to protect the soil from erosion if it is being cultivated. The Natural Resources Conservation Service is responsible for mapping the United States by soil type, through the Soil Survey Program.

Soiling Crop. (1) A green crop cut and fed to stock elsewhere, i.e., not at the field where grown. Such cut forage is sometimes called green meat. (2) Crops harvested when green and succulent condition and fed to animals shortly after cutting. It is neither dried nor stored for future feeding.

Species. (1) A natural group of plants composed of individuals similar in structure and physiology capable of producing similar fertile offspring; usually including several minor variations (subspecies). Different in structure and/or physiology from other such groups and normally do not interbreed with them. A sub component of a *genus.* (2) Basic classification unit of plants comprised of individuals that are recognized as distinct from other species, and which can freely interbreed among themselves. (3) Closely related groups of plants which are sub divisions of a genus. (4) Species represent the lowest and most important of the primary groupings used in classifying plants, animals, and microorganisms. While no single definition applies to all organisms, biologists rely principally on (1) morphological and genetic similarities and (2), for sexually reproducing organisms, the capability of interbreeding with one another but not other groups. If different species do interbreed, the offspring, if any, are often sterile. Biologists give species unique, binomial names: a generic name that includes closely related species, and a species-specific name. The horse, for example, is *Equus caballus;* the donkey or ass is *Equus asinus.* (Their offspring, the mule, is sterile.) As populations of organisms vary geographically and change over time (becoming extinct, or splitting or evolving into new species), species classifications are neither absolute nor immutable; where some biologists see variations within a species (and may designate subspecies), others may see separate species. About 1.5 to 2 million species have been named, but scientists estimate the total number of species could be 5 to 100 million, many of them probably undiscovered microorganisms. The Endangered Species Act (ESA) protects species designated as endangered or threatened with extinction; these protections prohibit taking endangered species and can include restrictions on habitat alterations, such as logging or water pollution. Because of the way "species" is defined in the ESA, policy debates have arisen over whether certain groups of organisms qualify for listing (e.g., northern goshawks and the Alexander Archipelago wolf).

Spinner. A tractor-drawn implement used for harvesting potatoes. Trailed and mounted types are in use. The main components are a digging share which lifts the row of potatoes and loosens the soil, and a revolving wheel set at right-angles to the row and bearing rotating tines. The tines strike the lifted row tinging the potatoes and soil to the side often against a ropenet, so that much of the soil passes through and the potatoes fall to the ground in a row. The spinner is particularly useful in wet soil conditions, but can cause damage to the tubers.

Splice grafting. (1) A grafting method where a long slanting cut is made at the base of the scion and a similar cut is made at the top of the stock and kept in close contact by tying together. (2) Method of grafting in which oblique cuts are made across both scion and stock of almost equal diameter. The cut surfaces of scion and stock are placed together and are tied after matching their surfaces.

Sporotrichosis. Sporotrichosis (Sporo for short) is a fungal disease present in sphagnum moss, soil, on rose thorns, and decomposing organic matter. No known cases have been documented occuring with peat moss. The fungus enters skin through cuts and scratches on fingers, arms or legs. Small pustules will develop on the skin. Pustules will swell and turn red. The disease can spread to the bones and other organs. Death can occur in extreme cases. If symptoms occur, see a doctor. Treatment can include, but is not limited to, potassium iodine solutions orally. Severe symptoms can be treated with amphotericin B. Check with a licensed physican. Possible problems can be prevented by wearing gloves and long sleeves when working with peat and sphagnum moss. Wash body parts thoroughly after working in the yard and garden.

Steerage hoe. A type of Hoe, rear-mounted on a tractor, requiring an additional operator to the tractor driver, who is able to steer or guide the implement to avoid crop damage. Steerage is effected either by a direct linkage mechanism from the tractor or by controlling the depth of ground wheels.

Subscription farming. Usually refers to a direct marketing arrangement between a local farmer and one or more nearby consumers, where the consumer pays, or agrees to pay, the farmer in advance for delivery of a variety of fresh farm products, as they are produced throughout the season. It provides a source of guaranteed, up-front income that the producer can use to finance planting and other operations.

Subsidy. A direct or indirect benefit granted by a government for the production or distribution (including export) of a good or to supplement other services. Generally, subsidies are thought to be production and trade distorting, resulting in an inefficient use of resources. Arguably, subsidies may be justified on grounds that they adjust for nonmarket considerations that are as important as market values. This term also is used to refer to federal reimbursements for meals served through child and elderly nutrition programs.

Succession. (1) A rotation of crops. (2) The sequence in which plants colonise a new habitat or bare ground, in which each plant community is superseded until a climax vegetation develops.

Successional cropping. (1) The growing of successive but different crops in the same field in the same season. (2) The planting of a crop in small areas over a period so that a little ripens at a time, ensuring a gradual continuous supply over a long period.

Succulent foods. Those feedingstuffs with a relatively high water content, approximately 70-90%. They are highly digestive but are low in oil, protein and fibre. They are valued particularly for winter feeding to balance dry fodders which tend to cause constipation, and especially for pregnant, sick or high yielding animals. Examples include the carbohydrate-containing tubers and root crops such as potatoes, carrots, mangels, swedes, and turnips etc. Various by-products with a low dietetic value are sometimes regarded as succulent foods, such as wet brewers' grains, wet sugar beet pulp, liquid skim milk, etc.

Succulent. (1) A fleshy plant that hold water in itself, usually in the leaves or stem. Thickened, juicy, fleshy tissues that are more or less soft in texture. (2) Plants with fleshy leaves and stems, usually tolerant to water stress.

Sucker. (1) An off-shoot that develops from an adventitious bud of the roots or lower stem of a plant. (2) Shoot growing in the axils of leaves (pineapple). (3) In rose, a shoot growing from the rootstock. (4) A vigorous shoot arising from a plant base or from below ground; also the adhering discs of a vine. A growth originating from the rootstock of a grafted plant (5) This is an extra shoot growing from the roots; remove it so it doesn't weaken the plant. (6) Many stems arising from the base of a tree or shrub and gradually spreading the diameter of its basal area. (7) On woody plants, new stems that èmerge from the roots. These can occur next to existing stems or many feet distant, depending on the species and how far the roots spread.

Sulphate of Ammonia. $(NH_4)_2SIO_4$ Ammonium sulphate, a nitroge-nous fertilizer occurring as a soluble salt in the form of white needle-like crystals once produced as a by-product of coal-gas manufacture but now synthesised. In common with other non-metallic ammonium salts, when applied to fields the sulphate radical combines with calcium in the soil and is leached out, depleting the soil of lime and increasing acidity. It contains about 21% nitrogen.

Sulphate potash. (K_2SO_4) Potassium sulphate, a fertilizer containing about 50% potash, occurring as a white crystalline powder, produced by the action of sulphuric acid on muriate of potash. Mainly used by fruit growers and in market gardens, and also on potatoes which produce tubers with a higher dry matter contents than when muriate of potash is applied.

Sulphur (S). A non-metallic chemical element, occurring as pale-yellow crystals in several forms, required by living organisms a constituent of protein (particularly keratin in animals) and some oils.

Sulphuric acid. An extremely corrosive, colourless, oily liquid, which chars organic matter and is sometimes used as a Contact herbicide, *e.g.*, to bum off diseased potato haulm. Also called oil of vitriol.

Summer annual. (1) A one year plant that sows in spring and flowers in summer or fall. (2) Germinate from seed in the spring and complete flowering and seed production by fall, followed by plant death, usually due to cold temperatures. Their growing season is from spring to fall. Examples: marigolds, squash, and crabgrass. These are also called warm season annuals.

Summer fallow. Cropland in semi-arid regions that is purposely kept out of production during a cropping season mainly to conserve moisture for the next season. It is common for wheat producers in semi-arid regions to rotate half their cropland to summer fallow each year.

Summer feeding. (1) The supplying of water to marshland in the summer, usually by impounding river water to prevent natural drainage and the lowering of the water table. (2) The feeding regime of cattle in the summer months involving outdoor graz-ing with or without supplementary rations.

Sun-curing. The process of drying in sun.

Sunflower. A composite plant the seeds of which are rich in edible oil, used for margarine, cooking oil and in medicine. The residue, after oil extraction, is used to produce cake or meal with a high Fibre content, sometimes used in Concentrates for cattle and sheep but not ted to pigs. The seed is sometimes ted to poultry, particularly during moulting.

Sweet potato whitefly. An insect pest of cotton, fruit and vegetable, and greenhouse crops. The Animal and Plant Health Inspection Service works cooperatively with producers to implement pest management strategies based on integrated pest management and biological control techniques.

Symbiosis. (1) An intimate physiological association of two or more species resulting in mutual benefit e.g. Rhizobium bacteria on the roots of legumes. (2) The phenomenon of two different organisms living to-gether for mutual benefit, *e.g.*, the association of nitrogen-fixing bacteria with leguminous plants in the root nodules. The bacteria obtain carbohydrates and other foods from the plants and fix atmospheric nitrogen (Nitrogen Fixation) which is made avail-able to the plants in various compounds. Lichens consist of fungi and algae living together symbolically.

Sympodial. (1) A zig zag or irregular form of growth. (2) *Orchids* with sympodial growth have a specialized lateral growth pattern in which the terminal *bud* dies. The growth continues by development of new shoots sprouting from or next to the ones of previous year (such as in the genus *Cattleya* or *Cymbidium*). The base of the *stem* of sympodial *epiphytes*, or in some species essentially the entire stem, may be thickened to form what is called a *pseudobulb*. These contain food reserves for drier periods. At their end appear one or two *leaves*, or sometimes four or more. Some sympodial *terrestrials*, such as Orchis and *Ophrys*, have two pseudobulbs between the *roots*. One is used as a food reserve for wintery periods, and provides for the development of the other pseudobulb, from which visible growth develops. In warm and humid climates, many terrestrial orchids do not need pseudobulbs. (3) Multiple upward branching from a commcn branch point, resembling an upward-turned hand.

Syrup feeds. By-products of the whisky distilling industry which are turned into animal feeds, particularly for dairy cattle. They comprise a blend of malt or maize draff and distiller's syrups and have been shown to contain high sources of energy and protein.

System of farming. The combination of products on a given farm and the methods or practices that are used in the production of these products is known as system of farming.

Systemic compound. A chemical which, when applied to foliage or the soil, is absorbed by a plant and moved in the sap to all parts of the plant. Insects sucking the sap of plants treated with systemic insecticides are poisoned. Translocated-herbicides are systemic compounds.

Systemic insecticide. Any insecticide that is applied to the soil in soluble form and absorbed by the plant's roots and which is designed to kill most of the sucking insects

Systemic pesticide/systemic herbicide. A chemical that is absorbed by a plant and is translocated in its vessels, either to kill feeding insects on the plant, or to kill the plant itself.

T

T Budding (Shield budding). (1) Budding where T shaped cut is made on the rootstock and the scion in the form a shield, is inserted under the two flaps of bark and the bud union is tied by exposing only the bud. (2) T budding, type of *grafting* in which a plant bud is inserted under the bark of the stock (usually not more than a year old). It is best done when the bark will peel easily and the buds are mature, as in spring, late summer, or early autumn. Budding is a standard means of propagating roses and most fruit trees in nurseries.

Take. (1) The taking possession of buildings and land by a farmer when the assumes a tenancy. Also called entry. (2) A crop which germinates or a successful graft are said to take.

Take-all. A disease of cereals due to a soil-borne fungus *(Gauman-nomyces graminis)* which infects the roots and stem base. causing dicoloration, stunted growth, premature ripening and the production of bleached ears containing little or no grain. One strain of the disease affects only wheat and barley, whilst another also affects oats but is mainly confined to the wetter areas in the North and West. Control is by rotation. Also called white heads.

Taking endangered or threatened species. "Taking," in layman's terms, means killing or removing a plant or animal of a species listed under the Endangered Species Act (ESA), or seriously damaging its chances of reproduction. Except under specified circumstances, taking is forbidden under the ESA. The definition of taking is one of the current issues in the ESA debate. Incidental take of a listed species is a taking which occurs in the course of some other legal activity, whether carried out by a federal or a nonfederal entity.

Talmadge-aiken plants. The approximately 250 meat and poultry plants in 10 states where USDA has contracted with state agency inspectors to conduct federal inspection activities. They are now formally known as Federal-State Cooperative Inspection Plants. Even though state employees are conducting the inspection in these plants, they are under the federal rather than state inspection programs.

Tank mixing. A relative new practice of mixing crop-protection chemicals (and also liquid fertilizers, adjuvant oils, trace elements and foliar feeds) before spraying on field crops, particularly cereals. The benefits include savings in time, labour, water and fuel, reduced wear and tear on machinery and less crop damage through wheelings. Manufacturers recommended mixes are cleared and approved under the pesticides safety precautions scheme and the Agricultural Chemicals Approvals Scheme.

Tankard mangel. A type of mangel which is tankard-shaped and classified with the intermediate types which have less of their swollen root above ground than the globe types.

Tannin. (1) Organic compounds present in some plants, chiefly glucosidal in nature, that cause an astringent taste. (2) Polyphenol compound in tea leaves. It is the second most valuable constituent of tea, after caffeine. (3) The browns in leaves after all colors have disappeared is caused by the pigment tannin. Tannin is present throughout the growing season but masked by the chlorophylls (greens), xanthophylls and carotenes (yellows and oranges), and anthocyanin (reds and purples). Tannins are used in leather refining processes.

Tapeworms. A class of parasitic flatworms which inhabit the small intestine of animals and are named from their dirty white tape-like appearance. They consist of a ribbon-like chain of segments anchored at one end and trailing through the gut, reaching several feet in length in some cases. They seldom require treatment in livestock.

Tapping task. The number of trees to be tapped by a tapper in a day.

Taproot. (1) Root originating from the radicle in dicotyledonous plants, growing vertically downward and providing anchorage to the plant. (2) The primary or thick main root growing straight down from a plant. (3) A *plant's* taproot is a straight tapering *root* that grows vertically down. It forms a center from which other roots sprout. Plants with taproots are difficult to transplant. The taproot is why *dandelions* are hard to uproot — the top is pulled, but the long taproot stays in the ground, and re-sprouts. A *taproot system* contrasts to a *fibrous root system*, with many branched roots. Most *trees* begin life with a taproot, but after one to a few years change to a wide-spreading fibrous root system with mainly horizontal surface roots and only a few vertical, deep anchoring roots. A typical mature tree 30-50 m tall has a root system that extends horizontally in all directions as far as the tree is tall or more, but well over 95% of the roots are in the top 50 cm depth of soil. Many taproots are modified to become storage organs. Some plants with taproots:

- Carrot
- Dandelion
- Kudzu
- Parsnip
- Patterson's Curse
- Poppy mallow
- Radish
- Turnip
- Welwitschia

Typical taproots:

- conical root: another word for taproot.
- *fusiform root :* tapers at the top and at the bottom, e.g. the taproot of a white *radish.*
- *napiform root :* is broader than it is long, e.g. the root of a *turnip.*

Taprooted: Having one main fleshy root that extends straight down into the soil. Examples include carrot and dandelion.

Targeting. A policy concept under which government farm program benefits would be directed toward specified groups of producers. One example of targeting might be to focus farm payments on small to medium-sized family owned and operated farms.

Tendril. (1) Slender prolongation of a stem or leaf that helps the plant to climb. (2) The twisting, clinging, slender growth on many vines, which allows the plant to attach themselves to a support or trellis; a twisting thread-like shoot used to support the plant. A climbing organ metamorphosed from a leaf, stipule or stem.

Tentiometer (soil moisture). Instrument used for measuring tension with which moisture is held in the soil. It consists of poorus clay cell filled with water which is placed in the soil. A pressure gauge is attached to the cell which registers the changing tensions as water moves through the cell towards the drier soil or into the cell from wet soil.

Terete. (1) Approximately cylindrical, usually tapering at one or both ends. (2) Cylindrical but usually slightly tapering at both ends, circular in cross section, and smooth-surfaced.

Terminal elevator. A large elevator (warehouse) facility with the capacity to transfer grain to rail cars, barges, or ships for transport to domestic or foreign markets. Terminal elevator markets are used as base locations for posted county prices.

Terminal market. A central site, often in a metropolitan area, that serves as an assembly and trading place for agricultural commodities. Terminal markets are usually at or near major transportation hubs.

Terminal. The end bud beyond which no further growth takes place normally until the following season. At the tip or distal end.

Terminator seeds. A descriptive term used by some for seeds that have been genetically engineered to produce a crop whose first generation produces sterile seeds, thus preventing a second generation from being grown from seeds saved from the first. This technology (currently 3 to 5 years from commercial application) was developed under a Cooperative Research and Development Agreement between the Agricultural Research Service and a private seed company. Supporters of the technology state that it is a way to build patent protection directly into high-value, genetically engineered crop varieties and thus recoup high research investment costs. Opponents are concerned that the technology could have harmful environmental and public health effects and argue that it would have an inequitable impact on farmers in developing countries who rely on saved seed for replanting and for developing locally adapted varieties.

Terrace gardening. (1) Gardening done on raised space of ground (terrace). (2) Gardening on the flat roofs or terraces of buildings.

Terrace. An embankment, ridge, or leveled strip constructed across sloping soils on the contour, or at right angle to the slope. The terrace intercepts surface runoff so that it can soak into the soil or flow slowly to a prepared outlet, decreasing rates of soil erosion.

Terrarium. (1) A glass or plastic-enclosed growing environment with high humidity. (2) A transparent enclosure, usually of glass, for providing a clean, humid environment to plants. (3) Terrarium, a miniature garden in an artificial environment, in which small plants and animals may be kept as ornament or for educational purposes. Fish bowls, small fish tanks, large bottles, and carboys are often employed as containers for terrariums; such vessels permit the necessary entrance of light for photosynthesis by the plants, and are suitable for display. The most important aspect of a terrarium is its relatively high humidity, maintained either by using a container with only a small aperture or by keeping the container covered. The humid atmosphere enables mosses and other sensitive woodland plants to flourish.

Teschen disease. A notifiable disease of pigs due to a virus, resulting in an excited condition with fever and paralysis, and often death. It was first recognised in Czechoslovakia.

Testicle. The reproductive gland of male animals occuring in pairs, producing sperm. Also called testis (plural, tests). A rope, chain or halter used to confine or restrain an animal within certain limits, e.g., by tying it to a post in the ground. (2) A similar ropelike restraint used as a safety measure, especially for young children and astronauts.

Tetrazolium test. A biochemical test in which seed viability is tested, not by germination, but by the red colour appearing when the seeds are soaked in 2, 3, 5, triphenyletzolium chloride (TTC).

Thinning of fruits. (1) Removal of a few young fruits from heavily bearing fruits with a view to provide greater advantage of space, light, water and food. It is usually practiced in papaya. (2) Removal of young fruits from a cluster to help the remaining ones to have more reserve food for development. (3) Removal of plants from thickly populated crop stand for maintaining optimum plant population. (4) Pruning branches for providing the remaining ones more space and reserve food for better growth. (5) Removing excess seedlings, to allow sufficient room for growth and better size and quality in the remaining plants. Removing stems or branches, to give the plant a more open structure.

Threshing machine. A machine once commonly used to thresh cereals and other seed crops. Most cereals crops are now harvested and threshed by combine harvester.

Threshing refraction. Components in wheat which differ from sound grain is called refraction.

Tonguted approach graft. Type of approach grafting wherein the two steins choosen are approximately of the same size. Two cuts are made to form a tongue on each of the two stems. The second cut is downward on the scion to provide a thin tongue on each piece. A closely fitting graft union can be obtained by interlocking these tongues tightly.

Fig.: Threshing machine

Trace element. A chemical element required by a plant or animal in very small quantities for its metabolism. Such elements are often essential for the functioning of enzymes, hormones and vitamins. They include iron, zinc, boron, manganese, cobalt, molybdenum and copper and a lack of these elements may cause a deficiency disease (e.g., Heart Rot in sugar beet and Pinning in sheep, due to insufficient boron and cobalt respectively). Deficiencies can also be included by the antagonistic action of other element which render otherwise satisfactory amounts unavailable to crops or animals. Also called micronutrient.

Tractor. A vehicle which has now almost entirely replaced the horse for pulling implements and trailers. The modern tractor is also able to operate various implements attached to its three point linkage and front loader arms, and to provide power to trailed and mounted machinery via the p.t.o. and hydraulic drive system. Some stationary machinery (e.g., sawbenches, Crushing mills, etc.) are also powered by tractors by belt pully. A wide range of tractors are available and are usually classified according to engine power as indicated below. Tractors have increased in size and power over the years so that models over 100 h.p. are much more common now than in the late 1160s. The 75 h.p. machine is now regarded as a 'common workhorse' on arable farms. Two-and four-wheel drive types exist, and the latter may have equal or unequal sized drive tyres.

Tuberous root. (1) A root, producing tubers e.g. tapioca (2) Tuberous roots are root tissue, unlike tubers, rhizomes, and corms which are stem tissue, and bulbs which are leaf tissue. There are no buds or "eyes" as in rhizomes or tubes. Tuberous roots are divided by separating the roots while maintaining a piece of the stem base in each division. Examples include: (i) Agapanthus; (ii) Canna and (iii) Clivia (Kafir Lily)

Turgor. Plant Physiology: the normal distention or rigidity of plant cells, resulting from the pressure exerted from within against the cell walls by the cell contents. The *cells* of *plants* are quite different from the cells of the other eukaryotic kingdom's *organisms*. Their distinctive features are: (a) A large central *vacuole* (enclosed by a membrane, the *tonoplast*), which maintains the cell's *turgor* and controls movement of molecules between the cytosol and sap. (b) A *cell wall* made up of *cellulose* and *protein*, and in many cases *lignin*, and deposited by the protoplast on the outside of the *cell membrane*. This contrasts with the cell walls of *fungi*, which are made of *chitin*, and *prokaryotes*, which are made of *peptidoglycan*. (c) The *plasmodesmata*, linking pores in the cell wall that allow each plant cell to communicate with other adjacent cells. This is different from the network of *hyphae* used by fungi. (d) *Plastids*, especially *chloroplasts* that contain *chlorophyll*, the pigment that gives plants their green color and allows them to perform *photosynthesis*. (e) Plants lack *centrioles* and *lysosomes* that are present in animal cells.

Like other eukaryotic cells, plant cells typically also have a *nucleus*, containing most of the cell's *DNA*, and *mitochondria*, the 'power plants' of the cell. However, some parts of a living plant's tissue are selectively dead so they can serve the plant without consuming vital nutrients. The three distinct types of plant cells are classified according to the structure of their cell walls and features of their *protoplast*. Plants will have a primary cell wall and sometimes a secondary wall as well. The protoplast These two major parts are what determines the function of each individual plant cell.

U

Ultimate wilting point. Moisture content of the soil at which all the leaves of plants growing in that soil are completely wilted and will not recover in an approximately saturated atmosphere without addition of water to soil. Lower end of wilting range.

Umbel. (1) A flower cluster in which all the individual flower stalks arise at one point, the cluster being flat shaped. (2) A flat-topped or rounded flower cluster in which the individual flower stalks arise from about the same point, as in the geranium, milkweed, onion, and chive. (3) Usually a flat-topped flower cluster, with pedicels and peduncles arising from a common point, resembling the supports of an umbrella.

Unarmed. Without either spines or prickles, though the leaves may have sharp teeth at the margins or tip like holly.

Unbroken Furrow. A continuous, smooth type of furrow slice such as that produced by a lea plough.

Undecorticated seeds. Seeds which have not been subjected to decortication.

Under drain. A type of drain constructed beneath the ground, e.g. mole drain, tile drain, etc., in a land drainage system, as distinct from an open ditch or surface drain. The water draining into such pipes is carried away by gravity flow and is usually discharged into a ditch. Before the advent of tile drains the ducts were often formed from locally avaiable materials, e.g., thorn or other brushwood faggots, stones, dried peat or turf tiles, or straw bundles.

Underbeam clearance. The measurement from the underside of the share to the underside of the beam of a plough.

Underground storage tank. For purposes of Subtitle I of the Resource Conservation and Recovery Act, this is any tank used to store petroleum or hazardous substances regulated under the Comprehensive Environmental Response, Compensation, and Liability Act, the volume of which is 10% or more beneath the surface of the ground. This does not include, among other things, any farm or residential tank of 1,100 gallons or less capacity used for storing motor fuel for noncommercial purposes, tanks used for storing heating oil for consumptive use on the premises, or septic tanks.

Undersow. To sow a grass seeds mixture with another crop (a Cover Crop), usually a cereal crop.

Understock. In connection with *grafting*, this is the portion of the root system of a vine onto which the scion, or shoot, is attached.

Understory trees. Small trees that grow well under larger ones. Common understory trees in a birch-poplar-spruce forest include ironwood, mountain maple, and serviceberry. Other common understory trees in a mixed maple-hardwood forest are witch hazel and black cherry.

Unexhausted manorial value. Refered to the amount of fertilizer, manure or lime remaining in the soil, and usefully available to a crop, after one or more crops have drawn upon it after its application.

Unfree water. Water which is held closely by soil colloids so that it is not readily available to plants. It freezes well below 0°C.

Ungulates. The term used for the herbivorous hoofed animals, e.g., cattle, deer, horses, pigs and sheep.

Unique farmland. Land, other than prime farmland, that has combined conditions to produce sustained high quality and high yields of specialty crops, such as citrus, nuts, fruits, and vegetables when properly managed.

Unisexual flowers. (1) Flowers having either pistils or anthers, but not both. (2) Withonly one sex (male or female) of flowers on a single plant.

Unit of fertilizer. 1% of 1 cwt (i.e., 1.121b). Thus a nitrogen fertilizer having 15% nitrogen in each cwt is said to have 15 units of nitrogen. A compound fertilizer having say 10,15 and 20 units of nitrogen, phosphate and potash, respectively, contains 10%, 15% and 20% respectively of each per cwt (shown as 1C : 15 : 20). With metrication fertilizer application is increasingly ex-pressed in kilograms per hectare.

Unit train. Generally refers to a string of freight cars that all carry the same commodity, frequently over long distances. Unit trains are widely used to haul such raw commodities as coal and grains, because they are less costly for railroads than mixed freight shipments. Unit trains have become a point of contention between agricultural shippers and the railroads, mainly because the shippers are increasingly being asked to fill longer unit trains (for example, 104 hoppers), which many local elevators are not equipped to handle.

Universal soil loss equation. A formula used to estimate erosion rates by considering climate, soils, and topographic conditions at a site, as well as any degree to which the use and management of the soil reduce erosion. It is being replaced by a revised universal soil loss equation.

Unkindly. A term applied (a) to soils which lack natural fertility or are cold, late, wet, etc., and (b) to crops or animals which lack thrift or are in poor condition.

Upland. (1) Cultivable land on high ground with good natural drainage. (2) Land at higher elevations than the alluvial plain or low stream terrace; all lands outside the riparian-wetland and aquatic zones. Used often in discussion of federal land management practices.

Urban horticulture. The study of stresses affecting city trees, gardens and landscapes.

V

Vacuum cooling. A method for extending shelf life of fruits and vegetables using a vacuum chamber where a low atmospheric pressure is maintained.

Vacuum silage. Silage prepared in a large, sealed, polythene 'bag'. The 'bag' is evacuated of air by a pump through a valve in the sheeting. This removal of air and the build up of carbon dioxide reduces respiration losses and aids fermentation, producing high quality silage.

Value-added agriculture. A concept that has gained currency in the small farm policy debate, in response to the concern that the farm value of the consumer food dollar continues to decrease (which, some small farm advocates contend, is due to the excessive profit-taking by processors and retailers). Value added agriculture might be any means to capture a larger share of the consumer food dollar by farmers. Examples include direct marketing; farmer ownership of processing facilities; and producing farm products with a higher intrinsic value (such as identity-preserved grains, organic produce, free-range chickens; etc.), for which buyers are willing to pay a higher price than for more traditional farm commodities.

Value-added products. In general, products that have increased in value because of processing; such products include wheat flour and soybean oil. Livestock are considered value added products because they have increased the value of pasture and feed grains going into them. The terms value-added and high-value are often used synonymously.

Valvate. (1) Applied to bud scales that meet along a definite, usually longitudinal, line without overlapping; the reverse of imbricate. (2) Scales that meet but do not overlap on a bud, typically in a pair.

Variegated. (1) Of more than one colour, often in patches, e.g., the leaves of certain plants. (2) Striped, margined or mottled with colors other than green, where green is normal; having markings or zones of tissue lacking green chlorophyll.

Variety. (1) A natural occurring population of plants of one species, different enough from that species to require a different name. Reproduces naturally from seed. (2) A naturally occurring group within a species that is distinct enough to warrant taxonomic recognition but not sufficiently distinct to be segregated as another species. (3) A sub- division or group of plants within a species which differs in some minor definable characteristics. (4) One of possibly many closely-related plant species. A naturally occurring variation of a wild species. (5) Subdivision of a species having a distinct though often inconspicuous difference, and breeding true to that difference. Also refers to clones.

Vascular bundle. (1) A discrete group of conducting vessels. (2) A minute spot within the leaf scar where the vessels were positioned.

Vascular. (1) Referring to the xyiem, phloem or both. (2) The vessels that conduct water or nutrients in plants. (3) Of, characterized by, or containing vessels that carry or circulate fluids, such as blood, lymph, or sap, through the body of an animal or plant.

Vase life. The length of time that a cut flower retains its aesthetic value after placing in a vase.

Vegetable dyes. Dyes prepared from various trees. They are known after the tree from which they are obtained e.g., santaline artocarpus and the cutch dyes prepared from the wood of *Pterocarpus Santalinus, Artocarpus integrifolia* and *Accacia catechu* respectively.

Vegetable garden, processing. Production of vegetables suitable for canning, dehydration or freezing, preferably around processing factories.

Vegetable. (1) A generally term applied to plants as distinct from animals. Normally applied to plants or parts of plants cultivated for human consumption or for stock-feeding, e.g., potatoes, carrots, cabbages, etc. Some fruits (e.g., tomatoes, cucumbers) and some seeds (e.g. peas, beans) are also considered as vegetables. Most vegetables are having useful amounts of vitamin C and minerals. Root vegetables have stored carbohydrates and seed vegetables are rich in protein. (2) Edible herbaceous plant or plant parts, commonly used for culinary purposes or as salads.

Vegetative controls. Nonpoint source pollution control practices that involve planting cover crops to reduce erosion and minimize loss of pollutants.

Vegetative propagation. The production of a complete plant by asexual methods like cutting, layering, budding and grafting

Vegetative reproduction. (1) Reproduction in plants not involving the flower or any sexual process, by the detachment of a part of the plant (e.g., rhizome, tuber, bulb, etc.), which then develops into a new plant. (2) Vegetative propagation

Vegetative. (1) Referring to non-reproductive structures or growth. (2) The leafy or green portions of a plant.

Vein(s). (1) A vascular tube in a leaf, typically referring to any within the blade except for the midrib. (2) The rib or branch of vessels in a plant leaf or other flat part. (3) The conducting tissue within a leaf and other plant parts.

Vermiculite. (1) A sterile mica product used in rooting/potting medium. (2) The mineral mica, heated to the point of expansion. A good addition to container potting mixes, vermiculite retains moisture and air within the soil.

Vernalisation. (1) Low temperature treatment of seeds and seedlings, found useful in some plants, to shorten the time required for flowering. (2) Preplant treatment sometimes accorded to seeds in which they are placed under conditions favourable for germination for a short period and then are exposed to condition where the germination process is arrested.

Vernation. The arrangement of leaves within a bud. *Vernation* (from *vernal,* since that is when leaves "spring forth" in Temperate regions) is the formation of new *leaves* or *fronds.* In plant anatomy, it is the arrangement of leaves in a bud. In *pine trees,* new leaves are short and encased in *sheaths.* Each leaf bundle consists of 2 to 5 needles. All the leaves on one section of branch grow in length together. In the *cabbage*, new leaves are folded over, each covered by the previous leaf. *Circinate vernation* is the name given to the manner in which new *fern* fronds emerge. As a new fern frond is formed, it is tightly curled so that the tender growing tip of the frond (and each subdivision of the frond) is protected within a coil. At this stage it is called a *crozier* (after the shepherd's crook) or *fiddlehead* (after the scrollwork at the top of a violin). As the lower parts of the frond expand and toughen up, they begin to *photosynthesize*, supporting the further growth and expansion of the frond. In the case of many fronds, such as that of the Australian tree fern in the picture at right, long hairs or scales provide additional protection to the growing tips before they are fully uncoiled. Circinate vernation may also be observed in the extension of *pinnae*, or leaflets, in the compound leaves of the *Cycads* (Division Cycadophyta).

Vestige. The remains of an exhausted or dead plant structure. A rudimentary or degenerate, usually nonfunctioning, structure that is the remnant of an organ or part that was fully developed or functioning in a preceding generation or an earlier stage of development.

Vetch. Vetches is a general term referring to many plants belonging to the legume or bean family. Many are used as ground covers (Crown vetch) or as a green manure crop to improve the soil (Hairy vetch). Hairy vetch, a true Vetch in the Vicia genus, is often sown in the fall and plowed under the following spring. The plant is an annual and will die out which isn't the best thing on banks. Plants don't respond to close mowings. As an interesting side note, the FAVA or FABA bean, made famous by the movie The Silence of the Lambs, is really a vetch, instead of a type of garden bean.

Veterinary biologics. Vaccines, antigens, antitoxins and other preparations made from living organisms (or genetically engineered) and intended for use in diagnosing, treating, or immunizing animals. Unlike some pharmaceutical products, such as antibiotics, most biologics leave no residues in animals. Veterinary biologics are regulated by the Animal and Plant Health Inspection Service, which licenses the facilities that produce them and conducts a program to ensure that animal vaccines and other veterinary biologics are safe, pure, potent, and effective.

Viability of seeds. It is represented by the germination percentage, which can be produced by a given number of seeds. A study of a viable seed is essential in successful propagation from seed. A reduction in seed viability may result from incomplete seed development on the plant, injuries during harvest, improper processing and storage or aging.

Vinca Blight. Sudden death or thinning of vinca groundcovers usually is caused by a canker disease that infects the stems, girdling the plants. Vinca blight can be minimized by regular yearly mowings in the spring, with the mower set on its highest settings. Rake and compost all clippings/cuttings. Fertilize and water to stimulate new growth.

Vine. (1) A slender stemmed plant that trails and climbs by means of attaching parts and stems. (2) A weak-stemmed plant that derives its support from climbing, twining, or creeping along a surface.

Vinegar. Liquid obtained by alcoholic and acetic fermentation of suitable materials containing sugar and starch. Vinegar, sour liquid consisting mainly of acetic acid and water, produced by the action of bacteria on dilute solutions of ethyl alcohol derived from previous yeast *fermentation*. The coloring and flavoring are characteristic of the alcoholic liquor (as cider, beer, wine, fermented fruit juices, solutions of barley malt, hydrolyzed cereals, starches, or sugars) from which the vinegar is made. Vinegar is used as a salad dressing, a preservative, a household remedy to allay irritations, a mild disinfectant, and, in cooking, as a fiber softener. Vinegar has been known from antiquity as a natural byproduct of wine; the name is derived from the French *vin aigre* [sour wine]. The manufacture as a separate industry began in France in the 17th cent. The wasteful, slow, or natural, process, a spontaneous fermentation in casks half full of beechwood shavings exposed to the atmosphere by bung holes, was superseded in the early 19th cent. by the quick, or generator, method. The generator used in present-day commercial manufacture is usually a tall, truncated cone or vertical, wood tank with a false bottom perforated to admit air that is generally forced through by a blower. The alcoholic solution is allowed to drip through a filling of hard-wood shavings or other material presenting a large surface area. Vinegar made by this method must be aged to remove a natural harshness. It is generally clarified, then pasteurized. Some vinegars are subjected to distillation which removes most of the flavorings other than acetic acid. In another process, the solution is aerated directly by a spinning rotor. The wood shavings are not needed in this case, and the process runs continuously. Acetic fermentation may be impeded by an excessive growth of mother of vinegar, a slimy mass of bacteria, or of the parasitic vinegar eel, a minute, threadlike worm.

Virgin land. Land which has never been cultivated. The Virgin Lands Campaign was an initiative by Nikita Khrushchev to open up vast tracts of unused (virgin) steppe in the northern *Kazakh SSR* and the *Altai* region of the *Russian SFSR*, started in *1954*. In the first year of the programme, 190,000 km² were *ploughed* up, and in *1955*, an extra 14 million were ploughed. With all this new land, a vast amount of people would need to be brought in from all over the *Soviet Union*: the *Komsomol* was charged with recruiting them. More than 300,000 people, mostly *Ukrainian*, arrived in the Virgin Lands to begin new lives as *farmers*. Hundreds of thousands of *soldiers*, *students* and *combine harvester* operators would join them; however, these people would stay for only a year's *harvest*. By the end of the mass *immigrations* to the Virgin Lands, Ukrainians outnumbered *Kazakhs* in many areas. For a brief time, Khrushchev inspired a *communist* zeal in the peoples of the Soviet Union, and concentrated that zeal on a task that, for an equally brief time, produced the expected results.

Vomitoxin. Deoxynivalenol (DON), also referred to as vomitoxin, is a naturally occurring mycotoxin produced by several species of Fusarium fungi. Wet and cool weather from flowering time to maturity promotes infection, resulting in scab or head blight in barley, wheat, oats, and rye. Wheat infected with scab has a tendency to have lighter weight kernels, some of which are removed during normal harvesting and cleaning operations. Vomitoxin does not represent a threat to public health among the general population. However, it can—in rare cases—produce acute temporary nausea and vomiting in humans and animals. Food and Drug Administration does not have an advisory level for vomitoxin in raw wheat intended for milling purposes, and relies on processors to reduce the level in finished products for human consumption to a level that does not exceed 1 part-per-million (ppm). Advisory levels also exist for animal feeds.

W

Wardian case. (1) A glass-topped enclosure, supplied with provision for bottom heat, used for plant propagation. (2) An ornate 19th century terrarium used for ferns and rare plants.

Warm season grass. (1) Grasses that make their major growth during the warmer part of the year. (2) Grass that grows and "greens up" in summer, sets seed in fall, and goes dormant (loses its green color and stops growing) during winter and spring. According to *Easy Lawns*, most warm-season grasses are *bunchgrasses* and will not tolerate any shade.

Warm season. Plants thrive in warm temperatures (65 to 90 degrees Fahrenheit daytime temperatures) and are intolerant of cool temperatures. Examples include corn, tomatoes and squash. Some warm season plants are sub-tropical and tropical plants grown as annuals in Colorado.

Wart disease. A notifiable disease of the potato, due to the fungus *Synchytrium endobioticum*, which causes the development of rough-surfaced, dark brown or black, tumorous growths on the tubers which, under favourable conditions, release motile spores which are able to infect young tubers, but which can persist indefinitely in the soil. Most varieties of potato grown novv-a-days are immune

Waste treatment pond. A shallow lagoon or similar storage facility, often man-made, used to treat liquid agricultural wastes, particularly liquid manure from livestock production farms, through the interaction of sunlight, wind, algae, and oxygen. Through natural biological processes, microscopic organisms consume wastes present in the water.

Water table. (1) The upper surface of ground water or that level below which the soil is saturated with water. (2) The surface at which ground water has settled in the ground and below which fissures and pores in the rock strata or soil are saturated with water. This surface is uneven and its position varies according to the amount of rain which has fallen. Where a water table rises to intersect the ground surface, a spring arises.

Watering the lawn. Cool season grasses require 1 inch of water per week to remain actively growing when temperatures are between 45 and 85 degrees F. If temperatures are between 85 and 90 degrees F., turf may require a 1-1/2 inches. Temperatures 90 degrees F. and higher signal 2 inches of moisture to keep grass actively growing. Less water amounts may force turf into a dormant state. During extended droughts, turf requires 1/2 inch every week to keep crowns alive. Amounts of 1/3 - 1/2 inch every three week keeps the root system alive.

Waterlogged soil. A soil which is saturated with water so that the pore space is completely filled with water, and conditions are unsuitable for plant growth since the roots are unable to obtain oxygen for respiration.

Watershed. (1) An area of land that collects and discharges water into a single stream or other outlet. Also called a catchment or drainage basin. (2) The ridge of high ground separating the catchment areas of two distinct river systems. (3) The total land area, regardless of size, above a given point on a waterway that contributes runoff water to the flow at that point. It is a major subdivision of a drainage basin. The United States is generally divided into 18 major drainage areas and 160 principal river drainage basins containing about 12,700 smaller watersheds.

Wattle. (1) The coloured fleshy skin beneath the throat of some birds Also the doddle of a goat. (2) A twig or flexible stick, woven with otherss to make a frame work for fences, roofs, hurdles, etc.

Weeding. Act of removing weeds from the infested area. Weeding chisel, a tool with a divided chisel-like end, for cutting the roots of large weeds under ground. — Weeding forceps, an instrument for taking up some sorts of plants in weeding. — Weeding fork, a strong, three-pronged fork, used in clearing ground of weeds; — called also *weeding iron*. — Weeding hook.

Weedkiller. A term for a herbicide. Herbicide, chemical compound that kills plants or inhibits their normal growth. A herbicide in a particular formulation and application can be described as selective or nonselective. In agriculture, selective herbicides are often used instead of tillage, or in combination with tillage and other agronomic practices, to control weeds without damaging crops. For these no-till or low-till systems, scientists are using biotechnology to develop crop varieties with increased tolerance for herbicides. Nonselective herbicides (e.g., paraquat) toxic to all plants, are used where complete control of plant growth is required. Contact herbicides kill only the parts of the plant they touch; systemic herbicides are absorbed by foliage or roots and translocated to other parts of the plant. Preemergence herbicides, mixed into the soil, will kill germinating seeds and small seedlings. Postemergence herbicides either hinder photosynthesis or inhibit growth. Early chemical herbicides were inorganic compounds. Herbicides such as ashes, common salts, and bittern have been used in agriculture since ancient times. Observation in 1896 that *Bordeaux mixture*, a fungicide, also provided control of certain weeds, led to the use of copper sulfate as a selective weed killer to control charlock (wild mustard) in cereals. By 1900, solutions of sulfuric acid, iron sulfate, copper nitrate, and ammonium and potassium salts were known to act as selective herbicides; soon thereafter sodium arsenite solutions became the standard herbicides, and they were used in large quantities until about 1960. Other inorganic herbicides include ammonium sulfamate, carbon bisulfide, sodium chlorate, sulfuric acid solutions, and formulations containing borate. Organic herbicides began to be produced in earnest with dinitrophenol compounds in 1932. A breakthrough occurred in the 1940s with 2,4-D (2,4-dichlorophenoxyacetic acid), a compound similar to plant hormones, which is a highly selective systemic herbicide when used in very small quantities. 2,4-D was quickly adopted to control broad-leaved weeds in corn, sorghum, small grains, and grass pastures, as well as in lawns and other ornamental turf. The phenoxyaliphatic acids and their derivatives, another major group of organic herbicides, succeeded because of their selectivity and ease of translocation. Other groups of organic herbicides include organic arsenicals, substituted amides and ureas, nitrogen heterocyclic acids, phenol derivatives, triazines, and sulfonylureas. In the 1960s and 1970s, a combination of 2,4-D and 2,4,5-T was widely used in Vietnam as a defoliant under the name *Agent Orange*. As a result of questions concerning the possible health effects of the use of Agent Orange, heightened awareness of possible ecological and health dangers attributable to herbicides has resulted in reevaluation of many compounds and has called indiscriminate use into question. Use of the dioxin-containing 2,4,5-T was prohibited in the United States in 1984. In 1975, Mexico, at the urging of the United States government, began spraying fields of marijuana with paraquat, which both eliminated the crop and raised fears of toxic side effects in marijuana users.

Weedwipers. Devices which smear herbicides on weeds which pro-trude above crops. They are usually tractor mounted but hand applicators are available. For annual and perennial weed control, pre crop sowing or planting and stubbles of all crops, for harvest management of cereals, oilseed rape, mustard, linseed, peas and field beans and for the destruction of grassland, and in top fruit for the control of weeds, and in non crop situations, aquatic situations and in forestry

Wetlands. Areas of predominantly hydric soils that can support a prevalence of water-loving plants, know as hydrophitic vegetation. Transitional between terrestrial and aquatic systems are wetlands typified by a water table at or near the surface, or the land is covered by shallow water at least part of the year. Types of wetlands are distinguished by water patterns (the frequency and length of flooding) and location in relation to upland areas and water bodies. Wetlands perform many functions including wildlife and fish habitat, storage and conveyance of flood waters, sediment and pollution control, and recreation. Under the swampbuster program, landowners may produce crops in these areas, but only if the water patterns, or hydrology, in the wetland area is not altered and any woody vegetation is not removed.

Wet-milling. A process in which feed material is steeped in water, with or without sulphur dioxide, to soften the seed kernel in order to help separate the kernel's various components. For example, wet-milling plants can separate a bushel of corn into more than 31 pounds of starch (which in turn can be converted into corn sweeteners or ethanol), 15 pounds of animal feed, and nearly 2 pounds of corn oil.

Wetting agent. (1) A substance which lowers the surface tension of water so that it spreads out rather than remains in droplet from when sprayed. Wetting agents are added to detergents to in-crease their efficiency and to sprays of insecticides, fungicides, herbicides or mineral sprays to improve the cover achieved when applied to leaf surfaces. (2) A substance added to a spray solution for improving its contact with plant surface.

Wheat bulb fly. A two-winged fly *(Hylemyia coarctaia)*, the larvae of which bore into the central shoots of wheat in the spring (and also winter barley and ray) causing them to turn yellow, and die if tillers have not developed. Control is by seed Dressing using chlorofenvinphos and carhophcnotion (Organophosphorous Compounds) and sometimes by spray treatment in the spring.

Whip graft. A graft in which a tongue cut on the Scion is fitted into a slit cut slopingly in the Stock. Also called a tongue graft. Method of grafting in which tongue shaped cuts are made on both stock and scion and are interlocked. Whip grafting, sometimes called tongue or bench grafting, is used primarily on species of plants that unite easily. It works best when the plant parts to be grafted are small. The best time to collect scions is when the plant is dormant. The scions should be about ¼ to 3/8 inch in diameter and approximately 1 foot in length. The best scions will contain a minimum of three buds. Store the scions in damp sawdust in a cool place, or seal them in a plastic bag and keep them in the refrigerator until the time of grafting. *Steps For Grafting:* (a) Cut off the understock diagonally at the desired point exposing a flat surface 1 to 2 inches long. The cut should be made by one drawing stroke of the knife. (b) Starting about two-thirds of the way from the heel of the "toe" of the cut, make a cleft running toward and through the center of the stock to form a thin "tongue." (c) Repeat the above process with the scion, being careful to make the cut in such a manner that the buds will point upward when attached to the stock. (d) The understock and scion are united by slipping the tongue of the scion inside the tongue of the understock until the scion is firmly placed. The cambiums of the understock and scion must make contact on one side of the union. It is best if the cambium layers make contact on both sides, but this is not necessary. (e) Secure the stock and scion together with raffia or waxed yarn, slipping the end under the first and last loop at each end of the tie. Rubber tape may be used in place of the raffia or waxed yarn. (f) Completely encase the union with wax or a tree wound compound, making certain that all cut surfaces are well covered.

Windbreak. (1) A living barrier that usually includes several rows of trees, and perhaps shrubs, located upwind of a farm, field, feedlot or other area and intended to reduce wind velocities. Windbreaks, also called shelterbelts, can reduce wind erosion, conserve energy or moisture, control snow accumulations, and provide shelter for livestock or wildlife. (2) A screen, particularly a line or clump of trees or a piece of woodland, providing protection from the wind to an area of land growing crops or livestock.

Woody plant. (1) A tree or shrub; any plant with lignified tissues. (2) Plants with woody fibre. (3) A woody plant is any *vascular plant* that has a *perennial* woody *stem* — that is, one that supports continued vegetative growth above ground from year to year. Many *annuals* appear to form woody stems in their first year, but nonetheless die back. A true woody stem contains *wood,* which is primarily composed of structured *cellulose* and *lignin,* and features a vascular, or circulatory, system. Most woody plants form new layers of woody tissue from year to year, each year over the layer of the preceding year, and so increase their stem diameter from year to year. Woody plants are usually either *trees, shrubs, or vines.*

Wraping. (1) The natural build-up of a saltmarsh by the deposition of silt. (2) The controlled flooding of agricultural land adjoining a river with silt-charged river water so as to build up fertile soil. (3) Covering fruits with suitable materials for improving the post harvest life.

X

Xanthophyll. A carotcnoid plant pigment (Carotene), yellowish in colour, present in chloroplasts and other plant parts where chlorophyll is absent. It assists in photosynthesis. The yellow in plant leaves during the fall is caused by the presence of xanthophyll. Xantophylls are present throughout the growing season, though their yellow is masked by the green of chlorophyll. Only when chlorophyll is destroyed through environmental changes, including autumn's cool temperature and lower light intensities, does the yellow appear. Xanthophyll intensity is based on the plant's genetics as well environmental conditions. Bright, sunny, cool autumn days will promote vivid yellow colors.

Xeriphyte. A plant naturally adapted to drought or low moisture conditions. The *general structure of the stem* is essentially a mirror of the root, with apical meristem at the tip, the three primary meristems just behind the apical meristem and the three ground tissues and more mature structures down lower, behind the three primary meristems. The other plant organ, the leaf, branches off from the stem—not from the root. The axillary buds are located in the leaf axils.

Xeriscape. (1) A landscape made of plants that require very little water or rain. (2) A trademark used for a landscaping method that employs drought-resistant plants in an effort to conserve resources, especially water.

Xeriscaping. The art of creating a landscape with the goal to reduce water usage is called xeriscaping. Plants requiring little or no watering, or those that can survive droughts, are chosen. Cultural practices such as soil preparation and mulching are also used extensively in xeriscaping. Xeriscaping DOES NOT mean using no water in the landscape. It implies the managed use of water; in other words, the effort to reduce the amount of water that homeowners need to supply in additional to typical rainfalls. Many plants tolerant of drought conditions are also tolerant of flood conditions. Included in this list are box elder and pecans. However, many drought-tolerant trees and shrubs may not be practical for home yards. Careful study and reading of xeriscaping books is essential.

Xerophyte. (1) A plant adapted to a very dry environment. (2) A xerophyte describes a *plant* that has structural (xeromorphic) and physiological adaptations which enable them to survive, or even thrive, in areas with very little free *moisture*. Xerophytes occur in all kind of environments, not simply those which immediately appear free of water such as *deserts*. *Cacti* and other succulents are typically found in deserts while *bromeliads* can be found in rain forests. Plants that live under arctic conditions may also have a need for xerophytic adaptations as water is not available for plant uptake when it is frozen. Many countries have xerophytes, including those which are wet such as the *United Kingdom* in *sand dunes* and sea shore strand lines. The adaptations are several, primarily these aim to limit water loss (including water storage), but also to obtain as much water as is possible from the environment.

Xylem. (1) Vascular tissue in plants which carries water and dissolved mineral salts from the roots, where they are taken into the various parts of the plant. It also provides mechanical support to plants, the cells being impregnated with lignin, and is located to the inside of the cambium. Secondary xylem comprises the bulk of the woody tissue in woody plants and is laid down by the process of secondary thickening. (2) Xylem vessels transport water throughout the plant, though essentially from the roots through the leaves. Xylem vessels live several months, before hardening off and becoming woody or wood. Xylem vessels transporting water within a woody plant are called sapwood. (3) Conducting vessels found throughout the plant. Xylem vessels transport water and minerals from the roots upward through the plant.

Y

Yeast. A single-celled ascomycete fungus, economically important in being capable of fermentation and an important source of vitamins.

Yellow rust. A fungal disease of barley and wheat, forming deep yellow pustules in closely parallel lines on (he infected parts of leaves and stems. The disease commences on volunteer plants and spreads to both spring and winter crops, particularly when conditions are cool and moist. Spores are able to survive the winter.

Yemma budding (chip budding). Budding where a single bud with a piece of wood is fitted to a suitable cut in the stock.

Yield monitoring. Collecting data on the amount of production at regular intervals combined with GPS readings. The resulting yield map is basic to decisions about fertilization, pest control, and other adjustments in a system of precision farming.

Yield of crop. Produce obtained from a crop grown in a unit area (one acre or one hactre). It depens upon many factors such as type of soil, cultural practice, use of manures and fertilizers, timely irrigation, season and variety etc.

Yield potential. Maximum production capability of a crop.

Z

Zero grazing. (1) A grazing system in which grass (or other fodder) is cut daily and then taken to cattle kept in a yard or in a small exercise paddock near the buildings. The system obviates prob-lems of poaching (particularly on heavy land), fouling, over-eating and wastage due to selective grazing. More grass can be provided from a given area by the system then by grazing. Also called mechanical grazing or soilage. (2) Zero-grazing means keeping animals in a stall, and bringing fodder to them instead of allowing them to graze outside. It is also sometimes called "cut-and-carry". The description below is about how to keep dairy cattle using zero-grazing. It is an intensive system that produces a lot of milk from a small amount of land. Zero-grazing can also be used with goats and sheep. They can be kept in a shed with a slatted wood or bamboo floor, raised about 1 m above the ground. The droppings fall through the slats into a pit beneath the shed. They can then be carried away to be used as fertilizer.

Zig-zag harrow. A type of Harrow in which the frames are zig-zag in shape, bearing short pointed tines, spaced so that each covers different ground. Used usually in the final stages of Seed Bed preparation, and also to remove tractor and drill wheel marks after sowing and to cover the seeds fully.

Zinc. Plant food element which is absorbed as zinc (Zn^{2+}) ion. It is considered essential in enzyme systems necessary for important reactions in the plant metabolism of life activity of plants. Zinc helps in the formation of some growth hormones, considered helpful in the reproduction of certain plants. Zinc aids in cell division, as well as enzyme and auxin productions/utilization. *Signs of Deficiency:* Small, thin, yellow leaves as well as low yields are symptoms of zinc deficiencies.

Zone. An area distinguished by a range of annual average minimum temperatures, used in describing plant hardiness. Regions that share similar climatic and rainfall conditions producing similar growing seasons.

Zoysia grass. Zoysia is fertilized in mid to late May with 2 pounds of nitrogen per 1000 square feet for an actively maintained lawn, and 1 pound nitrogen per 1000 square feet for low maintenance lawns. An additional 1 pound of nitrogen application of nitrogen can be applied in August. Zoysia is a warm season grass, thriving when temperatures are between 65 and 95 degrees F. There are several advantages to zoysia grass:

- Rapidly spreading
- Green during the summer
- Limited weed growth
- Few, if any, insect problems

- Slow growing, requiring fewer mowings.
- Extremely wear resistant when green

On the other hand, there are several disadvantages:

- Difficult to establish. Currently established by plugs, spaced six inches apart. Costly.
- Zoysia seed has a low germination percentage, making it difficult to seed.
- Will not grow in the shade.
- Invasive. Will grow under sidewalks and driveways, and neighbor's fences.
- Brown from mid October through mid May.
- ough on mowers. Mowers require frequent sharpening.
- Sensitive to "traffic" when dormant.
- Heavy thatch buildup, requires frequent dethatching or regulated burnings where permissible.

Zygote. (1) A fertilised ovum before it starts to divide and form an Embryo. (2) A *cell* that is the result of *fertilization*. That is, two *haploid* cells—usually (but not always) a *sperm* cell from a *male* and an *ovum* from a *female*—merge into a single *diploid* cell called the *zygote* (or *zygocyte*). Animal zygotes undergo *mitotic cell divisions* to become an *embryo*. Other organisms may undergo *meiotic* cell division at this time (for more information refer to *biological life cycles*). *Twins* and *multiple births* can be monozygotic (identical) or dizygotic (fraternal).

Zymogenous flora. (1) Organisms found in the soil in large numbers following the addition of readily decomposable organic matter. (2) So-called opportunistic organisms found in soils in large numbers immediately following addition of a readily decomposable organic substrate. Synonymous with copiotrophs.